Gideon Mantell and the Discovery of Dinosaurs

Gideon Mantell and the Discovery of Dinosaurs is a thorough but accessible biography – the first scholarly one – of a pioneering dinosaur hunter and paleontologist whose legendary achievements have been portrayed – not always accurately – in dozens of popular books about dinosaurs. How Gideon Algernon Mantell (1790–1852) actually discovered *Iguanodon*, the second dinosaur known to science and the first herbivorous one, is at last set right in this volume. It is not commonly known that Mantell also discovered seven other dinosaurs as well, proving at a time when majority opinion thought otherwise that they were not amphibious but walked on land – and sometimes on two legs. He devoted more than twenty-five years to restoring the original appearance of dinosaurs and almost single-handedly established in popular understanding the now-familiar idea that an Age of Reptiles had preceded the Age of Mammals. Honored by both the Royal and Geological societies for his scientific accomplishments, Mantell was closely associated with the outstanding naturalists of his day, including James Parkinson, William Buckland, Georges Cuvier, Robert Bakewell, Charles Lyell, Roderick Murchison, Michael Faraday, Charles Darwin, Louis Agassiz, and Richard Owen. He also corresponded for twenty years with Benjamin Silliman of Yale and had many interactions with other Americans.

Dennis R. Dean, Ph.D., a retired academic, is the author of *James Hutton and the History of Geology,* introductions and chapters in several books, important studies of Tennyson and Keats, and scholarly essays in such journals as *Isis, Annals of Science, Modern Geology,* and *Philological Quarterly.* He is presently the general editor of a history of the earth sciences reprint series.

Gideon Algernon Mantell (1790–1852). Engraved by W. T. Favey (ca. 1850) from a drawing by Pierre Senties and photographs by John Mayall (Sussex Archaeological Society).

Gideon Mantell and the Discovery of Dinosaurs

DENNIS R. DEAN

CAMBRIDGE
UNIVERSITY PRESS

CAMBRIDGE UNIVERSITY PRESS
Cambridge, New York, Melbourne, Madrid, Cape Town,
Singapore, São Paulo, Delhi, Mexico City

Cambridge University Press
The Edinburgh Building, Cambridge CB2 8RU, UK

Published in the United States of America by Cambridge University Press, New York

www.cambridge.org
Information on this title: www.cambridge.org/9780521420488

First published 1999

A catalogue record for this publication is available from the British Library

Library of Congress Cataloguing in Publication Data
Dean, Dennis R.
Gideon Mantell and the discovery of dinosaurs / Dennis R. Dean.
p.   cm.
Includes bibliographical references and index.
ISBN 0-521-42048-2 (hardbound)
1. Mantell, Gideon Algernon, 1790–1852. 2. Paleontologists –
England – Biography. 3. Iguanodon. I. Title.
QE22.M32D43   1999
560′.92 – dc21
[B]   98-16448
                                                              CIP

ISBN 978-0-521-42048-8 Hardback
ISBN 978-0-521-08817-6 Paperback

In Memoriam
Walter Baldock Durrant Mantell
1820–1895

Contents

Illustrations

Acknowledgments

My knowledge of Gideon Mantell's life prior to 1818 derives almost entirely from previously unutilized manuscripts and documents at the Alexander Turnbull Library, Wellington, New Zealand, together with a few newspaper items. Thereafter I extract many specifics from his private journal, the original of which is also at the Turnbull. E. Cecil Curwen's still helpful edition of that journal (1940) derived from an incomplete, sometimes faulty typescript commissioned by his father, of which Curwen published less than half. The final pages of Mantell's journal were not included in the typescript and came to light only in 1982, when I discovered them among his son Reginald's papers at the Turnbull. Mantell's revealing correspondence with Benjamin Silliman of Yale, begun in 1830, is another frequent source. Sidney Spokes diligently transcribed (but often failed to date) many of the same letters in his much-abridged 1927 biography of Mantell, the only previous one. Neither he nor Curwen had any real knowledge of the Mantell papers in New Zealand and neither was primarily interested in Gideon as a paleontologist. My version of his life is therefore significantly different from either of theirs, being fuller, more paleontological, and a good deal more reliable. Like all Mantell scholars since, however, I have benefited from their work.

I first visited Wellington, and the Alexander Turnbull Library (the national library of New Zealand), in July 1977 while en route home from a teaching assignment in Asia. Despite some preceding correspondence, I had not anticipated the astonishing richness of the Turnbull's Mantell Family Papers, an extensive but then relatively unknown resource for the study of British geology which I described in a short essay (1977), now superseded. Even so, it was not immediately apparent to me that this book had begun. My return to New Zealand some years later was funded by the National Science Foundation, a U.S. government agency. Without that grant (SES–8206469), this book would not have been possible. I wish very much to thank the Chief Librarians and their staffs at the Turnbull for courteous assistance then and since. The New Zealand Institute of Geological & Nuclear Sciences, the National Museum, and Victoria University generously permitted access to Mantell memorabilia in their possession. Among my individual New Zealand friends, Sir Charles (now deceased) and Lady Fleming were outstandingly gracious. Along with them, I gratefully remember Kerry Forsell and her family, George Henderson, Mike Daly, and the management and staff of the now defunct Carlton Hotel. John Yaldwyn, Director of the Museum when I was there, and A. P. Mason, of the Geological Society of New Zealand (Auckland), furthered my New Zealand interests most assiduously as this book was in its final stages of preparation.

Among (or in addition to) many British institutional staff members, I wish to thank Cyril Pike, the late Stephen Moore, Dr. Rendel Williams and his family, John Bleach, Joyce Crow, Thomas Bowen, Elizabeth Burdett, Norah Archer, Peggy Guido, J. B. Delair, Roy Porter, Charlotte MacKenzie, Jim Secord, John Thackray, Mary Sampson, Fiona Eldridge, the late Alan J. Charig, Ron Cleevely, William Edmonds, Rodney Steel, Brian Latimer, and Richard Stutely. Among the many persons who furnished useful information, I particularly wish to acknowledge Dr. Colin and Judith Brent, David Norman, Alan Shelley, the late Joan Eyles, Eric E. F. Smith, Patricia Gill, Dr. J. A. Edwards, Peter McKay, Alison McCann, Ian F. Lyle, M. D. Crane, Philippe Taquet, Mrs. A. D. Morris, Rowland C. Swift, Virginia Murray, Tom Sharpe, Hugh Torrens, John A. Cooper, W. A. S. Sarjeant, Garry J. Tee, Adrian Desmond, Christopher Hamel Cooke, R. J. G. Savage, and Richard L. Smith.

For researching Gideon Mantell, the most important British manuscript collections are those held by the British Museum (Natural History), Cambridge University Library, the Royal Society, the Geological Society, the East Sussex Record Office, the Sussex Archeological Society, the West Sussex Record Office, and Castle Museum, Norwich. John Murray, Ltd. and Lord Northampton own particularly significant private collections; that of the Eyles' has since moved to Bristol University. Further GM items can be found at the Fitzwilliam Museum, Cambridge; Edinburgh University Library, the British Library, the Royal College of Surgeons, University College, London, the Museum of the History of Science at Oxford, and the National Museum of Wales. The BL (Colindale) and Brighton Reference Library preserve relevant newspapers.

My research at Yale was funded in part by a grant from the National Endowment for the Humanities, a U.S. government agency. For aid while there I wish to thank John Ostrom, Barbara Narendra, and Judith Ann Schiff. Before coming to Yale I had already been well served by its library staff for some years. Though Yale is certainly the chief repository of Mantelliana in America, the American Philosophical Society, Philadelphia, holds more than fifty important letters. The Historical Society of Pennsylvania, the Academy of Natural Science, and Harvard University Library also provided useful manuscript material. Other contributing American institutions include the Library of Congress, the Newberry Library, the university libraries at Madison and Urbana, Northwestern University (Janet Ayers especially), the Folger Library, and the Smithsonian Institution. Through correspondence, I was also able to utilize important GM letters at the Bibliotheque Central du Museum National d' Histoire Naturelle, the Institut de France, and the Geological Society of Paris.

For permission to use materials in their possession, I wish to thank the institutions cited, whether in New Zealand, the United States, Britain, or France. In my quotations from nineteenth-century sources, as noted below, certain accidentals have been modernized for the reader's convenience, but my preferred geological and paleontological terminology is that of Mantell and his times. Similarly, my text and illustrations are devoted to conceptions of the dinosaur (and other prehistoric creatures) prevalent in the second quarter of the nineteenth century, with only brief indications where needed that some of them are no longer current.

For my illustrations I am indebted to the courtesy of a number of institutions and individuals, including (frontispiece) Joyce Crow and the Sussex Archaeological Society, Lewes; (map) Richard Allen Thompson of Evanston; (1.2) Margaret Kaiser and the National Library of Medicine, Bethesda, Maryland; (7.1a) Brian Latimer, Richard Stutely, and the Maidstone Museum, Kent; (9.2, 12.1) Mary Sampson and the Royal Society of London; (12.3) Judith Ann Schiff and Yale University; and (5.1, 6.2b, 9.1, 9.3, 9.4) Martin Pulsford and the Natural History Museum. Other photo work (map, 3.1, 4.1, 4.2a,c,d, 4.4, 6.1, 6.2a, 7.1b, 8.1, 8.2, 10.1, 10.2, 12.3a), was by CIMA Graphics of Evanston from originals or copies in my possession.

The remaining illustrations (1.1, 2.1, 3.2, 4.2b, 4.3, 6.3, 7.2, 7.3, and 12.2), many of them composites, were completed for me by staff members of the Alexander Turnbull Library. I wish to thank Marian Minson, Curator, Drawings & Prints; David Retter, Manuscripts Librarian; Mary Cobeldick, Reference Librarian, all of ATL; and Anna Sanderson, Photographer, National Library of New Zealand. Margaret Calder, Chief Librarian, Alexander Turnbull Library; Ian Keyes, Curator, Macrofossil Collection, Institute of Geological & Nuclear Sciences, Lower Hutt, NZ; and Dr. John Yaldwyn, Retired Director, Museum of New Zealand, Wellington, graciously provided access to materials under their supervision.

In commemoration of the subject of this biography, a Mantell Bicentenary Symposium (at which I had the honor of being the keynote speaker) was held at the University of Sussex, Falmer, in 1990. I therefore extend further gratitude to its organizers, its participants, and all others who helped to make the Symposium possible. The text of my address was published in the *Journal of Geological Education,* 38 (1990), 434–443.

At Cambridge University Press, I was ably served by Alex Holzman, history of science editor; Helen Wheeler, production editor; and Patricia Woodruff, copy editor. I owe special thanks also to my literary agent, Julian Bach, who brought CUP and me together.

Evanston, Illinois

Editorial Notes

The earlier years of Gideon Mantell's life were revolutionary ones, not only in politics, economics, and technology, but also in society, behavior, dress, outlook, understanding, and expression. Language especially was unstable. As literacy spread to lower economic classes not previously possessed of it, neoclassical regularities dissolved. In correspondence particularly, syntax became a good deal looser and more spontaneous. Punctuation and capitalization were erratic. Victorian grammarians and editors would eventually restore something like the formality we take for granted, but most of my sources for this book were written hurriedly in a sprawling, irregular idiom that now is difficult to read. When Gideon and others of his generation published such hastily composed material, they normally revised it; scholars will perceive that I have conservatively followed the earlier example. Throughout this book, therefore, quotations are meticulous but not invariably verbatim, having been enhanced where necessary with modern pointing and other stylistic incidentals.

The terms "geology" and "geologist" were uncommon anywhere before the 1790s. Then and later, fossils often appeared as "petrifactions" or "organic remains." The science of paleontology was not so called in England before the latter 1830s; an earlier term, "oryctology," dropped out of use soon after 1822. "Geology" then regularly included the study of fossils. "Dinosaur," coined in 1842, and "scientist," from a few years earlier, did not immediately gain acceptance. Names of formations, geological periods, and other now-familiar technicalities also emerged gradually. Since changes in language necessarily accompanied developments in science, a modern dictionary will not always reveal the nineteenth-century meanings of some words important to this book. My stratigraphic table and a few helpful definitions are therefore appended.

Classifying plants and animals binomially (into genera and species) had been usual since the mid-eighteenth century, but new discoveries were often identified only by genus – and frequently reidentified by others. Generic names (i.e., *Iguanodon*) were regularly capitalized but often not italicized. Specific names derived from proper nouns were also capitalized

(*Iguanodon Mantelli*). Now that various dinosaurs are well-known, however, some people speak no less familiarly of iguanodons than of cows. I therefore prefer a down style where possible. Contrarily, nineteenth-century writers often left important stratigraphical designations, like "Tertiary," uncapitalized. In the consistent modern usage I have followed, "Chalk" is the name of a formation; "chalk," of a rock.

Royalty and aristocracy throughout Europe ranked in part as follows: king and queen, prince and princess, duke and duchess, marquis (marquess in England) and marchioness, earl and countess, baron and baroness, and baronet or knight and dame. In England the king's sons were dukes as well as princes, multiple titles being common at all the higher levels. One addressed earls and other hereditary noblemen as "Lord Northampton"; a lifetime-only baronet or knight became "Sir Charles." The wife or daughter of a lord expected to be called "Lady," as did the wife of a baronet or knight. Among Gideon's generation, only servants and lesser family members were commonly addressed by their unadorned first names.

Nineteenth-century British money consisted of pence, shillings, pounds, and guineas, with twelve pence to a shilling and twenty shillings to a pound. The latter could be either a banknote or a coin ("sovereign"). If not spelled out, amounts are given here as £10/6/4 or as 2s/6p (though the normal abbreviation for pence was "d"). The guinea, no longer a coin itself but still commonly referred to, consisted of one pound and one shilling. £10/10/0, therefore, represented ten and a half pounds, ten pounds ten shillings, or ten guineas. In transatlantic exchanges the pound was worth about five American dollars.

Abbreviations

AJSA	*American Journal of Science and Arts* (ed. BS)
AJW	Alfred James Woodhouse
AMNH	*Annals and Magazine of Natural History*
Animalcules	GM, *Thoughts on Animalcules* (1846)
Anon.	Anonymous author
APS	American Philosophical Society, Philadelphia
AS	Adam Sedgwick (professor of geology at Cambridge University)
ATL	Alexander Turnbull Library (the national library of New Zealand, Wellington; repository of the Mantell Family Papers)
b.	born
BAAS	British Association for the Advancement of Science (founded 1831)
bapt.	baptized
BGaz	*Brighton Gazette* (weekly newspaper)
BGrd	*Brighton Guardian* (weekly newspaper)
BH	*Brighton Herald* (weekly newspaper)
BM	British Museum
BMNH	British Museum (Natural History); not then a separate institution
BS	Benjamin Silliman (of Yale)
Buckland, *Life*	Anna B. Gordon (daughter), *The Life and Correspondence of William Buckland, D.D., F.R.S.* (London, 1894)
CUL	Cambridge University Library

Castle Ashby	Compton Family Papers, Castle Ashby, Northamptonshire (private)
Charig	Alan Charig, *A New Look at the Dinosaurs* (1979)
CK	Charles König (of the BM)
CL	Charles Lyell
ca.	circa (at about this date)
Curwen	Edward Cecil Curwen or his edition of GM's J (1940)
d.	died
DAB	*Dictionary of American Biography*
DG	Davies Gilbert (formerly Giddy)
DNB	*Dictionary of National Biography* (British)
DRD	The author
DSB	*Dictionary of Scientific Biography*
EB	Etheldred Benett
EC	Elizabeth Cobbold
EMM	Ellen Maria Mantell (daughter)
ENPJ	*Edinburgh New Philosophical Journal*
esp.	especially
ESRO	East Sussex Record Office (Lewes)
FMC	Fitzwilliam Museum, Cambridge
Foss SD	GM, *The Fossils of the South Downs* (1822)
F. R. S.	Fellow of the Royal Society
G&M	WB, *Geology and Mineralogy* (1836)
GBG	George Bellas Greenough
GC	Georges Cuvier
Gent Sci	Jack Morrell and Arnold Thackray, *Gentlemen of Science* (early years of the BAAS; 1981)
Geol Exp	GM, Geological Expeditions (notebook, ATL)
Geol SE Engl	GM, *Geology of the South-East of England* (1833)
GFR	George Fleming Richardson
GM	Gideon Mantell
GS	Geological Society (of London; founded 1807)

Hist GS	Horace B. Woodward, *History of the Geological Society of London* (1907)
HMM	Hannah Matilda Mantell (daughter)
Horsfield	Thomas W. Horsfield, *History and Antiquities of Lewes* (1824, 1827)
Illus	GM, *Illustrations of the Geology of Sussex* (1827)
incl.	including
J	GM, journal (ATL)
JBP	Joseph Barclay Pentland
JD	James Douglas
JH	John Hawkins
JM	John Murray (publisher)
JP	James Parkinson
JS	James Sowerby
JSBNH	*Journal of the Society for the Bibliography of Natural History*
JSM	Johann S. Miller (formerly Müller)
LA	Louis Agassiz
LB	Lambart Brickenden
LGJ	*London Geological Journal*
Lit Gaz	*Literary Gazette*
LL.D.	Doctor of Laws (*Legum Doctor*)
LLJ	*Life, Letters, and Journals of Sir Charles Lyell*, ed. K. M. Lyell (1881)
Longman	Longman archives, University of Reading
LS	Linnaean Society (London)
MAL	Mark Anthony Lower or his *The Worthies of Sussex* (1865)
MAM	Mary Ann Mantell (wife)
Medals	GM, *Medals of Creation* (1844)
MM	Mantellian Museum
M.R.C.S.	Member of the Royal College of Surgeons
ms	manuscript

Murray	John Murray (Publishers) Ltd. archives (private)
Narrative	GM, *A Narrative of the Visit of . . . William IV . . . to . . . Lewes on the 22nd of October 1830* (1831)
n.d.	original not dated
ns	new series (of a periodical)
NZ	New Zealand
OM	Ordinary Minutes (of the GS)
Org Rem	JP, *Organic Remains of a Former World* (1804–1811)
Owen, *Life*	Rev. R. Owen (grandson), *The Life of RO* (1894)
Paris	Museum National d'Histoire Naturelle, Paris
Paris, Inst	Bibliotheque de l'Institut de France, Paris
Petrif	GM, *Petrifactions and Their Teachings* (1851)
Pict Atlas	GM, *A Pictorial Atlas of Fossil Remains* (1850)
PL	Pentland letters, ed. Sarjeant and Delair (1980)
PRO	Public Record Office (London)
Proc	*Proceedings* (of a learned society)
PT	*Philosophical Transactions* (RS, London)
QJGS	*Quarterly Journal* of the GS (1845ff.)
Ramble	GM, *A Day's Ramble in . . . Lewes* (1846)
RB	Robert Bakewell
RCS	Royal College of Surgeons (London)
Recherches	GC, *Recherches sur les ossemens fossiles* (1812)
Reg J	RNM, journal (ATL)
rev.	review
RIM	Roderick Impey Murchison
RNM	Reginald Neville Mantell (son)
RO	Richard Owen
RS	Royal Society (London)
SA	*Sussex Advertiser* (newspaper, Lewes)
SAS	Sussex Archaeological Society (Lewes)
SGM	Samuel George Morton (of Philadelphia)
Spokes	Sidney Spokes or his *Gideon Algernon Mantell* (biography, 1927)

SRI	Sussex Royal Institution
SSLI	Sussex Scientific and Literary Institution
St. Bart's	Saint Bartholomew's Hospital (London)
SW	Samuel Woodward (of Norwich)
SWA	*Sussex Weekly Advertiser* (newspaper, Lewes); later, *SA*
TH	Thomas Hawkins
TJB	Thomas James Birch
TL	Thomas Longman (publisher)
Town Book	*Town Book of Lewes*, ed. V. Smith (1972, 1976)
Trans	*Transactions* (of a learned society)
TW	Thomas Webster
UCL	University College London
WB	William Buckland (professor of geology and mineralogy at Oxford; Dean of Westminster)
WBDM	Walter Baldock Durrant Mantell (son)
WC	William Clift (of RCS)
WCW	William Crawford Williamson
WDC	William Daniel Conybeare
WHB	William Harding Bensted
WHF	William Henry Fitton
Wight	GM, *Geological Excursions Round the Isle of Wight* (1847)
Wilson	Leonard G. Wilson, *Charles Lyell* (1972)
Wonders	GM, *The Wonders of Geology* (1838)
WSRO	West Sussex Record Office (Chichester)
WW	William Whewell
Yale	Sterling Library, Yale University (Silliman Family Papers)
ZS	Zoological Society (London)

GM–BS	= writer and recipient of a letter
BS–GM, 3, 28 July 1830	= two separate letters
GM–BS, 2 & 26 Sept 1840	= one letter written on two separate days

Some missing dates have been supplied.

1

Castle Place

In the topography of southeast England, where Gideon Algernon Mantell (1790–1852) spent his life, strata are generally younger than elsewhere in Britain. Though lying atop older, the earliest exposed rocks date from near the very end of the Jurassic period, some 150 million years ago, when the region was mostly under water. Ammonites, belemnites, and (increasingly) bivalves were common marine species, often serving as food for plesiosaurs and ichthyosaurs, which had already developed to considerable size. Above them, in the air, gliding pterosaurs scanned the ocean for prey. Though far less impressive in wingspread, true birds had also begun to evolve. In freshwater environments, and even some marine ones, long-snouted crocodilians foraged shorelines. On land, dinosaurs were already dominant, including gigantic carnivores like *Megalosaurus*, huge plant-eaters like *Cetiosaurus*, and some early plated genera as well. Careful to stay out from under their immense feet, and usually no bigger than their toes, were our direct ancestors – early, diminutive mammals.

The Cretaceous or Chalk period began about 144 million years ago and lasted more than half that time. During it, the Weald (an extensive freshwater delta or floodplain in what are now Kent and Sussex) gradually submerged; sandstone, greensand, mudstone, and abundant chalk successively resulted. In fawn-colored Wealden sandstone, which Mantell was first to identify as a freshwater deposit, *Iguanodon* and his other dinosaurs were found. Before he discovered them, however, Gideon spent some years collecting marine fossils from the Chalk, a younger but much more prominent formation especially notable for its sponges, corals, ammonites, bivalves, and one kind of brachiopod (*Terebratula*). In decline by this time, the previously dominant ammonites had begun to uncoil their characteristically ribbed shells in a variety of bizarre ways (*Hamites, Scaphites, Turrilites*). Ichthyosaurs were dying out as well, having been superseded by various mosasaurs. Modern bony fishes, meanwhile, flourished – those of the tarpon group (*Osmeroides*) especially. Thus, plesiosaurs, pterosaurs, turtles, crocodiles, lizards, and in particular a great variety of dinosaurs dominated the food chain; fish, birds, mammals, insects, and flowering plants played important but lesser roles.

At the close of the Cretaceous period, sixty-five million years ago, dinosaurs of all sizes rapidly died out – just how suddenly or why, no one is sure – as did mosasaurs, plesiosaurs, pterosaurs, ammonites, belemnites, and other less familiar creatures. Correspondingly, the Age of Reptiles, or Mesozoic era, ended and the Age of Mammals, or Cenozoic era, began. Most of the now-vacant ecological niches formerly occupied by gigantic reptiles were eventually filled by mammals evolving toward larger size: mammoth, mastodon, giant sloth, tapir, and rhinoceros, for example. In a few unique environments, like the islands of New Zealand, mammals never appeared; their place was eventually taken by giant flightless birds – most famously the moa, some varieties of which probably survived into the eighteenth century A.D. before being hunted to extermination by man.

During the Age of Mammals also, significant orogenic movements bent European portions of the old sea bottom upward. Thus, Jurassic and Cretaceous strata in what eventually became southeastern England were steadily uplifted, forming a great bulging fold (or anticline), the top of which was subsequently sheered off by erosion to form a symmetrical topography of encased and inverted stratigraphic Us. Bordered at its farther edge by younger, more resistant chalk (the North and South Downs), and now a valley, the truncated Wealden Anticline became progressively older toward its vertical axis, around Winchelsea. Similar but downward warping also created the London, Hampshire, and Paris basins, with subsequent deposits of mammiferous sediments in all three. As part of the Age of Mammals also, Britain experienced glaciations, which contoured its landscape, scattered erratic boulders and thick gravels widely, and depressed subjacent crust, resilient uplift following. Perhaps no more than eighty-six hundred years ago, marine erosion (by a sea deepened with glacial melt) battered out the English Channel, severing Britain from continental Europe and assuring its historical identity as an island. Finally, the spectacular abrupt cliffs of coastal southeast England – at Dover and Brighton, for example, but also cleaving the Isle of Wight from Hampshire – are thought to be less than five thousand years old.[1]

1. See Wes Gibbons, *The Weald* (London, Boston, and Sydney, 1981); S. W. Wooldridge and Frederick Goldring, *The Weald* (London, 1953); David K. C. Jones, *The Geomorphology of the British Isles: Southeast and Southern England* (London and New York, 1981); J. G. C. Anderson and T. R. Owen, *The Structure of the British Isles,* second edition (Oxford, 1980); and, more generally, Rodney Steel and Anthony Harvey, eds., *The Encyclopedia of Prehistoric Life* (New York, 1979). For Lewes generally, see Thomas W. Horsfield, *The History and Antiquities of Lewes and Its Vicinity* (2 vols., Lewes, 1824, 1827), with contributions by GM; *The Town Book of Lewes, 1702–1837* (1972) and idem., *1837–1901* (1976), both

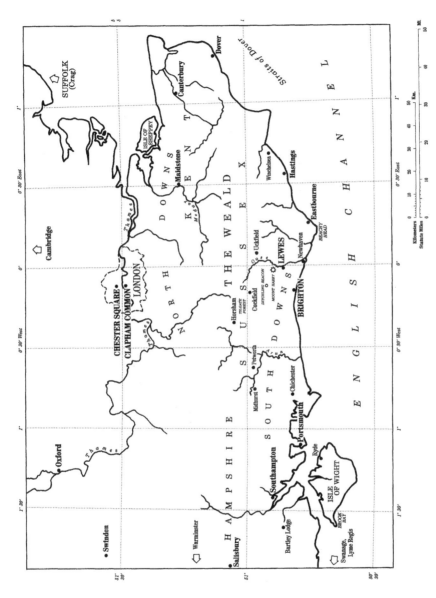

Map of southeast England

At this point we likewise come to a great divide, because the geological history related thus far, regarding both life and landforms in southeast England, was entirely unsuspected when Gideon Mantell was born in 1790 and known only imperfectly when he died in 1852; a surprising amount of the intervening progress, we shall see, is attributable to him.

As the attractive subject of dinosaurs and their discovery has already been granted a number of popular reprises, however, I wish to stress that this book consists throughout of original biographical research. The need is real, for while Gideon Mantell's role as a discoverer of saurian remains has been widely acknowledged, it has also been almost universally misinterpreted. Because so much of our earlier understanding derived from slipshod scholarship, I have insisted that my own account be resolutely factual throughout. In pursuing these researches, moreover, I soon came to see that the early history of dinosaur discoveries was rather different than had been previously supposed.

In that history, Gideon Mantell emerges now as one of several independent discoverers of *Megalosaurus* – a carnivore, and the first dinosaur of any kind to be described. He also discovered *Iguanodon,* the first herbivorous dinosaur; and *Hylaeosaurus,* the first armored dinosaur. Though not quite realizing that *Cetiosaurus* was a dinosaur, he was among the first to discover its huge remains and correctly identified them as reptilian. He then discovered *Pelorosaurus,* a gigantic brontosaurian; *Regnosaurus,* an enigmatic dinosaur whose more specific affinities are still being debated; and *Hypsilophodon,* which he thought plausibly enough to be a very young iguanodon. In addition, Mantell collected a great deal of important evidence pertaining to other archosaurs, including crocodiles, gavials, turtles, plesiosaurs, ichthyosaurs, and the first mosasaurian and pterosaurian remains discovered in England. No other individual contributed so much to our early knowledge of prehistoric saurians. And far more than anyone else, he impressed the Age of Reptiles upon contemporary minds.

Yet Gideon Mantell has been remembered thus far (in almost every popular book written about dinosaurs) only for his – or his wife's – accidental discovery of *Iguanodon,* a significantly distorted legend which I will examine later on. It is not commonly appreciated that he devoted some thirty years to his increasingly accurate reconstructions of *Iguanodon,* while discovering seven other dinosaurs as well. Gideon was the first person to collect dinosaur bones systematically and over a period of time with the specific intention of restoring the animals' original appearance. His work

ed. Verena Smith (Lewes: Sussex Record Society); together with two detailed local histories by Colin Brent: *Georgian Lewes* (1993) and, with William Rector, *Victorian Lewes* (1980). I am grateful to Dr. Brent for his contributions to this chapter.

in less popular areas of paleontology, moreover, is sometimes distorted even by specialists. In addition to saurians, Mantell discovered several dozen other prehistoric creatures. Among his further important finds were numerous mollusks, fishes, insects, sponges, plants, and foraminifera. Utilizing this previously overlooked supplemental information, he was the first researcher to place living dinosaurs within their real environment and, in so doing, raised seminal questions about prehistoric climates. As a pioneer geologist, furthermore, Gideon made fundamental contributions to stratigraphy, establishing the Weald of Sussex in particular as a classic locality of worldwide significance. Rather unexpectedly, he was likewise fundamental to study of the geology and flightless birds of New Zealand. These nonsaurian accomplishments of his deserve more scholarly recognition than they have received but are too numerous and diverse to be explored fully in this book. Though omitted with regret, they are at least alluded to in passing. Further information can then be found in my footnotes and in Gideon's own writings.[2]

Among the prominent researchers of his time, Mantell was unusually concerned to disseminate accurate scientific information throughout the public at large. He did so primarily in three related capacities: as the proprietor of an outstanding fossil collection, as the most popular geological lecturer in England, and as a foremost geological writer. In each case, he attracted both scientific and lay audiences of unprecedented size. Since Gideon was professionally a surgeon, however, doctoring routinely filled his days. Invariably on call, he was often summoned after dark as well. Much of his scientific work, consequently, took place around or beyond midnight, leaving him no more than a few hours' sleep. As if two such demanding careers – medicine and science – were not enough, Gideon was also a local and family historian; a productive archeologist and microscopist; a political activist and social climber; a minor poet and artist; and, in his valuable journal and extensive correspondence, an outspoken critic and chronicler of his times.[3]

---

2. Gideon Mantell's major books on earth science include *The Fossils of the South Downs* (1822), *Illustrations of the Geology of Sussex* (1827), *The Geology of the South-East of England* (1833), *The Wonders of Geology* (1838), *Medals of Creation* (1844), and *Petrifactions and Their Teachings* (1851), all of which are discussed below. An adequate survey of Mantell's contributions to every area of paleontology would necessarily include the work of his two sons, Walter and Reginald, for whom see later mentions in the present work.

3. See DRD, "A Bicentenary Retrospective on Gideon Algernon Mantell (1790–1852)," *Journal of Geological Education,* 38 (1990), 434–443, which is the text of my Ramsbottom Address, twice given in Sussex. A bibliography of Mantell by me, listing and remarking on his writings and publications in all areas, is available.

Origins

Norman French in origin, the name Mantell (pronounced "Mantle") refers to the common medieval cloak. According to Gideon's own notes (begun in 1814), the family of Mauntell, or de Mantell, came to England from Normandy and was still represented in Abbeville. As Mantell, it could be found in the Roll of Battle Abbey, a listing (by surname only) of the knights and squires who accompanied William the Conqueror in 1066 and won the Battle of Hastings. A William de Mantell then accompanied Richard the Lion-Hearted (d. 1199) to the Holy Land, as part of the dramatic but unsuccessful Third Crusade. Matthew Mantell was sheriff of Northamptonshire under King John in 1213. This line of Mantells remained in place through the Reformation, leaving their tombs and brasses in the chancel of Heyford Church, six miles west of Northampton. In 1541, however, John Mantell and some others ventured upon a "nocturnal frolic," poaching deer from Sir Nicholas Pelham's park in Sussex. Caught by three of Pelham's men, they murdered one and were subsequently executed, the greater portion of their estates being forfeited to the crown. As if to complete the ruin of his house, John's son Walter then joined Sir Thomas Wyatt's Kentish rebellion of 1554, which tried but failed to prevent Queen Mary's union with the powerful Catholic monarch Philip of Spain. Walter, his nephew, and Wyatt himself were executed the same year.

Gideon was highly conscious of this ancestry and the religious differences that had contributed so much to the social and economic degradation of his family. "I fully reciprocate your feelings as to the virtues of our blessed monarchs of the 17th century," he wrote ironically to his American friend Benjamin Silliman in December 1849.

> You forget that my family was despoiled of lands and fortune and station for their attempts in the previous century to maintain the Protestant faith. Sir Walter Mantell and his son [error for nephew] were executed in Kent in the reign of Queen Mary – see Foxe's *Book of Martyrs* – and in Charles II's time when the persecuted family took refuge in Lewes, their names are among the fined and imprisoned for being *convecticlers*.

Which is to say, they were Protestant dissenters who met clandestinely.[4]

---

4. That the Mantells pronounced their family name trochaically is evident from Thomas Mantell's and others' phonetic spellings of it, the puns of Horace Smith and others on it, and the placement of "Mantell" within metric lines of poetry. GM, "Memoirs of the Mantells" (ATL) is a collection of genealogical materials; see also William Berry, *County Genealogies: Pedigrees of the Families in the County of Sussex* (London, 1830), p. 20; William Berry, *County Genealogies . . . Kent* (London, 1830), pp. 332 and 185; Edward

Gideon traced his *direct* ancestry from the sixteenth-century John Mantell's brother Thomas, said to have been headborough of Lewes in 1562 and constable a decade later. His son and grandson, both named Thomas, were headboroughs of Lewes also. After two more generations, a further Thomas Mantell, born in 1716, married Gideon's grandmother, Susannah Austen (1716–1790). They had two daughters and five sons, including the future Reverend George Mantell (1757–1832) of Swindon and Thomas Mantell (1750–1807) of Lewes, Gideon's uncle and father, respectively. All eight of this latter Thomas Mantell's children (beginning with Sarah, his short-lived firstborn) would be named for biblical personages.

Gideon Algernon Mantell, the third son and fifth child, was prophetically named for an Old Testament patriarch who defeated the Midianites and subsequently became a judge of Israel; his middle name commemorated Algernon Sidney, a republican hero. He was born on 3 February 1790 in Lewes (now pronounced "Louis"), Sussex, a small provincial town of about five thousand persons with a long history of political and religious dissent. Gideon was baptized some time later by the Reverend George Barnard in an eventually Methodist chapel that Thomas Mantell, Gideon's father, had founded in 1788; no longer extant, it was then in St. Mary's Lane. Also on "Simmery-lane," in an unpretentious but comfortable home which paternal shrewdness and hard work (rather than formal education) had provided, Gideon lived with his father, Thomas; his mother, Sarah; his surviving older sister, Mary; his older brothers, Thomas and Samuel; and (in time) his younger siblings Joshua, Jemima, and Kezia.

As of February 1790 the French Revolution was less than a year old; when it became increasingly radical, concomitant British reaction harmed the Mantells a good deal. During his years in Lewes a generation before, from 1768 to 1774, Thomas Paine had fostered a political discussion group at the White Hart Inn (then the town's social center) on High Street, not far from St. Mary's Lane. That outspoken group of headstrong dissidents attracted William Lee the printer and a number of persons unknown. One of the latter was probably Gideon's father, who would have been eighteen to twenty-four. While definite connections between him and Paine's group are lacking, the Mantells and the Lees would thereafter be close friends and political allies for three generations. It is also entirely consistent with what little we know about Thomas Mantell's radical political beliefs to suppose that he came directly under Paine's influence during his formative years. As American revolutionary and author of *The*

Hasted, *The History of Kent* (4 vols., Canterbury, 1778–1799), III (1790), 97, 319–320; and Foxe's *Book of Martyrs*. Spokes's fuller genealogy supplements Berry but includes mistakes. GM–BS, 30 May 1842, and 14 Dec 1849 (Yale) discuss Mantell family history.

The Mantell Family

"Memoirs of the Mantells" (ATL) includes GM's birth and baptism; for Thomas Mantell and his Methodist chapel, see Horsfield, I, 306. The chapel was sold at Thomas Mantell's death in 1807.

THOMAS MANTELL (b. 1716) married Susannah Austen (1716–1790). Their five sons included Gideon's uncle and father.

GEORGE MANTELL (1747–1832), the Reverend, of Westbury and Swindon, son of the above and Gideon's uncle, married Martha Houston (1751–1821). Their son George (b. 1789), a Faringdon doctor and Gideon's cousin, had six sons and two daughters by his first wife, a former Miss Isles (d. 1832) and three daughters by his second, Rosetta Stacey.

THOMAS MANTELL (13 April 1750–1 July 1807), shoemaker of Lewes and Gideon's father, married Sarah Austen (25 Nov 1755–23 Dec 1828) of Peckham, Kent, in 1776. They had four sons and four daughters.

THOMAS AUSTEN MANTELL (1781, 1784, or 1786–30 Nov 1872), sheriff's officer, local politician, and auctioneer of Lewes, son of the above and Gideon's eldest brother, married Hannah (Ann) Groves (d. 1848) on 21 Dec 1809. Their six children, Gideon's nieces and nephews, included Sarah (b. 7 Oct 1810; bapt. 23 Mar 1812), Thomas (b. 26 Feb 1812; bapt. 23 Mar 1812), Charles, Ellen (d. 1849), George (all three baptized 20 June 1817), and Fanny (1821; bapt. 12 Jan 1822). TAM's age is given as ninety-one on his death certificate, as sixty-five in the census of 1851, and as eighty-seven in the census of 1871.

SAMUEL AUGUSTUS MANTELL (1789–23 Mar 1873), innkeeper and butcher of Lewes and Gideon's elder brother, married Henrietta (Harriet Mather) Kennard, ca. 1811. Their six children, also Gideon's nieces and nephews, included Arabella (b. 31 July 1817), Laura (b. 21 Oct 1818), Caroline (b. 9 June 1820), George (bapt. 7 July 1822), Thomas (b. 21 Sept 1824), and Samuel (b. 25 Jan 1828). An earlier Samuel was baptized on 26 Feb 1812. SAM's age is given as eighty-three on his death certificate, as fifty-five in the census of 1851, and as eighty-one in the census of 1871. In regarding him as older than Gideon, I am also following a genealogical memorandum in which GM identified himself as "3rd Son" (ATL).

JOSHUA MANTELL (3 Nov 1795–28 Mar 1865; *DNB*), surgeon of Newick and Gideon's younger brother, never married and was permanently institutionalized from 1836 onward. A hunchback deformed from birth, he was thought unlikely to survive and remained unbaptized until 8 Sept 1811. Joshua outlived Gideon by more than twelve years. Of Gideon's four sisters, SARAH (1779) died in infancy. MARY (Sept 1780–13 Apr 1854) wed her cousin Charles West, a London stationer, who was dead by 1833. JEMIMA (1798–unknown) married Edward Bevis, a widowed Brighton carpenter, on 4 Oct 1831. Surviving him, she then married a Mr. Fielder of London around May 1850. KEZIAH or Kezia Mantell (1800–30 May 1874), domestic servant of Bromley, never married and may have been defective. In 1800 her mother, Sarah Mantell, was approaching forty-five years of age. Keziah's birth year, like that of her older sister Jemima, is recorded only in the sometimes erroneous *International Genealogist's Index* (which lists Gideon's birthday as *8* Sept 1790, possibly his baptismal date). If all three are in fact baptismal dates, then Jemima may have been born in 1797 and Keziah in 1799. Keziah's death certificate gives her age as eighty-five.

GIDEON ALGERNON MANTELL (3 Feb 1790–10 Nov 1852; *DNB*), surgeon and fossil collector of Lewes, Brighton, Clapham Common, and London, married Mary Ann Woodhouse (9 July 1795–20 Oct 1869) of Maida Vale, London, on 4 May 1816. They had five children, the first of which died unnamed in 1817.

ELLEN MARIA MANTELL (30 May 1818–?1892), his oldest surviving child, was baptized on 3 Feb 1819, married the widowed publisher John William Parker (1792–1870) on 12 Feb 1848, and by him had three children, Walter, Edith, and Maud.

WALTER BALDOCK DURRANT MANTELL (11 Mar 1820–7 Sept 1895; new *DNB*), apprentice surgeon, New Zealand emigrant, naturalist, and preserver of his father's papers, married first Mary Sarah Prince (d. 1873) in 1869 and then, in 1876, Jane Hardwick (d. 1906). The widow of his son, Walter Godfrey Mantell, donated Gideon's and other family papers to the Alexander Turnbull Library in Wellington.

HANNAH MATILDA MANTELL (24 Nov 1822–12 Mar 1840), Gideon's younger daughter, his angelic invalid, was baptized on 1 July 1823 and died unmarried (of tuberculosis), as did

REGINALD NEVILLE MANTELL (11 Aug 1827–30 June 1857; new *DNB*), civil engineer (of cholera, in India).

═══

Rights of Man (1791) Paine certainly had his British admirers, but his religiously unacceptable *Age of Reason* (1794, 1796) managed to alienate most of them. The concurrent Reign of Terror in France, moreover, turned English opinion sharply away from even the most enlightened revolutionary views. "My poor father," Gideon remembered as an adult, "suffered greatly in fortune during the war with revolutionary France for his Whig principles."[5]

Because of his father's strongly held unorthodox beliefs in both religion and politics – Methodist and radical Whig, respectively – it was not possible for Gideon to enter the usual public schools. After learning to read,

---

5. GM–BS, 14 Dec 1849 (Yale). In 1796 Thomas Mantell voted for Justice William Green, an extreme Whig candidate, for Lewes Borough (poll book). For Paine, Lee, and Lewes, see John Keane, *Tom Paine: A Political Life* (Boston, 1995); Audrey Williamson, *Thomas Paine: His Life, Work, and Times* (London, 1973); David Freeman Hawke, *Paine* (New York, 1974); and the earlier recollections of Thomas "Clio" Rickman on which all accounts are based.

memorize, and recite biblical passages at home, he began his institutional education in 1796 at a dame school in nearby Fish Street, not far from his father's shop. A year later he moved across the river Ouse to John Button's Academy, or English Grammar School, opposite Cliffe Church, attending as a day scholar through 1802. The Cliffe section of Lewes, a crowded, insalubrious mercantile suburb, immediately adjoined Cliffe Hill, an Upper Chalk exposure of the South Downs. Button himself, a prominent Baptist Dissenter and openly radical Whig, was remembered years later as "a gentleman whose political sentiments were so accordant with those of Mr. Mantell the father that he was known to be on the Government black list."

As part of the sound, contemporary education he offered (based on English language and literature rather than classical), Button emphasized grammar, rhetoric, composition, penmanship, and daily oral recitations, which his better students were then required to perform in public. Here, Gideon's earlier training at home served him particularly well. Thus, on 22 June 1797, in the Assembly Room of the Star Inn, seven-year-old Master Mantell presented "The Parting of Hector and Andromache" from some version of Homer's *Iliad*. Similarly, on 20 December 1802, the young gentlemen of Mr. Button's Academy once again entertained "a full and respectable audience" of their friends with an evening of elocution at the Star Assembly Room. Master Mantell spoke that evening of his vile treatment on an imaginary "Trip to Paris" that included a particularly repulsive dinner, probably of snails. These schoolboy exercises, Gideon's earliest successes as a public speaker, were reported and commended in the Lee family's local newspaper.[6]

By 1801 Gideon had likewise become an author. His first book, "Sketches by G. A. Mantell, aged eleven, for his sister Jemima," in a limited edition of two copies, featured eight colored drawings by himself of Lewes Castle, Lewes Priory, and various other local sites, with accompanying calligraphy on facing pages. Again in 1801 he wrote in rhyme to his paternal uncle, the Reverend George Mantell, a Congregationalist minister in Westbury, Wiltshire, describing a characteristically tumultous general election at Lewes and its results. "To His Uncle," a verse epistle of fifty-six lines, then went on to describe Henry Shelley's victory fete, which included football, racing, bobbing for apples, and dancing.[7]

6. Benjamin Silliman remembered being told of Gideon's early Bible training (GM obituary, [*AJSA*, ns 15 (May 1853), 150]). The rest of this paragraph combines GM on himself in "Memoirs of the Mantells" and *Ramble* with *Gentleman's Magazine*'s comment on Button (GM obituary, 1852 [vol. 38, p. 645]) and information from Dr. Brent. *The Sussex Weekly Advertiser; or Lewes Journal*, 19 June 1797, p. 3 (adv.); 20 Dec 1802, p. 3.

7. Gideon's copies of "Sketches," "To His Uncle," and other childhood productions are preserved at ATL. A second copy of "Sketches" is in private hands.

By 1803 Gideon had gone to live with his uncle George, at Westbury. Remembered for his tremendous proportions (very tall, with gigantic shoulders and amazing strength), George Mantell was then pastor of an independent congregation at the Upper Meeting House. Here Gideon continued to be taught, partly by his uncle but often by himself. Not long afterward, both moved to Swindon, Wiltshire, then a market town of little more than a thousand persons. On 26 January 1804 the Newport Street Independent Chapel was founded in Swindon by the nonconformist Strange family, with Uncle George in charge. In association with the chapel, he also ran a Dissenting academy for boys, at which Gideon remained until 15 January 1805.[8]

Apprenticeship

On 3 February 1805 Gideon turned fifteen, the traditional age for a young man to begin a trade. Next day, therefore, he moved a short distance within Lewes from his family home in St. Mary's Lane to that of James Moore in High Street. A robust, ruddy-faced surgeon and general practitioner about forty years old, Moore had studied medicine under the illustrious John Hunter in London. Like his mentor, Moore was fond of natural history; besides collecting a few fossils and minerals, he enjoyed some local reputation as a botanist. In Lewes his medical practice became one of the more prominent, for in addition to his usual roles as surgeon, apothecary, and midwife, Moore also attended the poor of three nearby parishes (at twenty pounds per annum for each parish, with surgery and midwifery extra) and was ordnance surgeon to a troop of artillery stationed in barracks at nearby Ringmer. His own apprenticeship and all his professional career had been spent in Lewes, where he was a much respected figure.

Gideon was to sleep in a minimal room behind Moore's surgical office and to answer the night bell whenever it sounded. On his very first night, a messenger came through the snow and bitter cold to inform Moore he was needed as a midwife. Gideon dutifully answered the bell, roused his master, dressed, lit the outdoor lantern, and accompanied Moore to the stable, which was some distance from the house. The surgeon then saddled his own horse and rode off to attend his case. Less than an hour later

---

8. For these years, see J. B. Delair and DRD, "Gideon Mantell in Wiltshire," *Wiltshire Archaeological and Natural History Magazine,* 79 (1985), 219–224, which includes a previously unpublished map of Swindon by Gideon in 1804 (ATL). The map not only locates a Swindon quarry but even includes the homes of the evidently friendly quarrymen. In recollection, Gideon often regarded 1804 as the happiest year of his life, in part because of his companionable relationship with an otherwise unspecified Miss Strange.

Figure 1.1. Medical composite. *Top center:* Gideon Mantell's certificate of attendance at anatomical lectures (1811). *Top right:* Sketches of the skull, facing page
21, in his "Anatomy of the Bones, and the Circulation of the Blood" (1809). *Bottom right:* Tablet 2, showing head, neck, and spine, in his "Description of the
Muscles of the Human Body" (1810). *Bottom center right:* Tablet 6, showing the
hand, in his "Tablets of the Arteries" (1810). *Bottom center left:* Pages from his
"Notes on the Nervous System," a never-completed book. *Bottom left:* Articles of
agreement between Mantell and William Pearson (1838); and apprenticeship indenture, Mantell to James Moore (1805). *Top left:* Case notes on inflammatory
typhus, pp. 21–22 in Mantell's "Medical and Surgical Cases" (1814–1852). *Center
left:* Mantell's copy of *The Lancet,* 1 no. 15 (1839–1840). *Center right:* Mantell's
certificate from St. Bartholomew's Hospital, London (1811). All items are from
the collections of the Alexander Turnbull Library, Wellington, New Zealand.

the night bell roused Gideon for a second time, this summons being to a
village four miles away in the opposite direction. Not knowing what else
to do, he awakened Moore's wife, Ruth, who told him to dress and seek
out another surgeon in town. By this time the snow had turned to chill-
ing rain. It took Gideon two tries to find a surgeon who would agree to
go; afterward, he returned once more to bed, wet through and half perished from the cold. Moore himself returned just before dawn, but in consideration for his young apprentice stabled his own tired, sweating horse.

It was a night in the life of a young man that Charles Dickens might have immortalized; Mantell himself wrote of it later on.[9]

Whatever initial misgivings Gideon may have derived from such experiences, he soon consented to the arrangement with Moore. On 2 March, therefore, his indenture of apprenticeship was solemnly effected as follows:

> Gideon Mantell, son of Thomas Mantell of Lewes in the County of Sussex [shoemaker, but that word, or its equivalent, is blotted out in the surviving document], by and with the advice and consent of his Father . . . doth put himself Apprentice to James Moore of Lewes aforesaid Surgeon and Apothecary to learn his Art and with him after the manner of an Apprentice to serve from the fourth day of February last past during and until the full and term of Five Years.

In further traditional phrases Gideon promised to keep his master's secrets and not to waste his goods, nor commit fornication, nor contract matrimony. He was not to gamble away his master's goods at cards or dice, nor haunt taverns or playhouses, nor absent himself day or night unlawfully. Thomas Mantell agreed to pay Moore fifty pounds immediately and a further fifty pounds on 24 June 1806; he would also provide all of Gideon's clothing and laundry, as well as dinners for the first four years. Moore, in turn, undertook to teach Gideon his trade while furnishing him food, lodging, and medical care. These terms being mutually acceptable, the indenture was signed in turn by Gideon, his father, and Moore, who also gave Thomas a receipt for the fifty pounds he had just received.[10]

Thomas Mantell's signature at the bottom of this document was noticeably infirm, for though only fifty-five he had little more than two years to live. Gideon's father died on 11 July 1807, aged fifty-seven. In accordance with his will, made 7 October 1804 and proved on 15 December 1807, an auction of his real estate was held at the White Hart Inn on 24 December. Thomas Mantell's holdings included three houses, an orchard, and one vacant lot adjacent to the church of St. John's sub Castro; a large property in St. John's Street (then occupied by a paper mill company), including a choice house, stable, and adjoining garden; seven further properties in St. John's Street, all tenanted, together with a stable and two gardens; a building and a stable on Fish Street; another on the east side of the same street, occupied by his son Thomas; and three rental properties on the east side of St. Mary's Lane. Though the last were leasehold, Thomas had owned all the others. Spurred by his deep religious beliefs to practice

---

9. Anon., but GM, "Memoirs of the Life of a Country Surgeon" (London, 1848); GM's copy, ATL. Moore lived at No. 63 High Street, next door to William Lee and the *Sussex Weekly Advertiser.*
10. Indenture and receipt, 2 Mar 1805 (ATL).

both industry and thrift, as a businessman Gideon's father must have been not only formidable but extremely shrewd.[11]

Gideon's first duties as an apprentice under Moore included sweeping out the surgery, cleaning phials, and powdering and sifting drugs. He soon learned how to make pills and other pharmaceuticals. Within a few months Gideon was not only collecting empty phials door to door but compounding and delivering nearly all of Moore's medicines (his indenture, however, specifically excluded mercurial pills and a secret marine vegetable dentifrice). After his first year Gideon also kept all of Moore's accounts, wrote out his bills in a fine hand, bled noncritical patients, and extracted teeth. As Gideon's knowledge and competence increased, he was allowed to treat fractures and other routine cases. By the end of his fourth year with Moore, he had become an extremely valuable assistant.

Moore took a second apprentice at this time, which relieved Gideon of his more menial duties and even allowed him a vacation. Accordingly, on 31 July 1809, he set out with his close friend John Tilney and Tilney's dog, Coz, on a five-day walking tour to London that included its share of adventures – among them, a near brush with a press-gang. Besides other places, they visited the Isle of Sheppey, where Gideon avidly collected a number of fossils.[12]

As his apprentice days wound to a close, Gideon anticipated the formal medical education that would follow, using late-night and early-morning hours for study and composition. He began providently to teach himself a more systematic knowledge of human anatomy. The literary residue of this was a substantial holograph volume entitled "The Anatomy of the Bones, and the Circulation of the Blood," dated by him from Lewes in 1809 but actually begun in November 1808. It consisted of two series of drawings, captioned on facing pages, together with an elegant title page on which there was an urn inscribed "Gideon Mantell / De Ossibus Humani Corporis / Anatomia." Section one, "Of the Bones," included drawings of fetal skeletons one month, six weeks, three months, and four months old; front and back views of the adult skeleton; comparison of the adult and fetal skulls (with a paragraph on each of the major bones); and detailed sketches of all the body's bones, including parts of the skull, limbs, trunk, and spine – sixteen drawings in all. Section two, "Of the Circulation of the Blood," included discussion of the heart and veins, as well as three more drawings, "A General View of the Arterial and Venous System,"

11. Thomas Mantell's will is preserved on film at ESRO. His estate auction, announced in *SWA,* 5 Dec 1808, p. 3, revealed the extent of his holdings. He had apparently employed as many as twenty-three men (*Proc RS,* 6 [1853], 252).

12. GM, "Memoirs of the Life of a Country Surgeon" (note 9 above); GM, "Notes on a Pedestrian Excursion, Aug 1809" (ATL).

"Blood vessels, bones, and nerves of the arm" (both November 1808), and "The situation of the principal blood vessels from Eustachius" (February 1809). The whole project, 120 pages long, was completed 20 February 1809. As before, the drawings were exceptionally good, the captions lucid, and the handwriting elegant.[13]

"Although my apprenticeship was a most laborious servitude," Mantell was later to write,

> and comprehended many things which I ought not to have been required to perform, yet when I presented myself as a pupil to Mr. Abernethy, my knowledge of anatomy was equal to that possessed by most of the hospital's students of my own age, for the habit of copying drawings and making extracts of standard works had familiarized me with the elements and nomenclature of the science and given me a good theoretical acquaintance with its details, so that the lectures in osteology were readily comprehended, and by the frequent dissection of animals during my apprenticeship I had acquired a dexterity in the use of the scalpel which greatly facilitated the practice of human anatomy. In consequence I soon gained the especial notice, and ultimately the friendship, of my excellent and eminent teacher.[14]

John Abernethy (1764–1831) was then assistant surgeon at St. Bartholomew's Hospital in London and had been since 1787; like Gideon's master, James Moore, he once attended John Hunter's lectures. In 1788 Abernethy himself began lecturing on anatomy. Only three years later, special facilities to accommodate his teaching were built for him at St. Bart's. In 1793 he began to publish a collection of surgical and physiological essays and, after two papers in the *Philosophical Transactions,* was elected a fellow of the Royal Society in 1796. Abernethy moved to No. 14 Bedford Row in 1799 (staying there the rest of his life), married the next year, and had already fathered six children of an eventual nine when Gideon began to attend his lectures on 1 October 1810. By then Abernethy had also published his *Surgical Observations on the Constitutional Origin and Treatment of Local Diseases, and on Aneurism* (London, 1809), a book fundamental to Gideon's medical understanding.

Though founded in 1123, St. Bart's belonged to the eighteenth century. The Abernethy block, added in 1791, included a small operating theater. Whenever Abernethy lectured within it, this theater became so crowded that each ascending circle of students had its knees in the backs of those below them and were kneed in turn by those above. Precisely at 7 P.M.

---

13. GM, "The Anatomy of the Bones" (ATL, which also preserves two other similar productions by GM done at St. Bart's).
14. "Memoirs of the Life of a Country Surgeon," (note 9 above), p. 14.

Abernethy would appear, dressed in a black top coat and tippit cape, with black silk stockings and old-fashioned powdered hair. As one of his students remembered,

> His mode of entering the lecture-room was often irresistibly droll – his hands buried deep in his breeches-pockets, his body bent slouchingly forward, blowing or whistling, his eyes twinkling beneath their arches, and his lower jaw thrown considerably beneath the upper. Then he would cast himself into a chair, swing one of his legs over an arm of it, and commence his lecture in the most *outré* manner. The abruptness, however, never failed to command silence, and rivet attention.

Often, at the beginning of a session, he would enter the lecture room, look round at the crowd of new students assembled there, and proclaim solemnly: "Good God! what will become of you all?" His first lecture would begin by disparaging the then traditional distinction between the physician (internal diseases) and surgeon (external diseases, injuries, operations) because, he asserted, local disease can cause constitutional irritation and vice versa. In support of such broad principles, numerous cases were cited as examples. Abernethy expected his students to learn anatomy on their own, but shared anecdotal experience freely. As Sir Benjamin Brodie, himself a famous surgeon, remembered,

> He kept our attention so that it never flagged, and what he told us could never be forgotten. He did not tell us so much as some other lecturers, but what he did he told us well. His lectures were full of original thought, of luminous and almost poetical illustrations. . . . Like most of his pupils, I was led to look upon him as a being of a superior order, and I could conceive nothing better than to follow in his footsteps.[15]

On 18 April 1811, John Abernethy certified in writing that Mr. Gideon Mantell "hath diligently attended the anatomical lectures, those on the theory and practice of surgery, and the dissections, from the first of October to the present time." Accordingly, that day or the next, Mantell was awarded a certificate of attendance covering a period of six months. On the nineteenth he passed the necessary oral examinations and immediately received his diploma. Gideon was then enrolled by Edmond Balfour, sec-

15. My Abernethy material, including quotations, is borrowed from John L. Thornton, *John Abernethy* (London, 1953), pp. 73 and 78; with additions from Sir D'Arcy Power and H. J. Waring, *A Short History of St. Bartholomew's Hospital* (London, 1923), Abernethy's *Surgical Observations,* and hospitality at St. Bart's. See also G. MacIlwain, *Memoirs of John Abernethy, F.R.S.* (2 vols., London, 1853); W. F. Bynum, *Science and the Practice of Medicine in the Nineteenth Century* (Cambridge, 1994); and Owen H. Wangensteen and Sarah D. Wangensteen, *The Rise of Surgery* (Minneapolis, 1978).

Figure 1.2. George Cruikshank (1792–1878), "The Examination of a Young Surgeon" (London: M. Jones, 1 October 1811; courtesy National Library of Medicine; Bethesda, Maryland). A young surgeon, standing left, undergoes questioning ("Describe the organs of hearing") by members of the Court of Examiners of the Royal College of Surgeons (established 1800), who are seated round a semicircular table; skulls and bones decorate the Master's chair, behind which hang two skeletons. For an exhaustive description, see Mary Dorothy George, *Catalogue of Political and Personal Satires Preserved in the Department of Prints and Drawings in the British Museum, IX (1811–1819)*, (London: British Museum, 1949), pp. 45–46. Gideon Mantell underwent just such an examination in 1811.

retary of the Royal College, and had officially become a surgeon. Four days later, on the twenty-third, he also received a certificate of attendance from the Lying-in Charity for Married Women at Their Own Habitations (instituted 1757) and was officially a midwife as well.[16]

Having thereby attained his educational goals, Mantell hoped to obtain an assistantship under some London practitioner. As no such offer was forthcoming, however, he returned crestfallen to Lewes and entered into *partnership* with James Moore, his former master. During the ten years they were together, annual receipts trebled from £250 to £750. Gideon himself attended between two and three hundred cases of midwifery every year and was often up for six or seven nights in a row, his only sleep being an occasional hour's nap in his clothes. He also attended the brutal military floggings at Ringmer until they stopped.[17]

16. Both documents at ATL.
17. "Memoirs of the Life of a Country Surgeon," (note 9 above), esp. p. 16; GM, untitled letter on flogging, *The Times*, 8 Dec 1835, p. 7 [J, 6 Dec 1835]; and GM's medical records (ATL and ESRO).

By 1814, apparently, Moore and Mantell were renting (from Thomas Read Kemp, a real estate speculator) No. 3 Castle Place, one of four newly built townhouses adjacent to Castle Bank on High Street. On 8 June 1815 the architect Amon Wilds paid £830 for all four and then sold No. 3 to Moore for £780 on 31 August. By May 1816 Mantell was renting the same house from Moore, whose health had begun to fail. On 25 March 1818, therefore, he and Gideon dissolved their medical partnership. In exchange for unilateral access to their now lucrative practice, Gideon had only to pay Moore £81 per annum for seven years and £40 annual rent. That June he hired an assistant of his own. With every prospect of a comfortable future ahead of him, Gideon then bought No. 3 outright from Moore in November, paying £725. Prosperous and respectable, he bought No. 2, the other middle house, from Wilds for £600 the following May. Under Gideon's direction, Wilds then joined the two together behind a common facade, thereby creating the double-sized "Castle Place" from which all of Gideon's earlier publications were inscribed.[18]

18. Deeds at ESRO; GM's 25 Mar 1818 agreement with Moore (ATL); J (25 Mar 1818). As Mantell subsequently explained to Silliman, "My house is on the high street, but a little garden at the back joins the bank on which the castle stands" (GM–BS, 20 July 1832 [Yale]). See also Chapter 2 below.

2

Oryctology

Throughout his adult years, after he had become famous, Gideon Mantell accounted for his paleontological interests in various ways, but most of the stories told by and about him can be reconciled into a consistent development. In 1798, while a pupil at John Button's Academy in Lewes, eight-year-old Gideon read some of the currently fashionable "improving" literature for children, much of which had Evangelical overtones. Fifty-three years later he specifically recalled having been decisively influenced by a story in one of these books, "Eyes and No Eyes; or, The Art of Seeing" by Anna Letitia Barbauld and her naturalist brother, John Aikin. In this simple but effective tale, two boys, Robert and William, have just returned from a walk across Broom Heath. Totally oblivious to natural history, Robert thought the walk exceedingly dull. William, on the other hand, curious about plants and much else, easily found objects to please him, including a probable Roman camp. Thus, some persons are alert to the world about them whereas others are not, and Robert was admonished to follow William's example. Gideon freely resolved to do the same.[1]

Another anecdote (together with further evidence) suggests that Mantell began to collect fossils around Lewes when he was twelve or thirteen. "While yet a mere youth," the story goes,

> he was walking one summer evening with a friend on the banks of a stream communicating with the Ouse, when his observant eye rested upon an object that had rolled down from a marly bank which at that particular spot overhangs the stream. He dragged it from the water and examined it with great attention. "What is it?" was the natural inquiry of his friend. "I think, Warren," he replied, "that it is what

---

1. First published during the 1780s, "Eyes and No Eyes" proved remarkably enduring. I have it from Aikin and Barbauld, *Evenings at Home* (London, 1860), pp. 278–284. Gideon acknowledged his indebtedness to the story in a lecture of 1851 (*BH*, 16 Aug, p. 3; *BGrd*, 20 Aug 1851, p. 7) while reminiscing in Brighton at the opening of the short-lived Mantellian Academy of Science. See also *Wight*, p. 21.

they call a fossil." The "curiosity," which proved to be a fine specimen of the ammonite, was borne home in triumph by the two friends;
and from that moment young Mantell became a geologist.

If Gideon himself related some version of this incident to Mark Anthony
Lower, the Sussex historian, he must have begun collecting fossils at Lewes
by 1802 or 1803. Warren was Stewart Warren Lee, son of the local newspaper editor.[2]

Fossils certainly interested Mantell in 1804 at Swindon. As he wrote nine
years afterward to a correspondent,

> I passed some of my earlier days at an academy there, and well re-
> member the deep impression made on my imagination by the won-
> derful stories of petrified snakes and crocodiles told me by the quarry-
> men. Although I was then too young to understand the nature of
> fossils, curiosity tempted me to pick up a few specimens, and they are
> now in my cabinet.

Those he selected must have been ammonites, the "petrified serpents"
young Gideon saw being broken up to resurface local roads. On asking
his Uncle George how these shells that look like coiled snakes came to be
in the rocks of Swindon he was told, reasonably enough for 1804, that
they had been left by the biblical Deluge ("at that time the resting-point
of ignorance," as Mantell recalled almost half a century later). His curiosity piqued, Gideon next consulted the *Encyclopaedia Britannica* (1797),
which accurately reflected a minimal understanding of fossilization and
offered nothing more than an elementary classification of specimens. Ammonites, it affirmed, were "petrified serpents," and other figured bodies,
only *"lusus natura"* – in other words, natural deceptions and not fossils
at all (247). Of course, very few persons in 1797 realized how much past
life had differed from present, and no one knew anything about dinosaurs,
which were yet to be identified.[3]

Whatever the precise facts may be, young Gideon Mantell was certainly
collecting fossils by 1803 and further developed his interest in them at
Swindon the next year. On returning to Lewes, he was encouraged by several adult friends, including Thomas Woollgar, who gave him some specimens and had the Lewes Library Society (which Woollgar founded) purchase significant books about fossils; William Constable then impressed

2. Mark Anthony Lower, *The Worthies of Sussex* (Lewes, 1865), p. 158. GM later claimed
   that one of the specimens he gave Sowerby had been collected personally "in 1802 or 1803"
   (*Mineral Conchology*, II, 64; see below).
3. GM–EB, 3 Dec 1813 (ATL); *Medals*, II, 927. *Pict Atlas*, pp. 14–16 (from Parkinson). *BH*,
   16 Aug 1851, p. 3 (Deluge quote). *Encyclopaedia Britannica*, third edition (1797), XIV,
   245–249; J (22 June 1832).

Gideon with a mammoth's tooth brought from America. Throughout his busy apprentice years with Moore (who collected a few as well), Mantell continued to acquire fossils and was actively in the field by 1809 when, on his walking tour with John Tilney, he picked up numerous specimens on the Isle of Sheppey. Gideon also took a small collection of fossils with him to St. Bart's in the fall of 1810 and while there bought further specimens from Joseph Stutchbury, a London dealer.[4]

On completing his medical education in the spring of 1811, Mantell enlarged his geological interests. The decisive experience that transformed him from a casual hobbyist to a serious original investigator of fossil life was his meeting later the same year at Hoxton Square with James Parkinson (1755–1824), a surgeon of radical political sympathies closely associated with St. Bart's. Now fifty-six and the father of six children, Parkinson had devoted much of the earlier 1790s to pamphleteering on behalf of parliamentary reform and universal suffrage. From John Hunter and other influences, Parkinson gained an enthusiasm for fossils, which (never a field collector) he began to acquire through purchase around 1798. By 1801 he was soliciting contributions of fossils toward his projected book, a monumental work eventually completed in three folio volumes. According to Gideon's subsequent recollection, the last volume had just appeared when Parkinson agreed to see him. A remarkably ambitious attempt to record and depict all known British fossils (which at the time included no successfully identified large vertebrates older than the mammoth), *Organic Remains of a Former World* (1804, 1808, 1811) became a primary reference for Gideon at the Lewes Library Society in 1812 and would influence him all his life.

Unfortunately, Parkinson's beautifully illustrated volumes of British fossils were too expensive for a just-graduated student like Mantell to buy. Probably in November 1811, however, their author made himself and his famous collection available. Visiting Parkinson at home, Gideon then found him "rather below the middle stature, with an energetic, intelligent, and pleasing expression of countenance, and of mild and courteous manners; readily imparting information, either on his favorite science or on professional subjects." Still actively in practice, Parkinson had also written several important medical essays and would eventually publish one in 1817 on the disease that bears his name. "He kindly showed and explained

---

4. Thomas Woollgar (1761–1821), linen draper of Lewes, retired from business around 1800, devoting himself to natural history, local history, and the Lewes Library Society. Mantell remembered him fondly in *Foss SD* (1822), pp. 115, 167, 191, 197–198; and *Ramble* (1846), p. 120. Constable was attested more vaguely (*BH,* 16 Aug 1851, p. 3). Gideon's postmortem ("Case of a Distorted Spine," 1854) recalled his fossils at St. Bart's; see also Joseph Stutchbury-GM, May 1812; and GM–WBDM, 25 Sept 1851 (both ATL).

to me the principal objects in his cabinets," Mantell remembered, "and pointed out every source of information on fossil remains." Throughout the next year, Gideon spent whatever time he could reading books that Parkinson had recommended to him and began in earnest to form a comprehensive collection of Sussex fossils.[5]

In 1812 the young surgeon compiled another private book, his first geological one. Called "A Systematical Arrangement of Secondary Fossils" by G. A. Mantell, M.R.C.S., it included a five-part abridgment of Parkinson's *Organic Remains,* extracts from the work of Georges Cuvier and William Martin, and a descriptive catalog of his own rapidly augmenting collection. The abridged remarks corresponded with five systematically arranged specimen cases, lettered A through E respectively. Additional parts related fossils with the strata in which they were contained, according to the stratigraphical discoveries of William Smith, and then recapitulated Gideon's collection, which included (with additions) 38 fossil plants, 75 zoophytes, 44 echinoderms, 203 shells, 46 fishes, and 7 arthropods (including crabs and trilobites, both of which he classified as insects). As various entries demonstrate, Mantell was digging regularly at some favorite sites, notably Hamsey, two miles north of Lewes; and Castle Hill, eight miles south, on the coast at Newhaven. Yet in four important taxonomic categories – reptiles, amphibians, birds, and mammals – he had virtually no specimens to list.[6]

---

5. *Pict Atlas,* pp. 13–14. For Parkinson, see *DSB* (Patsy A. Gerstner); R. J. Cleevely and J. Cooper, "James Parkinson (1755–1824): A Significant English 18th Century Doctor and Fossil Collector," *Tertiary Research,* 8 (4) (July 1987), 133–145; and John C. Thackray, "James Parkinson's *Organic Remains of a Former World* (1804–1811)," *Journal of the Society for the Bibliography of Natural History,* 4 (1976), 451–466. According to Thackray, the third volume appeared at the end of October, with a full set costing eight guineas and sixpence. An objective account of the author's own symptoms, *Essay on the Shaking Palsy* (London, 1817) first described what we call Parkinson's disease.

6. Individual entries in GM's "Systematical Arrangement" (ATL) reveal a number of significant personal relationships, as specimens were given him by various well-wishers. Under "Vegetable Fossils," for example, Mantell listed bituminous wood from Norwich, presented by his friend George Chassereau; fossil wood from Jamaica and other specimens (George Edward Woodhouse, of Oxford Street, London); and siliceous wood from Swindon (Thomas Strange). George Edward Woodhouse, Jr., of Maida Hill, Paddington, contributed a number of fossil zoophytes, mostly corals from Worcestershire. Other specimens from Worcestershire and Maida Hill were gifts from Mary Ann Woodhouse, Gideon's future wife, who was then seventeen. (Since he spelled her name as "Marianne," Gideon had probably not known her long.) Mantell received still further additions from John Tilney, now in London; James Woodhouse, the brother of George, Jr., and Mary Ann; James Moore, his medical partner; and Arthur Lee of the *Sussex Weekly Advertiser.* Nor did Gideon omit a large number of specimens collected by himself, including those from the Isle of Sheppey in 1809.

The final and conceptually most significant portion of Mantell's catalog, "Fossils considered in connection with some strata in which they are contained," derived almost verbatim from Parkinson. "According to the actual observations of Mr. [William] Smith," it began, "the following are the upper strata which have been discovered in this island, disposed in the order in which they occur" (442). But this colloquial enumeration of nineteen poorly defined layers by no means intimated how bafflingly complex more specialized attempts at stratigraphical correlation would soon become. William Smith (1769–1839), in some respects the father of English geology (as he was later called), became a surveyor of canal routes and other public works in western England. By 1799 he had grasped and applied some principles fundamental to modern geology: that rock materials deposited as sediments form strata; that these strata form recognizable and predictable sequences over wide areas; and that individual strata can be identified and correlated on the basis of unique fossils found within them. Smith's discoveries (together with similar fieldwork by Cuvier and Brongniart in France) effectively began the science of stratigraphy, heightened British awareness of fossils momentously, and made additional knowledge of their stratigraphic occurrence extremely desirable. Smith was slow to publish, however, and the results of his work were not generally known until after the restoration of peace in 1815. It is therefore remarkable that Gideon, happening on Parkinson's preliminary account of Smith's work in 1812, should have recognized its significance so fully as he did.[7]

Reaching Out

The next stage in Mantell's geological career, as he saw, was to become part of the greater London scientific community through personal contacts, publications, and memberships in the various learned societies. The

---

7. Parkinson, *Organic Remains*, III (1811), Chap. 32; pp. 440–455 esp. Parkinson also noticed Smith's ideas in *Trans GS* that year (1:324–354), but may have learned of him only recently (Thackray, p. 45). He specifically notes another publication of 1811 (by John Farey) as his source. See also Joan M. Eyles, "William Smith: Some Aspects of His Life and Work," pp. 142–158 in Cecil J. Schneer, ed., *Toward a History of Geology* (Cambridge, Mass. and London, 1969); and *DSB*. Two brief, preliminary versions of Georges Cuvier and Alexandre Brongniart, "Essai sur la geographie mineralogique des environs de Paris" appeared in 1808, followed by a useful translation (*Philosophical Magazine*, 35 [1810], 36–58). A much fuller version was then published in 1811, both as a memoir of the Imperial French Institute and as a separate publication (Paris: Baudouin, 1811). It would reappear once more in 1812 as part of Cuvier's four-volume *Recherches*. See "Georges Cuvier" (Franck Bourdier) and "Alexandre Brongniart" (M. J. S. Rudwick) in *DSB*. I discuss Cuvier and Brongniart more fully in Chapter 4.

earliest acquaintances important to these ambitions were William and Arthur Lee, of the *Sussex Weekly Advertiser,* who published Gideon's first geological essay, "On the Extraneous Fossils found in the Neighbourhood of Lewes" in March and April 1812 and his second, "Of the strata in the vicinity of Lewes" in March 1813. These were his earliest formal attempts to explain the fossils and geology of Sussex.

At the beginning of his essay "On the Extraneous Fossils" Mantell compared, as he often would, paleontological remains with archeological ones. "To a mind alive to contemplation," he wrote, "the mouldering relics of former times afford an inexhaustible source of amusement and instruction." "If," Gideon continued, "the puny efforts of man can excite such emotions, with what reverence and astonishment must the works of the Creator impress our minds!" And yet how few realize that "the very hills which surround them may be regarded as vast tumuli in which 'organic remains of a former world' are entombed." While understanding of fossils languished till recently, he explained, Parkinson, Cuvier, and Martin have since created a genuine science. As no one had hitherto described the abundant fossil remains nearby, Mantell attempted to do so. Though admitting to little knowledge, time, and opportunity, he nonetheless mentioned an impressive series of genera and locations, citing specimens from his own collection. By so doing Gideon hoped to interest others in the delightful study of oryctology (as paleontology was then called), a science "replete with stupendous proofs of the power and wisdom of the Almighty."

"Of the strata," written and published in March 1813, proved to be more sophisticated. By then, for example, Mantell had become aware of James Sowerby's attempt to classify and depict all of Britain's fossil shells in a serially produced work entitled *Mineral Conchology,* begun in 1812. Thus, "the assiduity and perseverance exhibited by the most eminent naturalists in their oryctological pursuits," he prophesied optimistically,

> render it probable that the period is not far distant when a more accurate knowledge will be obtained of the natural history of those animals and vegetables which existed in the earliest ages of our planet, and whose interesting remains, preserved in the mineral kingdom, afford the most important information respecting the nature of those astonishing changes which this globe has suffered since its original formation.

Nothing could be more useful to this progress, he believed, than close descriptions of the fossils to be found in specific areas.

Having by now read the great memoir by Cuvier and Brongniart on the environs of Paris, however, Mantell noted that many fossils mentioned there occurred also in the vicinity of Lewes. Similarly, he had traced a stratum of marl from Hamsey to Offham and elsewhere in Sussex, confirming the identity of its various outcroppings by finding the same species of ammonite within them, "for it is now an established geological fact," he declared presciently, "that some fossils are peculiar to, and are only discovered in, particular strata," as Smith and the two French geologists had observed. Gideon next listed the two major strata of Lewes (Chalk and Marl, but he was overlooking a good deal), together with fossils characteristic of each. "This slight sketch of the strata and Secondary fossils of Lewes," he concluded,

> affords many interesting and important facts. The evident displacement which some of the strata have sustained, and the complete removal of others; the discovery of innumerable remains of animals, differing in their forms and organization from any that are now known to exist; the regularity observable in the confinement of certain fossils to particular strata, and the agreement which exists between our strata and those of France are phenomena which impress the mind with astonishment.

He also highlighted remarkably preserved fossil alcyonia, delicate creatures formerly inhabiting the ocean's floor that must surely have existed where we now find their well-preserved remains and could hardly have survived the violence of a debacle. A comparison of this newspaper essay with its predecessor a year earlier demonstrates the rapidity with which Gideon's geological understanding was maturing. He was already in the vanguard of his times.[8]

On to London

William and Arthur Lee also published Mantell's first medical paper, a letter on vaccination (6 September 1813). They were thus central in turning him from a private author into a public one. In 1813, moreover, not long after Gideon's return to Lewes from St. Bart's, William Lee had decisively

8. GM, "On the Extraneous Fossils found in the Neighbourhood of Lewes," *SWA*, 23 Mar 1812, p. 4; 6 Apr 1812, p. 4. GM, untitled letter, "Of the strata in the vicinity of Lewes," *SWA*, 8 Mar 1813, p. 4; 15 Mar 1813, p. 4. Both copied into Letterbook (ATL). GM referred to Parkinson's *Organic Remains* and *Trans GS* paper; Cuvier and Brongniart, "Essai sur la geographie mineralogique des environs de Paris" (Paris, 1811), together with a series of papers on vertebrate paleontology by Cuvier; William Martin, *Petrificata Derbiensia* (Wigan, 1809) and *Outlines of an Attempt to Establish a Knowledge of Extraneous Fossils on Scientific Principles* (Macclesfield, 1809). The latter is dedicated to A. B. Lambert.

augmented Mantell's introduction to the scientific establishment by recommending him to the Reverend James Douglas (1753–1819), who wrote
William Lee on 18 March 1813 to praise the *Sussex Weekly Advertiser*
for inserting Gideon's paper on strata three days before. Douglas' present
geological outlook, strongly influenced by Cuvier's, was catastrophic: the
earth's history, for him, consisted of a series of distinct epochs separated
from one another by episodes of violent change. A petrified forest at Felp-
ham, Sussex, for example, seemingly proved that the earth had undergone
a series of "various and successive revolutions previous to the last general
catastrophe of the Mosaic deluge." His letter reaffirmed all of these significant opinions.

Lee having been the intermediary, on 10 April 1813 Douglas wrote directly to Mantell from London. Having sent A. B. Lambert, vice-president
of the Linnaean Society, a copy of Gideon's newspaper essay "Of the
strata," Douglas thought Lambert would welcome a more formal paper
on the same subject. He also believed such an offering would gain its author introductions to some of the most famous naturalists in Britain, for
oryctology and geology had together become very popular. Understandably flattered, Mantell replied the next day; professional responsibilities
did not permit him to visit London immediately, but he would call on both
Douglas and Lambert when he could. Douglas then agreed to provide a
letter of introduction to Lambert and the Linnaean Society. He was
pleased to learn a week later that James Sowerby had just published one
of Gideon's specimens.[9]

Sowerby (1757–1822), who had already compiled Baconian catalogs of
British botanical specimens, began in 1812 to issue ad seriatim *The Mineral Conchology of Great Britain,* "or colored figures and descriptions of
testaceous animals or shells which have been preserved at various times
and depths in the earth" (7 vols., London, 1812–1845; incomplete). More
specialized than Parkinson's *Organic Remains,* this was the first modern
effort to name and illustrate fossil shells from throughout Britain. Without attempting any form of systematic arrangement, however, Sowerby
described specimens as they came to him. Though defective in that respect,
Mineral Conchology greatly stimulated the collection of fossil shells, established many new species, standardized nomenclature, and led to more
sophisticated studies. During the earlier years of his project, Sowerby was

9. GM, untitled letter on vaccination, *SWA,* 6 Sept 1813, p. 4. For JD, author of *A Disser-
tation on the Antiquity of the Earth* (London, 1785) and *Nemia Britannia; or, a Sepulchral
History of Great Britain* (London, 1793; barrow openings), see *DNB* and R. Jessup, *Man
of Many Talents . . . James Douglas, 1753–1819* (London, 1975). JD–Wm. Lee, *SWA,*
22 Mar 1813; also in GM's Letterbook. JD–GM, 10 Apr 1813; GM–JD, 11 Apr 1813;
JD–GM, 17 Apr 1813; GM–JD, 23 Apr 1813 (all ATL). Sowerby appeared on 1 April.

Figure 2.1. Geological composite. *Top row, left to right:* Gideon Mantell's copy of his "Outlines of the Natural History of the Environs of Lewes" (Lewes: John Baxter, Sussex Press, 1824), open at the title page to show the lithographed frontispiece "View of the Western Keep of Lewes Castle" drawn by F. Pollard; Mantell's "Outlines of the Mineral Geography of the Environs of Lewes" (1817), open at the title page to show the author's sketch; James Sowerby, *The Mineral Conchology of Great Britain* (London: Benjamin Meredith, 1812ff.), open to show Plate 55, *Ammonites Mantelli* (collection of the New Zealand Institute of Geological & Nuclear Sciences, Wellington). *Center row, left to right:* Mantell's "Systematical Arrangement of Secondary Fossils," 1812, showing the title page facing Mantell's watercolor sketch of "A fossil crab from Malta" (qMS–1303). Mantell's copy of James Parkinson's *Organic Remains of a Former World* (London: Sherwood, Neely & Jones, 1811), vol. 1, open to show George Edward Woodhouse the elder's presentation letter of 26 November 1812 to Mantell; with (below) Mantell's "Geological Extracts and Observations" (1813–1845), open at article one (MS–1503). Mantell's "Sketches of the Most Remarkable Fossils in Knorr's *Monumens des Catastrophes*" (1817), open at the title page (qMS–1302). *Bottom row, left to right:* Watercolor sketch of *Turrilites* by Mantell in GM–EB, 12 September 1814 (MS–Papers–0083–010A). Mantell's copy of *The Gleaner's Portfolio, or Provincial Magazine,* vol. 1, no. 1, August 1818 (collected edition, Lewes: Sussex Press, 1819), open at p. 8 to show his article, "A Sketch of the Geological Structure of the South Eastern part of Sussex." Watercolor sketch of *Alcyonia* (later *Ventriculites*) in GM–EB, 16 March 1814 (MS–Papers–0083–010A). Except as noted, all items are from collections of the Alexander Turnbull Library, Wellington, New Zealand.

considered the foremost British authority on fossil conchology. Many of the specimens he described were loaned or donated to him by energetic locals, including Etheldred Benett, Elizabeth Cobbold, and Gideon Mantell.

Sowerby and Mantell were actively exchanging letters by January 1813. On 6 February, the latter sent his finest specimens of *Scaphites, Turrilites,* and *Ammonites* from the marl pit at Hamsey, just north of Lewes. These being the rarest and best specimens available to him, he hoped they would be acceptable. Sowerby replied gratefully that the turrilites were larger than any he had earlier possessed and the scaphites "very perfect." Thus encouraged, Gideon corresponded frequently with Sowerby for the next several years, but almost exclusively about the fine specimens he was providing, many of which appeared eventually in *Mineral Conchology*. One that Sowerby called *Ammonites Mantelli* has ever since commemorated "the indefatigable G. A. Mantell, Esq.," who found a number of examples at Ringmer, near Lewes.[10]

In London

Through his friend John Tilney, who had married one of their girls, Gideon now came to know various members of the Woodhouse family in London. James Woodhouse, the grandfather, had been a poetical shoemaker, the author of several published works. George Edward Woodhouse, the father, established himself as a successful linen draper at 151 Oxford Street. Together with his wife, Mary Ann, he had four children. Of these, George Edward Woodhouse the younger became an attorney, at 11 King's Bench Walk. James worked with his father's drapery firm until December 1822, embarked for New Holland, then returned in 1828 to become a linen draper once again, at 124 Edgeware Road. Mary Ann, born 9 July 1795, and her younger sister, Hannah, would both be associated with Gideon Mantell.

George Edward Woodhouse the elder was also the first of Mantell's patients whom we know by name, and the young surgeon must have come

---

10. For Sowerby, see esp. R. J. Cleevely, "The Sowerbys, the *Mineral Conchology*, and Their Fossil Collection," *JSBNH,* 6 (1974), 418–481; and *DSB*. GM–JS, 6 Feb 1813; JS–GM, n.d.; GM–JS, 3 Mar 1813; and subsequent letters, many of which mention John Tilney, who served as go-between (ATL, originals and Letterbook). GM–JS, 9 Jan 1813, and a dozen other privately owned letters were graciously made available to me by the late Joan M. Eyles; still others are at BMNH. JS, *Mineral Conchology*, I (begun 1812; pub. 1815), 51, 81, 119, 169, 171–172, etc. In all, according to Cleevely, GM contributed seventeen specimens to Volume 1, seven to 2, four to 3, three to 4, four to 5, and thirty-four to 6. *Ammonites Mantelli* (I, 119) is now the genus *Mantelliceras*. Sowerby's collection of fossil shells was acquired by the BM in 1860.

all the way from Lewes by coach to see him on repeated visits throughout the autumn of 1812. The only surviving evidence of their professional
relationship, however, is this letter:

> Maida Hill, Paddington
> 26 November 1812

> My dear Friend,
>   I beg your acceptance of this Work as a small token of my gratitude
> and esteem, and as a trifling acknowledgment of the peculiar profes-
> sional exertions and kind attention which I have experienced from you
> during my illness. I am in too weak a state to add more, and am,

> Your sincere and affectionate friend,
> George Edward Woodhouse, Senior

The "Work" in question, costing almost nine pounds, comprised all three
volumes of Parkinson's *Organic Remains,* a marvelous gift that Gideon
treasured for the rest of his life.[11]

Woodhouse, unfortunately, did not recover, dying in May 1813. That
same month, Mantell went to London – probably for the Woodhouse funeral, but with sufficient additional time to meet Sowerby, Douglas, and
the affable, supportive Lambert. A letter of introduction from Douglas
affirmed that Gideon wished to join the Linnaean Society. With Sowerby
as cosponsor, Lambert obligingly proposed him at the next meeting, en-
countering no opposition. Gideon's election to membership would therefore take place routinely that November or December. Meanwhile, Lambert further assisted Mantell by introducing him through correspondence
to another of Sowerby's contributors – Miss Etheldred Benett (1776–
1845) of Norton House, Warminster – Lambert's sister-in-law, and a
woman known both at London and Oxford for her extensive collection
and knowledge of Wiltshire fossils.[12]

On 7 December 1813, thanks to the generous sponsorship of Lambert,
Sowerby, and Douglas, Mantell was elected a fellow of the Linnaean

---

11. Compiled from business directories and subsequent entries in GM's journal. Tilney prob-
    ably married a daughter of Charlotte Vollor, who was George Edward Woodhouse the
    elder's sister and his partner's wife. (In a journal entry of 21 May 1821 Gideon referred
    to his wife and Mrs. Tilney as cousins.) Woodhouse wrote his letter of 26 Nov 1812 to
    Gideon on the front flyleaf of Parkinson, I (ATL). His will is at PRO, London.

12. Lambert–GM, 17 June 1813; LS–GM, 7 Dec 1813; and misc. JD, JS (ATL). GM–EB, 30
    June 1813; EB–GM, 15 July 1813; GM–EB, 30 July 1813; EB–GM, 16 Nov 1813; GM–
    EB, 3 Dec 1813; EB–GM, 5 Jan 1814; GM–EB, 16 Mar 1814; EB–GM, 13 May 1814,
    etc. (ATL). See also Sarah E. Nash, "The Collections and Life History of Etheldred Benett
    (1776–1845)," *Wiltshire Archaeological and Natural History Magazine,* 83 (1990), 163–
    169. Etheldred Benett is often considered the first woman geologist.

Society. He attended the necessary meeting on 1 February 1814 and was officially admitted then. While in town this time Gideon met George Bellas Greenough (1778–1855), founding president of the Geological Society of London, and gained another very useful correspondent. After studying mineralogy years before in Germany, Greenough traveled extensively and soon amassed an extensive collection of geological specimens, which he displayed at his home in Parliament Street. On 16 March, Mantell described his first visit to their mutual friend for Miss Benett; the Wiltshire fossils in Greenough's collection seemed to him *very* analogous to those of Sussex. Illustrating his letter with eight beautifully colored sketches, Gideon then went on to discuss *Alcyonia,* a puzzling fossil zoophyte that had interested him since 1812. As Miss Benett now learned, he would offer his new colleagues a paper on this curious fossil.

Conventionally presented in epistolary form as a letter of 20 May 1814 to A. B. Lambert, Gideon Mantell's "Description of a fossil Alcyonium" was read before the Linnaean Society on 7 June. The author believed his proposed new species, *Alcyonium chonoides* (a funnel-shaped sponge hitherto identified as a coral), peculiar to the Upper Chalk; he had found many specimens of it at Bridgewick chalk pit, near Lewes. The living animal, Mantell suggested, "possessed great powers of contraction and expansion, which enabled it to assume various dissimilar forms," from a vertical funnel to an almost horizontal disk (402). As a result, its fossils had often been assigned to more than one species. Following a sophisticated physiological argument, Gideon analyzed some probable *A. chonoides* specimens depicted by Parkinson as separate species. When contracted, he observed, *A. chonoides* greatly resembled a modern-day Tahitian sponge. "Although I dare not flatter myself that the preceding observations will add much to oryctological science," Gideon concluded, "yet, as it is of the first importance that we should be extremely accurate in our reference of fossils to their prototypes, it is humbly presumed that an attempt to prove the identity of specimens which had formerly been considered as distinct species will not be thought wholly uninteresting" (406–407).

After the reading, Mantell sent a copy of his finished paper to Lambert but heard nothing regarding it for months. In October another went to Miss Benett, who assured Gideon that his researches were interesting and worthwhile. In May 1815 she then took this copy with her to London, showing it to Sir Joseph Banks and other influential members of the scientific community. Through her efforts, Mantell's paper (which included a postscript dated 28 September 1815) was in print by late October, though not published till the Society's meeting in November. For all the agonizing delay, Gideon then achieved recognition as the author of an unquestionably respectable scientific paper on one of the most characteristic and controversial fossils to be found in the vicinity of Lewes.[13]

Now twenty-five, and a man of some importance in his hometown, Gideon Mantell had necessarily become one of its most eligible bachelors. While none of his deeply personal correspondence with Mary Ann Woodhouse the younger has been preserved – he destroyed it later on – Gideon surely visited often at No. 2 Maida Hill between 1812 and 1816. During those years Mary Ann changed from a girl of seventeen to a young woman of twenty. Once licensed to marry, she and he did so two days later (4 May 1816) at St. Marylebone Church, part of which was still under construction. Since Mary Ann had not quite reached twenty-one, her mother's consent as guardian was required. Gideon, however, was fully twenty-six, the same age at which his father married. The newlyweds then withdrew to Lewes, establishing their home at Castle Place, amid the ominous chill and gloom of the century's coldest, most overcast summer.[14]

Stratigraphy

Though Sowerby and others had encouraged him to remain primarily a collector of shells and sponges, this old-fashioned outlook soon failed to satisfy Mantell, who had already been exposed through Parkinson and the French geologists to the more historical study of stratigraphy. From 1812 onward, therefore, Gideon regularly sought information and specimens pertinent to the various layers of the earth. In September 1813, for example, he began a notebook on the topic with a stratigraphic section of Highgate Tunnel from *European Magazine*. A far more significant paper, elucidating the geological structure of Sussex, then appeared in the Geological Society's *Transactions:* Thomas Webster's "On the Freshwater Formations in the Isle of Wight, with some Observations on the Strata over the Chalk in the south-east part of England" (1814). Like Mantell, Webster had been strongly influenced by Cuvier and Brongniart's pioneering

---

13. Nomination/election papers, LS. GM–EB, 16 Mar 1814 (ATL). For Greenough, see *DNB* and *Hist GS*. GBG–GM, 11 July 1814 is the first surviving letter. As Greenough then departed London for extended travels, regular correspondence between them did not begin till Feb 1815 (ATL and UCL). GM, "Description of a fossil Alcyonium from the Chalk strata near Lewes," *Trans LS*, 11 (1815), 401–407; ms versions ATL, LS. JP, *Organic Remains*, II, 145. See also JS–GM, 22 Dec 1813, 21 Feb 1816 (ATL).

14. GM's marriage license is on film at the Greater London Record Office, with two further documents at PRO. Because the immense 1815 eruption of Mount Tamboro in the Dutch East Indies affected climates worldwide, 1816 was famous throughout Europe and America as "The Year without a Summer."

study of the geological environs of Paris. (He was also indebted to Greenough, who had given him a good deal of information about the extent of the Chalk strata, and to Sir Henry Englefield, for sponsoring his researches throughout the Isle of Wight.) Gideon read Webster's impressive paper soon after it was published and carefully transcribed excerpts from it into the same notebook. He would spend much of the next decade working out their implications. Later on, Mantell materially amended Webster's observations and analyses, but he continued to cite them respectfully for more than thirty years.[15]

G. B. Greenough particularly stimulated Gideon's stratigraphic investigations at this time. For some years he had been at work on a geological map of England – like Smith's, which was about to appear. Unlike Smith, however, Greenough utilized information derived from correspondence. On 1 February 1815, therefore, Mantell sent an involved attempt to answer Greenough's most recent queries about strata in the vicinity of Lewes. Gideon now recognized five major layers; in ascending order (as described), they were Green Sand, Blue Marl, Chalk Marl, Hard Chalk, and Upper or Flinty Chalk. Greenough suggested in response that the two marl layers were probably aspects of a single stratum. In November and December, Mantell discussed the two marl problem in letters to Etheldred Benett, who eventually agreed with Greenough.[16]

Not surprisingly, Gideon's marriage temporarily interrupted his scientific endeavors. Thus, he wrote James Sowerby on 4 September 1816 that domestic circumstances had made him "a truant to geology" until recently, but he was now able to send specimens from a recent fossil hunt. On the first of October, Mantell sent fifty-seven such specimens to Miss Benett, writing out identifications of them all. Gideon also renewed his lapsed correspondence with Greenough. "Compelled by peculiar circumstances to abandon geological pursuits for a time," he explained lamely, "I can only express my regret at the vexatious occurrence," by which he presumably meant the ten-month interruption rather than the marriage.[17]

In actuality, Mantell was never that much distracted from geology, for by July 1816 he had met John Hawkins (1758–1841), an older man pri-

---

15. Stratigraphy. GM, Geological Extracts and Observations (ATL), with about forty entries made between 1813 and 1845. Webster's paper appeared in *Trans GS*, 2 (1814), 161–254; GM excerpted pp. 176, 178, and 190. For Webster, see *DSB* (John Challinor) and Wilson, pp. 96–103.

16. GM–GBG, 1 Feb 1815; GBG–GM, 3 Feb 1815; GM–EB, 20 Nov and 7 Dec 1815; EB–GM, 20 Aug 1816 (all ATL). Eventually, Blue Chalk Marl became Gault; and Chalk Marl, part of the long-controverted Upper Greensand (*Petrif*, p. 4).

17. On his marriage: GM–JS, 4 Sept 1816; GM–EB, 1 Oct 1816; GM–GBG, 16 Oct 1816 (all ATL). The Marriage Act of 1753 required rites by the established church for all except Quakers and Jews.

marily interested in Cornish mines who had studied under the influential German teacher Abraham Gottlob Werner at Freiberg. Though generally more concerned with economic minerals than with fossils, Hawkins (now living at Bignor Park, Sussex) was currently writing a memoir about the geology of his present neighborhood for a work of regional history. He therefore welcomed proffered assistance from Gideon, who no longer regarded his own knowledge of fossils as entirely callow.

"You will be glad to hear," Hawkins wrote a friend on 3 August,

> that I have met with a person in this country engaged on similar pursuits who has promised to assist me in ascertaining the names of these fossil bodies or in imposing new denominations. He is the author of a paper lately printed in the Geological Transactions [error for Linnaean] – a Mr. *Mantel*, a young surgeon settled at Lewes who has entered upon the pursuit with great zeal and in a truly scientific way.

Mantell, Hawkins continued, has formed a wonderful collection of everything found in his neighborhood and showed him numbers of Sowerby's *Mineral Conchology* as well. As he had done with Etheldred Benett and others, Gideon sent Hawkins a fine hamper of local fossils in September, together with a section of the strata around Lewes and remarks thereon. Hawkins then replied with information about western Sussex, emphasizing its Malm Rock.[18]

Mantell's most stimulating correspondent, however, continued to be Greenough, who replied amiably enough when Gideon renewed contact with him in October. Just returned from an extensive geological tour throughout Europe, Greenough was once again deeply involved with his forthcoming geological map of England. He continued to request geological information of any kind regarding Sussex and thought that a catalog of the Secondary fossils found near Lewes, as proposed by Gideon in a letter of 4 March 1815, would be entirely appropriate, for it was only by collecting an immense stock of accurate local information that one could hope to generalize successfully.[19]

18. JH-Samuel Lysons, 3 Aug 1816 (in Francis W. Steer, ed., *The Letters of John Hawkins and Samuel and Daniel Lysons, 1812–1830* [Chichester, 1966], pp. 32–33, quoting p. 32; other letters are also of interest). GM–JH, 20 Sept 1816; JH–GM, 27 Oct 1816 (both ATL). James Dallaway, *A History of the Western Division of the County of Sussex*, I (London, "1815"), includes "Observations on the Geological Phaenomena," pp. cxlv–cl, by John Hawkins, who acknowledges his indebtedness to Mantell (cl).

19. GM–GBG, 4 Mar 1815, 16 Oct 1816 (UCL); GBG–GM, 13 Nov 1816, 2 Apr 1817 (ATL). Greenough's map, begun so early as 1808, was finally published on 1 May 1820 (though dated 1 Nov 1819). His pervasive skepticism – as manifested, for example, in GBG, *A Critical Examination of the First Principles of Geology* (London, 1819) – lastingly influenced GM (*Petrif*, p. 6).

Besides Greenough, other correspondents also urged Mantell to write more extensively on his local fossils. Thus, James Sowerby advised him in October 1816 to publish a catalog of them. "I would furnish what plates I have to illustrate it," he offered generously, "and engrave any you might desire to add." Gideon responded on New Year's Eve, admitting to grander authorial ambitions but fearing he could not "attach sufficient interest to a work of that kind so as to make it prudent to risk publication." Sowerby reassured him five days later and again offered Mantell his plates, pointing out that this arrangement had worked successfully in the case of a recent book on the minerals and fossils of Scarborough. Gideon needed no further urging. Writing on his birthday – 3 February 1817 – and once more on the twenty-fourth, he sent first Sowerby and then Greenough a detailed outline of Sussex geology as he expected to present it. A further copy of the outline, transcribed by Mary Ann Mantell with illustrations by Gideon, was completed in March for Etheldred Benett, who thought Mantell's listing of Sussex strata far more accurate than Smith's "very erroneous" one of the previous year.[20]

"Outlines of the Mineral Geography of the Environs of Lewes" by G. A. Mantell (Lewes, 1817) was, in the Benett copy, a bound ledger-sized manuscript with seventeen pages of text and eight plates, including a fine watercolor landscape of Lewes.[21] Elaborating on each layer, Gideon now proposed the following stratigraphical divisions:

Strata above the Chalk

   a. Alluvial deposition
   b. Series of strata above the Chalk at Castle Hill near Newhaven

20. JS–GM, 9 Oct 1816, 4 Jan 1817; GM–JS, 31 Dec 1816 (all ATL). Some copies of [Frederick Kendall], *A Descriptive Catalogue of the Minerals and Fossil Organic Remains of Scarborough and the Vicinity* (Scarborough, 1816) were extra-illustrated with as many as forty-eight Sowerby plates. GM–JS, 3 Feb 1817; GM–GBG, 24 Feb 1817; EB–GM, 31 Mar 1817 (all ATL). EB's disparaging reference is to William Smith, *Strata Identified by Organized Fossils* (London, 1816).

    Sometime early in April, Mary Ann bore an unnamed first child, who died a few days later (GM–GBG, 19 Apr 1817; GM–JH, 28 Apr 1817; GM–EB, 26 Apr 1817 [all ATL]). Weak and depressed, she remained an invalid for several months, during which she and Gideon completed four projects: the Benett copy of Gideon's "Outlines"; "Sketches of the Most Remarkable Fossils Represented in Knorr's *Monumens des Catastrophes*" – an exceptionally beautiful manuscript; "Figures and Descriptions of Monastic Pavements, from the Ruins of Lewes Priory"; and "Original Correspondence on Subjects Connected with Geology, Natural History, and Antiquarian Researches" (cited by the present author as Letterbook). The latter includes seventy-four letters by GM or to him between April 1812 and 28 July 1817, the last few being added subsequently.

21. GBG's copy of "Outlines" is at GS; all others, ATL.

Chalk Formation

    a. Upper or Flinty Chalk
    b. Lower or Hard Chalk
    c. Chalk Marl
    d. Blue Marl

Strata below the Chalk

    a. Green Sand Formation
    b. Blue Clay or Oak Tree Soil
    c. Ferruginous Sand

This more sophisticated sequence revealed Gideon's continuing indebtedness to William Smith.

The next year, for a short-lived local periodical, *The Gleaner's Port-Folio, or Provincial Magazine,* Mantell wrote "A Sketch of the Geological Structure of the South-Eastern Part of Sussex," portions of which appeared in the first two numbers. By reason of "the great displacement which some of the strata have sustained, and the entire disintegration and removal of others," Sussex geology was unusually complicated; readers were therefore cautioned that the stratigraphic succession he now attempted should be regarded as tentative. Beginning with the oldest stratum, Gideon listed (8–9):

1. Ferruginous Sand, with subordinate beds of sandstone, etc.
2. Blue Clay, or Oak Tree Soil (including Sussex Marble)
3. Brick Earth.
4. Green Sand.
5. Chalk formation, comprising the Blue Marl, Chalk Marl, Lower or Hard Chalk, and Upper or Flinty Chalk.
6. Plastic Clay (as at Castle Hill, near Newhaven).
7. Blue Clay of the Levels.
8. Alluvium.

Ferruginous Sand occurred extensively toward the north but contained few fossils. The subordinate beds, however (later to be Gideon's Wealden Series), included fossil shells, plants, wood, *the teeth of fishes and alligators (or alligator-like animals), and bones of the latter.* This was the earliest indication in any of his writings that he had discovered the remains of dinosaurs.[22]

---

22. GM, "A Sketch of the Geological Structure of the South Eastern Part of Sussex," *Provincial Magazine,* pp. 8–11 (Aug 1818), 68–71 (Sept 1818) [incomplete]. Reissued, 1819.

3

Fossils of the South Downs

On 20 November 1817 Gideon Mantell sent Etheldred Benett a box of fossils and a brief catalog. Having been unusually fortunate in his own field endeavors the previous summer, he could easily afford such a show of generosity. "My collection of Chalk fossils," Mantell advised her concomitantly, "is now so extensive and contains so many rarities that I intend, if possible, to publish a work on the subject in the course of next year and notwithstanding the disadvantages of my present situation for such an undertaking." Mrs. Mantell, he continued, wished to engrave the plates and with Gideon's letter was sending an impression of her first attempt. "The next," he said cruelly, "will, I hope, exhibit considerable improvement."

Miss Benett replied in December that a volume describing Gideon's extensive collection of Chalk fossils could not fail to be worthwhile. As for Mary Ann's plate, "I am much pleased with her first attempt at etching," Etheldred commented tactfully, "and a little practice to enable her to work stronger and bolder appears to me all that is wanting to make them a great ornament to your work." In subsequent letters, she at first declined and then later accepted (but never fulfilled) Mantell's invitation to contribute a listing of Wiltshire fossils while strongly endorsing his basic plan, the soundness of which he had begun to doubt. "A theoretical work in the present stage of the science of geology would only puzzle a cause already sufficiently intricate," she assured him on 21 April, "but a work such as yours – which will contain so many facts, as well as plates of unpublished fossils from the places in which they are found in the highest preservation – must always be valuable." She repeatedly urged Gideon to persevere, and, giddy with the prospect of scientific fame, he eventually found the necessary courage.[1]

1. GM–EB, 20 Nov 1817 (also GM–GBG, 11 Nov 1817), EB–GM, 14 Dec 1817; GM–EB, 31 Jan 1818; EB–GM, 23 Mar 1818 (all ATL); GM–GBG, 1 Apr 1818 (UCL); EB–GM, 21 Apr 1818 (ATL). EB, "A Catalogue of the Organic Remains of the County of Wilts" (Warminster, 1831), cites GM, *Illus* frequently; it was reprinted in Sir Richard Colt Hoare, *The History of Modern Wiltshire* (6 vols., London, 1822–1844), III ("1830" but later),

By September 1818 Mantell was resolutely at work on what would become his first published book, *The Fossils of the South Downs* (1822). It began slowly, for he noted around 21 January 1819 that his time had been taken up almost entirely by his medical practice and a harsh winter; he had "consequently done but little to the drawings" for his intended tome. On 25 February, however, Gideon finished those of fossil plants from Castle Hill that would eventually become Plate VIII. The next evening was one of several devoted to sketches (Plates X–XV) of *Ventriculites,* the *Alcyonium chonoides* of his Linnaean Society paper, now renamed. On 24 April he sent six of Mary Ann's engravings to Elizabeth Cobbold, who replied a week later with enthusiastic praise. Still, it was difficult to make headway for a time, and on 23 June Gideon wrote sadly to James Parkinson: "Circumstances of a domestic nature have so much interfered with the progress of my intended work that I am quite in despair of its being completed till all my discoveries are anticipated by more able and industrious collectors."

In a journal entry for the same day (his first in more than six weeks) Mantell explained.

> A long interruption to my diary has taken place, the consequence of bustle and unremitting exertion in some domestic arrangements. I have purchased the house adjoining mine on the west side, and have made a way into the drawing room and front bedroom from the passage in my own residence. The whole of my collection is removed into this new drawing room, in which I am now writing. The masons, carpenters, and painters who have been employed in this alteration have not yet entirely finished their labors.

Among these extensive alterations, Gideon eliminated several walls, one staircase, and a front door. It was then necessary to redesign the High Street facade, which acquired an elegant Ionic entrance, a false window directly above it, grill work, a rooftop balustrade, and, as its most conspicuous feature, two-story pilasters at either side with ammonite volutes on top. Such volutes were the punning trademark of Gideon's architect, *Amon* Wilds, who used them elsewhere as well; at Castle Place, they referred also to Mantell's fossils. In addition to his rebuilding project, Gideon was further busied during these months by his medical practice, guests, correspondence, and incoming specimens.

Progress toward his book resumed in August, when Mantell provisionally allotted his plates. "I find much difficulty in selecting subjects for

pp. 117–126. A presentation copy of 1831 to WDC with holograph corrections by EB is at the Wiltshire Archaeological and Natural History Museum, Devizes. There is also an unpublished manuscript of hers ("Sketches of Fossil Alcyonia" [1816]) at GS.

my work," he then advised Greenough. "I shall restrict the number of plates to thirty or perhaps thirty-five – and if it should take, shall publish (in the form of an appendix) engravings of the remainder." About twenty plates had already been engraved. Toward completing the others, on the twenty-seventh Gideon drove with Mary Ann to the summit of Mount Harry, west of Lewes. Armed with a camera lucida (patented by Wollaston in 1807), he outlined the surrounding hills, then finished a geological analysis of them that evening; these became his fine plates II and III. On 21 September John Webb Woollgar brought him a copy of William Smith's just-published section of the Sussex strata. Sarah Godlee presented her sketches of Rottingdean strata (Plate V) eight days later. In mid-October (his earliest opportunity), Mantell walked under the cliffs from Brighton to Rottingdean, ascertaining for himself the correctness of her analysis. By 24 November, all but six of the necessary drawings had been completed.[2]

Some further delays were attributable to the birth of Gideon's first son, Walter Baldock Durrant Mantell, on 11 March 1820 and a comic episode in April when the proud father learned for the first time of his having been elected a fellow of the Geological Society two years before! (Owing to a secretarial mix-up, he had never been notified.) He finished the last drawing on 9 May and on the twenty-fourth helped Mary Ann with her engraving (probably Plate IV). On 4 June the last copper plates arrived from London. When, on 23 July, Mantell dispatched his colored geological map of lower Sussex (Plate I) to be lithographed, the forty-two full-page plates were done. Of these, the first seven consisted of his map, views, and sections; all the others depicted specimens (more than 350 in all), with as many as 34 on the same plate.

On 31 August, Gideon began writing his text, parts of which required additional fieldwork, often with Mary Ann. Throughout the last week of September, for example, they rode out together on a series of excursions (later reported more generally in Gideon's Chapter 5) to delimit the unfossiliferous Ferruginous Sand formation. Mantell then researched "The

2. J (ca. 21 Jan, 25, 26 Feb, 24 Apr 1819); GM–EC, 24 Apr 1819; EC–GM, 2 May 1819; GM–JP, 23 June 1819 (all ATL). J (23 June 1819). Mantell bought No. 2 Castle Place on 15 May 1819 (ESRO). For Amon Henry Wilds (1762–1833) and his trademark, see Clifford Musgrove, *Life in Brighton* (Hamden, Ct., 1970), pp. 178–179. GM–GBG, 23 Aug 1819 (UCL); J (27 Aug, 21, 29 Sept 1819). William Smith, "Strata in Sussex" (1 May 1819), a long foldout with landscape and strata beneath letterpress, offered a cross-section of the Weald; Hamsey and Castle Hill were mentioned. GM had previously received Buckland's stratigraphic table (EB–GM, 21 April 1818 [ATL]). J (15 Oct, 24 Nov 1819). John H. Hammond and Jill Austin, *The Camera Lucida in Art and Science* (Bristol, 1987), "Part One: The Sketching Instrument."

Figure 3.1. *Foss SD*, Plate III. (top) The Weald, looking west from the summit of Mount Harry (640 feet) on 27 August 1819, Ditchling Beacon (812 feet, at left) to Newborough Beacon. Rounded hills and shallow valleys were regularly cited as evidence of a recent deluge. (bottom) Section of lithology, from Malling Hill (east of Lewes) northeastward. The uniqueness of each layer suggested that the history of the earth consisted of successive geological revolutions.

Rocks" at Uckfield on 18 October with Lupton Relfe (his publisher, who had just married Hannah Woodhouse) and, ten days later, braved Terrible Down with James Woodhouse. On these and other excursions, Gideon continued to refine his stratigraphy.[3]

Five hundred copies of his lithographed geological map arrived from London on 3 December; with paper and coloring, they had cost thirteen pounds. Three days later Mantell sent John Hawkins a spare one and reported the progress of his book: "All the plates for my work are finished or nearly so," he confirmed, "and I have written a considerable part of the text, but when it will be finished I have not the least idea, for I am so much the creature of surrounding circumstances as to have but little command over my time." In this letter and another, Gideon discussed technical points of local stratigraphy with Hawkins in great earnest. By year's end, as part of his usual self-assessment, the aspiring author could record comfortably that his book was "in a state of forwardness."

Throughout January 1821 Mantell the surgeon was so overwhelmed with his customary annual billing that no time remained for authorship. Once accounts were out of the way, however, he began to make rapid progress. The first week in March found Gideon busily employed in his profession while writing industriously at night. On the sixth he received a number of plates, proofs, and drawings from London and less than a week later was drafting a prospectus (which, delayed by as many as half a dozen cases of midwifery per day, did not reach his printer, John Baxter of Lewes, until the twenty-fourth). Issued on 2 April, it identified Mantell's forthcoming book as *The Fossils of the South Downs; or, Outlines of the Geology of the South-Eastern Division of Sussex.* But the subtitle and stratigraphical sequence would change prior to publication.

Gideon immediately dispatched copies of this announcement, and sample engravings, to a variety of scientific friends, virtually all of whom helped to circulate his prospectus and were eventually among his subscribers. Advertisements for *The Fossils of the South Downs* appeared in the *Sussex Weekly Advertiser* on 6 May 1821 and in the *Brighton Herald* on the twelfth, both of them listing the King and other prominent supporters. Mantell then sent clippings far and wide. On 9 June he wrote more fully to his influential scientific neighbor (and potential patron) Davies Gilbert:

---

3. J's record of the birth no longer exists, having been torn out by WBDM. TW–GM, 16 Apr 1820 (ATL): As secretary of the GS, Webster was "not at all aware that you had received no notice of your election." Mantell had been admitted on 15 May 1818, but TW was then extremely ill, his duties being performed by others. J (4, 9, 24 May, 4 June, 23 July, 31 Aug, 23, 28, 29 Sept, 1, 8, 18, 28 Oct 1820).

From the number of engravings, I find that the expenses of publication will amount to £600. Two hundred copies will therefore be required to cover the expenses, and of these ninety are already subscribed for. My list contains the names of His Majesty (four copies), the professors of mineralogy and geology of both universities, the earls of Egremont and Northampton, the bishops of Durham and Chichester, etc., but there are yet many noblemen and gentlemen in this part of Sussex who generally patronize such undertakings that still withhold their support from mine.

Naming several, Mantell obviously hoped Gilbert would solicit them. Specially colored copies, he announced, were also available at six guineas (twice the uncolored price), three having already been sold. Gideon was proceeding "tolerably fast" with the manuscript and thought the whole work might be got through the press that winter if enough support materialized. Accordingly, on 1 July he issued a second prospectus, listing his present subscribers, announcing a goal of *four* hundred in all, and promoting the colored edition. Even so, his text was far from written.[4]

Lyell

For the next several months Mantell scarcely mentioned his announced book, probably because it had become clear that subscriptions alone would be grossly inadequate to cover expenses already incurred. As part of his promotional efforts he had begun a not very satisfactory correspondence with William Buckland (1784–1856), the influential Oxford geologist, who responded to Gideon's polite, legible letters with a characteristic series of short, hasty scrawls. In return for his patience, however, Mantell was enabled to meet young Charles Lyell (1797–1875), the colleague who, after Cuvier, would influence him more than any other.

"Mr. Lyell of Bartley Lodge, Hampshire, called on me," Gideon recorded on 4 October; "he is a pupil of Professor Buckland of Oxford, and enthusiastically devoted to geology." Lyell, the scion of Scottish gentry, spent his childhood at Bartley Lodge, near Southampton. From 1810 to 1815 he attended school at Midhurst, Sussex. Following a visit to Scotland, Lyell matriculated at Oxford in February 1816. After reading his

4. J (3 Dec 1820). GM–JH, 6, 21 Dec 1820 (WSRO). J (31 Dec 1820; 3, 19 Jan, ca. 5, 6, 12, 24 Mar, 2 Apr 1821); *Foss SD*, first prospectus (J. Baxter, Lewes, 2 Apr 1821 [ATL]); J(2, 8, 11, 19, 20 Apr, 6 May 1821); replies, incl. DG–GM, 6 Apr, JSM–GM, 19 Apr, J. M. Cripps–GM, 7 June 1821 (all ATL); *Foss SD*, "Subscribers," pp. xi–xiv; GM–W. Wood, 7 June 1821 (ATL); GM–DG, 9 June 1821 (ESRO); also GM–JH, 23 May 1821 (WSRO). *Foss SD*, second prospectus (J. Baxter, Lewes, 1 July 1821 [ATL]).

father's copy of Robert Bakewell's *Introduction to Geology,* Lyell was stimulated to enroll in Buckland's course on mineralogy (May 1817). That same year he visited James Sowerby in London and Dr. Joseph Arnold in Yarmouth, both of whom shared their fossil collections with him. Now seriously interested in geology, Lyell augmented his knowledge by further reading, personal observations, and extensive travels throughout England, Scotland, and the Continent. As of February 1820 he also began to study for the law. On 2 October 1821 Lyell rode from Bartley Lodge to Midhurst, his old school, and while geologizing there was referred to Gideon.

"About twenty years or more ago," Mantell recalled in June 1841,

> one beautiful summer evening, a young Scotchman called at Castle Place (Lewes) and announced himself as a Mr. Lyell, who was fond of geology, had been attending Jameson's lectures at Edinburgh, had visited his former alma mater, Midhurst Grammar School, in the west of Sussex; and rambling about the neighborhood found some laborers quarrying a stone which they called "whin." As this term is Scottish for trap [i.e., basalt], the young traveller was much puzzled to know how such a rock appeared in the south of England and, upon inquiring of one of the laborers why the stone is so called, was referred to "a monstrous clever mon as lived in Lewes, a doctor who knowed all about them things and got kurosities out of the chalk pits to make physic with."

Formerly a quarryman at Lewes, the laborer had been one of Gideon's collectors.

Lyell then elected to ride a further twenty-five miles over the South Downs to Castle Place, where he arrived at sunset. Introducing himself as a pupil of Buckland's (*not* Jameson's), Lyell flattered Mantell with honest appreciation, relating in particular how Dr. Arnold had referred constantly to his *Alcyonium* paper of 1815 while showing some fossils. After tea with him and Mary Ann, Lyell then saw Gideon's collections. Thoroughly delighted with each other, the two men talked geology until midnight; afterward, Mantell stayed up to write his journal entry (subsequently emphasized by a pointing finger in the margin) and a letter – mostly about Lyell – to John Hawkins, who lived only twelve miles from Midhurst and would be able to resolve the problem of Sussex whin. On Friday the fifth, Lyell went by himself to geologize at Newhaven. Following a pleasant excursion, he returned to Castle Place in time for tea with Gideon at 6 P.M., and they sat up talking again till one. Later that Saturday, when Lyell had to leave, Mantell eagerly accompanied him on horseback as far as Ditchling Beacon, a viewpoint north of Brighton, showing Lyell geological features of interest along the way. Before parting, the two men had become lifelong friends.[5]

The First Book

Gideon worked diligently on his manuscript throughout November and December. He must have completed a good deal of it by then because when pages 197–198 went to the printer old Thomas Woollgar (died 22 December 1821) was still alive. Mantell therefore wrote proudly on New Year's Eve of having made considerable progress on the verbalizing of his book, which he hoped now to complete in the spring. "The subscribers to it," he added wishfully, "are more numerous and respectable than I could have expected." Neither their number nor their quality sufficiently allayed his fears, however. "I begin this new year with considerable apprehension," he recorded on 1 January; "before the close of it (should my life be spared) I shall in all probability appear before the world as an author, and experience all the vexations and anxieties inseparable from a first literary attempt." There were then no further journal entries on any subject until propitious 1 May, when he noted: "This day I have written the last page of my *Illustrations of the Geology of Sussex* – the dedication [to Davies Gilbert]. Since my last entry I have completed the manuscript and corrected all the sheets as they passed through the press, receiving a sheet per post almost daily and returning it corrected the same evening; the printing has taken up three months." He received the first completed copy from London five days later.[6]

5. The meeting with Lyell. GM–WB letters are at GS; WB–GM, at ATL. J (4–6 Oct 1821); GM–BS, 14 June 1841 (Yale), emended; GM–JH, 4 Oct 1821 (WSRO); Lyell, *LLJ*, I, 377; Wilson, pp. 92–94. The Sussex "whin" was actually chert (*Foss SD*, p. 71n; Bakewell, 1815, p. 491). Lyell's mentor, the Reverend William Buckland, had been appointed reader in mineralogy at Oxford in 1813 and then reader in geology six years later. On the latter occasion, 15 May 1819, Buckland preached an inaugural lecture entitled "Vindiciae Geologicae; or the Connexion of Geology with Theology Explained" (published 1820). In it he attempted to establish that geology *confirmed* natural religion (the evidence of God in nature), as well as the biblical narratives of the Creation and the Flood. See J. M. Edmonds, "Vindiciae Geologicae," *Archives of Natural History*, 18 (1991), 255–268.

6. J (31 Dec 1821; 1 Jan, 1, 6 May 1822); *Foss SD*, passim. Mantell's dedicatee, Davies Gilbert (born Davies Giddy in Cornwall, 1767), befriended Thomas Paine in London and was elected to the Royal Society that same year (1791), having been sponsored by his geological neighbor and fellow radical John Hawkins, later of Bignor Park, Sussex. In 1808 Giddy married, changed his name, and became an early, prominent member of the Geological Society of London (founded 1807). Active as well in the Linnaean Society and the Society of Antiquaries, he was also founder-president of the Royal Geological Society of Cornwall (1814), holding office for life. In 1827 he would become president of the Royal

Figure 3.2. *Foss SD* composite. *Top left:* Gideon Mantell's own copy of his *Fossils of the South Downs* (London: Lupton Relfe, 1822), open to show p. 196, his smaller geological map of Sussex and Kent. *Top center:* The original woodblock for that map (collection of the New Zealand Institute of Geological & Nuclear Sciences, Wellington). *Top right:* Mantell's undated watercolor sketch, Strata at Castle Hill near Newhaven (before 1818, probably; E–295–098). *Bottom left:* Prospectus and printer's bill for *Fossils of the South Downs* (MS–Papers–0083–120). *Bottom center:* Four fossils depicted in the book (collection of the New Zealand Institute of Geological & Nuclear Sciences). *Top left: Conulus subrotundus,* illustrated on Plate XVII, Figs. 15 & 18; described, p. 191. *Top right:* Belemnite from the Chalk at Brighton, Plate XVI, Fig. 1; described, p. 201. *Bottom left: Hamites attenuatus,* Plate XIX, Figs. 29 & 30; described, p. 93. *Bottom right: Conulus albogalerus, var. acuta,* Plate XVII, Figs. 16 & 19; described, p. 190. *Bottom right:* Mary Ann Mantell's extra-illustrated copy of *The Fossils of the South Downs,* open to show the normal frontispiece and title page. Except as noted, all items are from the collections of the Alexander Turnbull Library, Wellington, New Zealand.

Society, and in that capacity select William Buckland and other authors of the Bridgewater Treatises (natural theology). See MAL, *The Worthies of Sussex* (Lewes, 1865), pp. 212–215; *DNB;* and A. C. Todd, *Beyond the Blaze: A Biography of Davies Gilbert* (Truro, 1967).

Published on 6 May 1822 by Lupton Relfe in an edition of five hundred copies (only twenty-five of which, issued in August, were fully colored), Mantell's book was a royal octavo of 328 pages and 42 plates. As his first and in several respects most important major publication, this one underlies all the others. Though officially *The Fossils of the South Downs; or, Illustrations of the Geology of Sussex,* it would sometimes be cited (as above) by subtitle, which had changed since his prospectus of April 1821. Awkwardly enough, Gideon subsequently published a separate book entitled *Illustrations of the Geology of Sussex* (1827), which he then represented as a continuation of *The Fossils of the South Downs.* Not surprisingly, flustered scholars have occasionally interchanged the two.

What became Chapter 1 of *The Fossils of the South Downs,* an anonymous "Preliminary Essay," began as a formal letter to Mantell (probably received in May 1821, just after his first prospectus had come out) from the Reverend Henry Hoper of Portslade, a fossil collector himself, "On the Correspondence Between the Mosaic Account of the Creation, and the Geological Structure of the Earth." Both geology and Genesis, for him, affirmed the immense antiquity of the earth, and while many strata seemed to have formed gradually, there was ample support for the biblical record (as he conceived it) of sequential, violent revolutions. Both geology and the Bible agreed, furthermore, that the earth's crust – including its granite – was formed from water and that life in the past differed greatly from today's. At least one stage of that past life perished in a worldwide body of water, an event Hoper readily equated with the Deluge. The present life and surface features of the earth, therefore, were not of any great antiquity. In general, Hoper's well-written essay accorded with the geological orthodoxy of his time.[7]

As was characteristic of lay understanding, however, Hoper failed to acknowledge the fuller significance of stratigraphy, which did not straightforwardly confront the public until 1822. William Smith, we know, had grasped the essence of stratigraphic correlation during the 1790s; his work was then publicized, more than a decade later, by several followers. Parkinson's summary of Smith, in 1811, stimulated Gideon to begin his own study of stratigraphy the next year. He was then influenced by other pioneers: Cuvier and Brongniart, Thomas Webster, John Farey, William Phillips, Greenough, Hawkins, and Buckland most importantly. In *The Fossils of the South Downs,* Gideon also utilized William D. Conybeare and William Phillips' important *Outlines of the Geology of England and Wales* (1822), which though not yet published he had seen in proof.[8]

---

7. *Foss SD,* p. ix (note also p. 205). J (6 Sept, 26 Dec 1822, 28 July 1823).

8. As we saw in Chapter 2, Mantell had been interested in stratigraphy since 1811. His notebook on the topic, "Geological Extracts and Observations" (ATL), included about forty

Geology and Evolution in 1822

- There was no geological time scale.
- There were no geological periods.
- The vast extent of geological time was unrecognized.
- Individual strata had only occasionally been grouped into formations.
- International correlations of strata being extremely rare, there were no world-wide "systems."
- Uppermost layers of gravel and pebbles were regularly attributed to the Noachian Deluge.
- Other changes of strata (and most landforms) were also explained catastrophically.
- Thus, geological changes of almost all kinds were assumed to be rapid, violent, and devastating. But the nature of the catastrophe was often left unspecified.
- That fossils represented forms of past life was no longer doubted. Each stratum, moreover, had characteristic fossils, often unique to itself.
- The extinction of species was accepted but seldom emphasized.
- There was no prevalent theory regarding extinction.
- How species originated was equally unclear, but the already proposed idea that one kind of animal or plant might evolve into another was almost universally rejected. Animals and plants had no ancestors. Mankind had no predecessors.
- The great majority of known fossils were shells. Other invertebrates remained poorly known. Trilobites, for example, were still considered insects.
- Vertebrate paleontology had scarcely begun. Only a few recently extinct large mammals and a few exceptionally well-preserved Jurassic reptiles were known; interpretations of the latter were still controversial. Amphibians and, more importantly, dinosaurs, had yet to be discovered. Even the abundant fossil fishes had not yet been closely studied. Almost nothing was known about either fossils or strata older than what we now call the Mesozoic.
- *The Fossils of the South Downs* was a major contribution to the science of its day.

---

entries made between 1813 and 1845, several of the early ones being from Thomas Webster's important paper on the Isle of Wight (*Trans GS,* 2 [1814], 161–254). John Farey contributed stratigraphical analyses to successive volumes of James Sowerby's *Mineral Conchology* (I, 1815). William Phillips' *A Selection of Facts* (London, 1818) included an important table of strata compiled by Buckland. Conybeare and Phillips' *Outlines* (June 1822), revising the latter's *Selection,* is particularly notable for Conybeare's additions. Among these, WDC cites GM's LS paper on alcyonia (76), includes *Ammonites Mantelli* (124), and twice alludes to GM's forthcoming book (135, 155), which in fact appeared first. Mantell saw *Outlines* in proof and was thus able to avail himself of its contents: WDC–GM, 14 Apr 1821 (ATL); JP–GM, 28 June 1821 (ATL, in GM's copy of *Org Rem,* II); JSM–GM, 12 Nov 1821 (ATL), sending proofs. Gideon then emended the beginning and ending of his Chapter 5 to include the new information (see his note, p. 25).

Chapter 2 of Mantell's book, the first by himself, briefly described south-eastern Sussex, the geological structure of which he then adumbrated in Chapter 3 (pp. 22–23), proposing the following divisions:

SECONDARY FORMATIONS

I. Green Sand Formations
    i. Iron Sand
    ii. Tilgate Beds
    iii. Weald, or Oak Tree Clay
    iv. Green Sand
II. Chalk Formations
    v. Blue Chalk Marl
    vi. Grey Chalk Marl
    vii. Lower Chalk
    viii. Upper Chalk

TERTIARY FORMATIONS

III. Formations Above the Chalk
    ix. Druid Sandstone
    x. Plastic Clay
    xi. London Clay
IV. Alluvial Formations
    xii. Diluvium
    xiii. Alluvium ("the effect of causes still in action")

The order of this succession, Mantell asserted, always remained the same; while the geographical extent of each bed was not entirely known, a higher-numbered bed would never be found underneath a lower. This list, with its new emphasis on "formations" (see his Chapter 4), was more complex than any Gideon had hitherto proposed; it also added, for the first time, his Tilgate Beds, Druid Sandstone, London Clay, and Diluvium. Mantell's explicit proposal that only the Alluvium had been laid down under present conditions was equally new, though not original. Subsequent chapters, ranging in length from a single paragraph to numerous pages, then discussed each of the thirteen principal divisions in turn.[9]

In particular, Tilgate Forest (Chapter 6) was full of interesting remains.

9. As Mantell soon realized, his Iron Sand and Tilgate Beds are identical. Both were later called the Hastings Sands; together with the Weald Clay, they form what P. J. Martin (in 1828) was first to call the Wealden Formation (Lower Cretaceous). Gideon's Green Sand is now the Lower Greensand and his marl layers are Lower and Upper Gault. Lower and Upper Chalk are still so called. All three of Mantell's "Formations Above the Chalk" are Eocene; only London Clay remains current. Diluvium is Pleistocene and glacial; Alluvium is Recent. Building on Wernerian antecedents, Cuvier and Brongniart (1812) originated the modern concept of formations, which Gideon first utilized in his unpublished stratigraphical manuscript of 1817.

On contemporary maps this region lay irregularly between Horsham, Crawley, and Cuckfield in northern Sussex. Now Mantell had become the first to use its name as a geological designation. (In actuality, however, his famous Tilgate Forest consisted of two adjacent quarries at Whiteman's Green, just north of Cuckfield.) Strata there included various sandstones, limestones, and calcareous slates, all lying on blue clay. Unfortunately, Gideon had discovered their rich fossil content too late for comprehensive treatment in the present book. "When the idea of this work first suggested itself," he explained, "my information concerning these deposits was too imperfect for publication, and their description was in consequence excluded from the original plan. The organic remains I then possessed were but few and uninteresting, and the plates were devoted to other objects before more illustrative specimens occurred." Should circumstances permit, however, he would "hereafter lay before the public delineations of several extraordinary fossils recently discovered in these strata" (42n). Those described in *The Fossils of the South Downs* included plants (evidently tropical), shells, fishes, tortoises, and some extremely curious vertebrae, teeth, and bones of *unknown but obviously gigantic reptiles.*

Mantell's twenty-first and final chapter, "Concluding Observations," apparently written in April 1822, featured a geological map and section of Sussex, updated summaries of each formation, and some unusually catastrophic assertions. The separation of the Isle of Wight from the mainland, and of England from Europe; Diluvian strata and fossils; and the present form of the earth's surface, with its rounded hills, all served to convince Gideon that "our globe has been overwhelmed, at a comparatively recent period, by the waters of a transient deluge," as Cuvier, Greenough, and Buckland all agreed (303). Eight numbered conclusions then followed, including Gideon's explicit recognition that the Tilgate beds were of freshwater origin.[10]

In subsequent opinion, this final discovery (of the freshwater origin of the Wealden strata) emerged as Mantell's most significant contribution toward understanding the geology of Sussex. According to W. H. Fitton's geological sketch of Hastings (1833), for example, "It was not until the appearance of Dr. Mantell's *Illustrations of the Geology of Sussex* in 1822 that the full value of the evidence which this district affords was made to appear." In that "excellent work" Gideon proved his extraordinary Tilgate fossils "must have originated in a lake or estuary" during a much warmer

---

10. *Wight,* pp. 268–269n; *Wonders,* sixth edition, p. 366n; and *Petrif,* pp. 206–207 are insistent recollections. The whole question of saltwater and freshwater strata grew out of Brongniart's Paris Basin studies, with subsequent English contributions. In *Foss SD,* GM's only clear recognition of the freshwater origin of the Weald is on pp. 303–304.

time. Similarly, John Phillips would later write: "Until the appearance of Dr. Mantell's works on the geology of Sussex, the peculiar relations of the sandstones and clays of the interior of Kent, Sussex, and Hampshire were entirely misunderstood. No-one supposed that these immense strata were altogether of a peculiar type, and interpolated amid the rest of the marine formations as a local freshwater deposit." Further commendation of the same kind appeared throughout the nineteenth century until William Topley's frequently carping citations of Gideon in *The Geology of the Weald* (1875) grudgingly confirmed *The Fossils of the South Downs* in its status as one of the classic works of British regional geology, a reputation it has since maintained.[11]

The book's contributions to paleontology are harder to trace in that many discoveries were renamed or alternatively classified by later workers, as happened with much of the nomenclature proposed at this early time. Mantell himself claimed (on pages 307–308) to have announced four new genera, all of them zoophytic. Of these, *Ventriculites* and *Marsupites* are still considered valid; two others, *Choanites* and *Spongus* (both created in collaboration with Charles König), have since been modified. Gideon also announced new species: fifteen zoophytes; four echinites; twelve ammonites, three hamites, two scaphites, one each of belemnite and turrilite, and nine further univalves; seven new species of inocerami, three pectens, five terebratulae, and nine more bivalves; as well as one new crustacean and four fishes. Thus, he proposed seventy-six new species in all – surely an outstanding contribution, and all the more so when one realizes that Gideon made no claims for the valuable and now historic reptilian fossils already in his possession.[12]

---

11. William Henry Fitton, *A Geological Sketch of the Vicinity of Hastings* (London, 1833), pp. 13–14; John Phillips, "Geology," VI, 631, in *Encyclopedia Metropolitana* (26 vols., London, 1817–1845). Conybeare and Phillips, pp. 140, 155, had supposed the Wealden formation marine. Among later works, William Topley, *The Geology of the Weald* (London, 1875) cites GM almost fifty times. For other references, see the histories of geology by Karl Alfred von Zittel (1901), Horace B. Woodward (1911), and John Challinor (*Annals of Science*, 26 [1970], 177–234; pp. 226–228).

12. In fifteen instances, Gideon Mantell utilized his Adamic prerogative as a discoverer of new creatures to name appropriate species in honor of his scientific colleagues: Charles König, Davies Gilbert, James Parkinson, James Sowerby, the late William Martin, John Marten Cripps (who had secured the King's patronage for *South Downs*), the late Reverend Joseph Townsend (an earlier investigator of Chalk strata), Miss Benett (*Ventriculites,* which had been renamed from *Alcyonium* by JSM), J. S. Miller (*Marsupites,* which, like *Ventriculites,* was already mentioned in his book of 1821), the late Thomas Woollgar, the Reverend Henry Hoper, Cuvier, Brongniart, Thomas Webster, and William Elford Leach. See esp. W. E. Leach–GM, 1 May 1820; and JSM–GM, 27 Nov 1820 (both ATL).

Reception

Though they lacked Mantell's specialized knowledge of either geology or paleontology, British reviewers uniformly praised his efforts. The earliest notice, appearing in the *Sussex Weekly Advertiser* on 27 May, was predictably warm. Three periodical reviews followed, all of them quite favorable. Thus, in June, *Monthly Magazine* called *South Downs* a "splendid" work, excellent not only for its admirably drawn and engraved plates but also for the quality of its text. Each fossil, it held, had been outstandingly described; Cuvier himself could not have done better. Two months later, *Literary Chronicle and Weekly Review* lauded the book as a valuable contribution toward the rapid progress of geology, "and though the study is one which might not be considered as very attractive to a lady, yet in the present instance Mrs. Mantell, either from love of the science or of her husband, has most ably seconded his efforts by engraving the numerous plates which he had designed." It gave *South Downs* a hearty commendation. In the last, longest, and most prestigious review, *Philosophical Magazine* (for September) likewise alluded to an infant science of geology, which must continue to make gains when embraced with such ardor as *South Downs* displayed. These notices may have been only marginally prescient, but all of them were positive, reputable, and at least potentially advantageous.[13]

More professionally, *The Fossils of the South Downs* had already been mentioned by J. S. Miller in *Natural History of the Crinoidea,* by Conybeare in the *Transactions* of the Geological Society (both 1821), and by Conybeare and Phillips in *Outlines of the Geology of England and Wales* (1822). Another favorable notice written prior to its publication appeared belatedly in Parkinson's *Outlines of Oryctology* (July 1822). Mantell's book would then be quoted and cited profusely in Thomas Horsfield's *History of Lewes* (1824), as well as in subsequent editions of Cuvier and Brongniart on the environs of Paris and of Cuvier on fossil animals. De la Beche twice mentioned it in his 1824 translation of a paper on the Chalk by Brongniart. Buckland cited *The Fossils of the South Downs* repeatedly in his paper on *Megalosaurus* in 1824, his coprolite paper in 1829, and his Bridgewater Treatise of 1836. Lyell did so as well, naming Gideon's work in each volume of his famous *Principles of Geology* (1830–1833). Helped by its author's tireless promotion, and the loyalty of his friends, *South Downs* readily attracted the professional notice it genuinely deserved.

13. J (6 May 1822). Reviews: *SWA,* 27 May 1822, p. 4; *Monthly Magazine,* 53 (June 1822), 446–447; *Literary Chronicle and Weekly Review* no. 172 (31 Aug 1822), p. 551; *Philosophical Magazine and Journal,* 60 (Sept 1822), pp. 211–215.

As a contribution to science, *The Fossils of the South Downs* unquestionably succeeded, but it failed to achieve either the popular or the social acceptance for which its author had hoped. At year's end, Mantell
noted ruefully, "the publication of my work on the geology of Sussex, although attended with many flattering circumstances, has not yet procured
me that introduction to the first circles in this neighborhood which I had
been led to expect. . . . I have published my work with eclat, though without any profit; in fact, at present with loss." That serious works of science were unprofitable had become current wisdom. Thus, as G. B. Greenough wrote Mantell on 29 May 1822, he was sorry but not surprised
to hear that the sale of *South Downs* was unlikely to repay its expenses,
Gideon's being "the fate of almost every author who works for the advancement of science." Greenough's maxim was only too true.[14]

14. J (31 Dec 1822); GBG–GM, 29 May 1822 (ATL). Despite his own exertions, a promo-
    tional broadside by Lupton Relfe citing the favorable reviews, and private attempts by
    friends, Mantell's book was, as he had feared, too specialized and too expensive to at-
    tract the lay public of Byron and Scott. Though 400 subscriptions, as the second prospec-
    tus had called for, might have made *South Downs* financially viable, there were never
    more than 152. Desperate both to avoid crushing financial losses and to insure publi-
    cation, on 1 February 1822 Gideon had even offered *South Downs* to John Murray.
    When Murray of course refused, *South Downs* achieved publication with Lupton Relfe
    only because its enormous costs were glumly underwritten in large part by George Edward
    Woodhouse, Gideon's more successful brother-in-law. Thus, according to an undated
    memorandum preserved among Mantell's papers (ATL), the book's account as of spring
    1822 stood as follows: printing costs, £424/15/7; revenue from subscriptions, £113/
    12/0. Largely for his sister's sake, Woodhouse generously supplied most of the £311/3/7
    still wanting with earnings from his law practice. It was no way to make money.

4

Iguanodon

Insofar as Gideon Mantell is known to the general public at all, he is remembered primarily for an unlikely story in which his wife, Mary Ann, supposedly discovered an odd-looking fossil tooth that Gideon immediately recognized as saurian; from it and other unpromising relics he then fashioned *Iguanodon*, thereby creating everyone's favorite prehistoric animals, the dinosaurs. Through popular imagination and inadequate scholarship, this specious tale has acquired additional dimensions and thoroughly undeserved credence. Serious researchers, on the other hand, have questioned even the appropriateness of Mantell's assumed priority, as he was neither the first to find dinosaur remains nor the originator of their presumably saurian identification. Nor was *Iguanodon* the first dinosaur he found. What actually happened can no longer be fully recovered, but certain ascertainable facts may restrict our conjectures.[1]

Historical Background

Once it had become fully established, during the latter 1790s, that prehistoric *elephants* substantially unlike present ones had existed – that species, in other words, became extinct – it followed that life on earth was, in some important sense, developmental or at least successive. The study of past life, through fossils, then became increasingly popular. For a time, the prehistoric animal that most captured European imaginations was the mammoth, particularly after 1799 when a frozen one was found substantially intact in Siberia. Thus, the first vertebrate fossil we can associate with Gideon Mantell was a proboscidian tooth shown him during his

---

1. Versions of the Iguanodon legend appear in Spokes (pp. 18–24), and various books by E. M. Colbert, W. E. Swinton, Björn Kurtén, and others. J. N. Wilford (*The Riddle of the Dinosaur*, 1985) saw an early draft of my argument, which I first presented before the Geological Society of America on 31 Oct 1983. A portion of this chapter was also published in *Modern Geology* (1993) and later in book form.

apprentice years by William Constable. In 1811, James Parkinson reviewed almost everything then known about vertebrate fossils in the third volume of his *Organic Remains,* most of which derived from papers by Cuvier and other Continental writers. Parkinson's was unquestionably the first substantial discussion of fossil bones Gideon ever read; from it he learned Cuvier's opinion that there had been a time of reptiles on earth, before the time of mammals. In 1812 Mantell acquired a copy of Thomas Ashe's *Memoirs of Mammoth, and Various Other Extraordinary Stupendous Bones, or Incognita, or Non-Descript Animals* (Liverpool, 1806). As its title suggests, this sixty-page popular guide to mammalian remains in the Liverpool Museum emphasized their gigantism and mystery. It was the first book on fossil bones that Gideon actually owned, with Parkinson soon to be the second.[2]

The modern study of vertebrate paleontology (not yet so called) began almost concurrently, when Georges Cuvier (1769–1832), the brilliant comparative anatomist of Paris, republished some of his earlier work in four substantial volumes as *Recherches sur les ossemens fossiles de quadrupèdes (Researches on the Fossil Bones of Quadrupeds;* Paris, 1812). The first volume included a newly written preliminary discourse of 120 pages, a paper on the Egyptian ibis, and (coauthored with Alexandre Brongniart, the primary contributor) a geological description of the environs of Paris that had appeared previously. A pioneering attempt to ascertain the strata and included fossils of a geological region, Cuvier and Brongniart's "Description" not only stimulated but strongly influenced stratigraphical endeavors in England. Its revelation of alternating freshwater and saltwater strata in the Paris basin, moreover, helped to convince many British followers that geological history consisted largely of periodic revolutions, in which land or sea had precipitously imposed upon the other.[3]

Volume II detailed living and fossil rhinoceri, hippopotami, and tapirs,

---

2. Gideon remembered William Constable in GM–SGM, 12, 19 Dec 1832 (APS) and in a lecture of 1851 (*BH,* 16 Aug 51, p. 3). Parkinson, III (1811), begins with a preface acknowledging Lamarck and Cuvier and upholding the still-controversial concept of extinction. Gideon's copy of Ashe, inscribed by him "1812," is at ATL.

3. Georges Cuvier (1769–1832), the son of French-speaking Montbeliard Protestants, attended German universities. Six years' employment as a private tutor in Normandy brought his well-developed interests in anatomy and natural history to maturity. From 1795 onward, he was a very prominent naturalist and civil servant in Paris, working primarily at the Museum d'Histoire Naturelle. Alexandre Brongniart (1770–1847), a native Parisian, pursued mining, ceramics, and teaching as well as research on fossil and living animals. The two men collaborated freely. See Chapter 2, note 7 above. As is already evident, the fact that England and France were at war till 1814 in no way hindered admiration for Parisian science in London. Cuvier and Brongniart originated the concepts of geological basins and (in its modern sense) formations.

with an extended section following on living and fossil elephants. Cuvier's investigations of the latter, more than a decade earlier, had established the reality of extinct species. Mammoth and mastodon, therefore, were the first prehistoric creatures to achieve general public recognition. A skeleton exhibited in London so impressed everyone with its dimensions that "mammoth" became a fashionable adjective. For Cuvier, fossil elephants also raised fundamental problems about the distribution of species, the relation of fossil elephants to living ones, and the cause of their extinction (which he took to be a transient, overwhelming flood).

Volume III, devoted to fossil bones from the vicinity of Paris, included lengthy systematic discussions of *Paleotherium* and *Anoplotherium* (two large fossil mammals), with much on the techniques of comparative anatomy. Here, as in Volume II, Cuvier emphasized the importance of teeth. "The first thing to do in studying a fossil animal," he counseled, "is to investigate the form of its molars, determining whether it be a carnivore or an herbivore; if the latter, one can then ascertain to which order of herbivores it most nearly belongs" (2). This methodological opinion, like many of Cuvier's views, had great influence in England and was fundamental to Gideon's discovery of *Iguanodon*.

The especially important final volume of Cuvier's *Recherches* dealt with ruminants, ungulates, carnivores, and fossil reptiles. Among these further prehistoric mammals, Cuvier anatomized the well-known "Irish elk," various other fossil deer, the aurochs and further bovines, horses, pigs, bears, dogs, cats, sloths (*Megalonyx* and *Megatherium*), dugongs, and seals. While calling attention to bone-bearing deposits in a number of countries, he particularly emphasized the ossiferous caverns of Germany. Remains of tropical animals found in such caverns, Cuvier stressed, must have belonged to individuals once living nearby. Obvious as that sounds, it meant either that the climate of Europe or the animals' habitat had changed greatly. Cuvier's remark, therefore, underlined the problem of prehistoric environments – a study in which Gideon Mantell would play an important role.

The thirteenth memoir in Volume IV surveyed fossil bones from oviparous quadrupeds, a category including crocodiles and lizards (sometimes grouped together as "saurians") as well as fossil turtles. Living African crocodiles were distinguishable from the longer-snouted gavials of India and the alligators of North and South America. Several fossil crocodiles had also been found, including two with gavial-like snouts discovered in the cliffs of Honfleur and Havre. Other specimens found elsewhere, Cuvier argued, probably belonged to one or the other of these species. (But a skeleton published in 1719 by William Stukeley possibly differed, and the supposed crocodiles found in Thuringia were actually lizards.) Finally, he pointed out, all these fossil crocodiles occurred in strata significantly older than those containing the remains of *Paleotherium* and *Anoplotherium*.

In his fourteenth memoir, Cuvier described a well-known fossil uncovered at Maastricht, the Netherlands, in 1766. Though previous researchers had thought the creature a marine mammal, a fish, or possibly a crocodile, Cuvier identified it as being anatomically a lizard. If so, this particular lizard had been remarkably large, for its skeleton stretched over twenty-three feet. Cuvier agreed, however, that the Maastricht animal must have been marine. Thus, he not only hypothesized a lizard as large as a crocodile but the only one known to inhabit salt water. Cuvier's fifteenth memoir went on to discuss the handful of other fossil reptiles then known, including another spectacular conjecture: the pterodactyl. First described in 1784, this anatomical puzzle had been supposed a bird, a bat, an amphibian, or a new kind of mammal. In 1809, however, Cuvier identified it as a flying *reptile*. The last memoir in Volume IV then concerned fossil turtles. Though Cuvier failed to recognize *Ichthyosaurus, Plesiosaurus,* or any creature now regarded as dinosaurian – and himself strongly emphasized fossil mammals – he nonetheless demonstrated the reality of prehistoric reptiles, while creating the methodology and manifesting the daring through which dinosaurs would be discovered. Cuvier was also the first naturalist anywhere to realize that an age of reptiles had preceded the dominance of mammals.[4]

4. In 1800 an essay of Brongniart's on the classification of reptiles established four subgroups: chelonians (turtles), ophidians (snakes), batrachians (amphibians), and saurians (lizards and crocodiles). Cuvier and most of his followers readily accepted this scheme. As a result, true amphibians ("batrachians") and reptiles with amphibious habits ("amphibians") were long confused; similarly, lizards and crocodiles amalgamated to some extent. Cuvier believed, moreover, that major classes of life were invariably linked by intermediates, a concept Gideon utilized in 1848 when comparing his seemingly mammalian *Iguanodon* with a seemingly reptilian sloth. Again, in 1851, he considered *Telerpeton* intermediary between amphibians and lizards (but in either case a reptile). Pierre Latreille had proposed in 1804 that amphibians be considered a separate class; Richard Owen established the order *Crocodilia* in 1831.

Like a number of major concepts, the Age of Reptiles idea developed gradually. Cuvier's chief stimulus was his French predecessor Buffon, who had supposed (in *Epochs de la nature,* a comprehensive earth-history of 1778) that terrestrial life *began* with the large mammals about fifteen thousand years ago. Having observed reptilian fossils in older strata, Cuvier pointed out the discrepancy ("Discours Préliminaire"; I, 68), as Parkinson and other writers did subsequently. Yet Cuvier's thirteenth memoir, and popularizations based on *Recherches* as a whole, failed to emphasize this earlier era. Cuvier's more important affirmation of the Age of Reptiles appeared only in the final volume of his second edition (1824, p. 10), in which he had been heavily influenced by the discoveries of British researchers, including Conybeare, Buckland, and especially Mantell, who was unquestionably responsible for expanding and popularizing the concept. See also GC–WB, 20 June 1824 (Paris).

As *Essay on the Theory of the Earth* (translated by Robert Kerr, with additions by Robert Jameson; Edinburgh, 1813), Cuvier's preliminary discourse had a separate literary existence. Achieving five editions (the last in 1827), it soon became a major force in British popular geology. However he was read, whether in French or English, Cuvier effectively advocated the primacy of fossils over other kinds of geological evidence. The remains of quadrupeds, he argued, were especially significant. More obviously even than shells, such fossils proved that geological changes in the past had been sudden, revolutionary, and more powerful than any present-day agency could effect. It was apparent from mammalian evidence, for example, that the last of Europe's major catastrophes had been an immense flood. In England this was often taken to mean that geology specifically supported the biblical Deluge of Noah, but Cuvier himself never defended such an interpretation and did not attempt to explain how any of his revolutions were caused. Between 1812 and 1832, however, while his thought predominated in England, Cuvier deeply influenced Gideon Mantell and almost everyone else who was contributing to the development of geology.[5]

Though dinosaurs were still unidentified, the existence of prehistoric reptiles was known. So long before as 1605, in fact, Richard Verstegan had depicted fossil bones of "fishes" that were actually plesiosaurian. Ichthyosaurian fragments appeared in the first book devoted exclusively to British fossils – Edward Lhwyd's of 1699 – identified as the remains of fish. In 1719 William Stukeley published "An Account of the Almost Entire Sceleton of a Large Animal in a Very Hard Stone." The first articulated skeleton of its kind to be found, this well-preserved plesiosaur (also wrongly identified by Cuvier ninety years later) was thought to be a crocodile. In 1758 an actual fossil crocodile was recovered, only to be misidentified in turn. Similarly, Cosimo Alessandro Colloni described the pterodactyl in 1784 as if it were marine. A suitable methodology for identifying fossil bones did not yet exist.

The first prehistoric reptile to attract significant attention as a new, gigantic species was the Maastricht animal. Described in 1786 as if a whale, it was transformed a decade later into a crocodile, then reidentified once more by Cuvier in 1808 as a saurian. Further publicized by Parkinson (1811), Cuvier (1812), Mantell (1821 and 1822), and Buckland (1824), it

5. He also published second (1821–1824) and third (1825) editions of *Recherches*. Though differing from the first edition in numerous respects, they are almost identical with each other textually. The third added an excellent frontispiece portrait of Cuvier and a rewritten "Discours Préliminaire" (196 pp.); beyond that, the same sheets were reissued. In his publications after 1824, GM regularly cited Cuvier's second edition, which he owned; his copy is now at Yale.

THE AGE OF REPTILES

Figure 4.1. "The Age of Reptiles," a primarily Jurassic but environmentally incorrect frontispiece by G. H. Nibbs to G. F. Richardson, *Geology for Beginners*
(2nd edn., 1843). Pterodactyls glide overhead; one has been caught by a plesiosaur. Behind, an ichthyosaur crunches GM's salmon-like *Osmeroides;* other
ichthyosaurs and some ammonites are in the water. In the foreground depths are
further ammonites, crinoids, alcyonia (GM's *Ventriculites*), corals, sea anemones,
and other marine creatures. Crocodiles, turtles, and various plants adorn the
banks. None of these creatures is a dinosaur.

finally emerged as *Mosasaurus*. By the early nineteenth century, many isolated specimens of both *Ichthyosaurus* and *Plesiosaurus* had been found
and illustrated, but the creatures themselves remained elusive and were
named only in 1818 and 1821, respectively. Cuvier had redescribed and
named *Ptero-dactyle* in 1809.[6] Between 1808 and 1824, then, largely because of Cuvier, the earliest valid interpretations of fossil reptiles appeared.
    The first person in England to become famous for discovering reptilian

6. Cuvier himself is the primary historian, on whom Zittel and others have drawn. For
   nondinosaurian prehistoric reptiles, see Dr. Peter Wellnhofer, *The Illustrated Encyclopedia of Prehistoric Flying Reptiles* (New York, 1996), and H. G. Seeley, *Dragons of the
   Air: An Account of Extinct Flying Reptiles* ([1901] New York, 1967) for pterosaurs; S. R.
   Howe, T. Sharpe, and H. S. Torrens, "Ichthyosaurs: A History of Fossil 'Sea-Dragons,'"
   (Cardiff, 1981) for ichthyosaurs; and George Cumberland, "Some Account of the Order
   in Which the Fossil Saurians Were Discovered," *Quarterly Journal of Science, Literature,
   and Art,* 27 (1829), 345–349, more generally. Eric Buffetaut, *A Short History of Vertebrate Palaeontology* (London, 1987) is a well-written modern survey. In 1819, Cuvier renamed "Ptero-dactyle" as "Pterodactylus," which is still the official designation.

fossils was Mary Anning (1799–1847) of Lyme Regis, though her father and mother – Richard and Mary – had found and sold significant specimens earlier. After Richard, a cabinetmaker, died of tuberculosis in 1810, the Annings supported themselves almost exclusively by selling superb large fossils to well-heeled collectors or savants. While both her mother, Mary, and brother, Joseph, were astute collectors also, the Anning legend has settled on young Mary as its heroine. From age ten onward, apparently, she roamed the sea cliffs of Lyme herself, with her family, or while accompanying famous geologists like Buckland, who were astonished by her intuitive discoveries as she brought to light the nearly complete examples of *Ichthyosaurus* and *Plesiosaurus* from which modern identifications of both derived. Thus, in 1811 young Mary extracted the decisive ichthyosaur about which Everard Home (one of Gideon's medical examiners the same year) wrote the first of six pretentious papers, all of them mistaken. Mary also found the nearly complete plesiosaur that William Conybeare and Henry De la Beche analyzed in their important paper of 1821, a pterosaur of which Buckland wrote in 1828, and important cephalopods as well, before dying of breast cancer at age forty-eight. Though she was never a member of it, her portrait hangs appropriately in chambers at the Geological Society of London.[7]

Mary's most generous customer, Lt. Col. Thomas James Birch (who changed his last name to Bosvile in 1824), bought many of her best specimens, including an ichthyosaur wrongly described by Home in 1819 as *Proteo-saurus*. On 2 July of that year Birch, from Lincolnshire but vacationing at Brighton, visited Gideon Mantell at Castle Place to see his fossils. On entering, however, Birch at once perceived that more time would be required and therefore delayed his perusal until the fifth, when he and Gideon spent the evening together in geological chitchat. Birch dropped by again at 4 P.M. on the thirteenth to bring his new friend a fish's head from Southerham quarry. He geologized the Sussex coast by himself, then wrote Mantell from Brighton on the sixteenth, recalling his pleasure at seeing Gideon's collection, narrating his own efforts at Newhaven and Brighton cliffs, and promising to convey plates of fossils from Gideon's forthcoming *South Downs* to Home in London.

On 16 December, after a western tour that had included Etheldred Benett and her collection, Birch wrote Mantell from London to present him with

7. Basic are W. D. Lang, "Mary Anning, of Lyme, Collector and Vendor of Fossils, 1799–1847," *Natural History Magazine*, 5 (1925), 64–81; and Lang, "Mary Anning (1799–1847) and the Pioneer Geologists of Lyme," *Dorset Natural History and Archaeological Society Proceedings*, 60 (1939), 142–164. H. S. Torrens, "Mary Anning (1799–1847) of Lyme," *British Journal for the History of Science*, 28 (1995), 257–284, is a recent updating. Mary pronounced her own name "Annin."

an engraving of *Proteo-saurus,* Home's paper having just appeared in print, together with two smaller ones illustrating the head and body of this "extraordinary animal." Gideon replied the next day, thanking Birch for the engraving and hoping to obtain a similar fossil for his own collection. In his next letter of 5 March 1820 Birch obligingly promised to send "some fine things from the Blue Lias" but then announced unexpectedly that he was going to sell his collection, "for the benefit of the poor woman and her son and daughter at Lyme, who have in truth found almost *all* the fine things which have been submitted to scientific investigation." Not having uncovered any major fossils for nearly a year, the Annings (whom Birch had visited the previous summer) were desperately selling furniture in order to pay their rent. Discovering the family's plight, Birch responded with incredible generosity by arranging to auction on their behalf specimens the Annings had found, then sold him at high prices. "I may never again possess what I am about to part with," Birch predicted correctly, "yet in doing it I shall have the satisfaction of knowing that the money will be well applied."

The auction of Birch's fossils, held at Bullock's in London on 15 May 1820 with Gideon Mantell in attendance, realized more than four hundred pounds overall, but the major specimen on which Home had based his paper of 1819 failed to attract a sufficient bid. (Eventually purchased by the Hunterian Museum of the Royal College of Surgeons, it survived there until the air raid and fire of May 1941.) How much sale money eventually reached the Annings through the annuity purchased for them is unknown, but, as James Sowerby wrote, Birch's "generous method of disposing of his collection will long be remembered." Probably because of it, on 4 June 1824, "Thomas Birch Esq., now Thomas James Bosvile Esq., was admitted as a member of the [Geological] Society." Birch further assisted paleontological science by subscribing to J. S. Miller's *Natural History of the Crinoidea* (1821), Gideon's *South Downs* (1822), and Cuvier's later editions. Before dying in 1829, he also donated specimens to the Bristol Institution (of which Miller was head) and to Oxford.[8]

Everard Home, we know, had his troubles in attempting to classify even

8. Birch's identity was established by H. S. Torrens in "Colonel Birch (c. 1768–1829)," *Newsletter of the Geological Curators Group,* 2 (1979), 405–412; and idem., *The Geological Curator,* 2 (1980), 561–562 (the same periodical renamed). See also "Ichthyosaurs" (note 6 above). My account of Birch and Mantell derives also from journal entries; letters (TJB–GM, 16 July, 16 Dec; GM–TJB, 17 Dec 1819; TJB–GM, 5 Mar 1820 [quoted]; EB–GM, 20 Sept 1819; G. Cumberland–GM, 8 Apr 1820 [re the annuity], all ATL; EB–GBG, 28 Feb 1820, UCL); J. Sowerby, *Mineral Conchology,* III (1820), 120; GS (OM), 4 June 1824; and GM, "A Few Notes on the Prices of Fossils," *London Geological Journal,* 1 (1846), 13–17.

relatively complete specimens of *Ichthyosaurus*. In his first paper of 1814, Home detected both fish-like and crocodilian affinities in the unnamed creature. His second paper, of 1816 (stimulated by another Anning discovery), affirmed that the animal was some kind of extinct fish. With still more specimens in 1818, Home then realized he was examining an air-breather. His fourth paper, based on the Birch specimen and read in March 1819, specified a saurian, which (from supposed affinity with the salamander, *Proteus*) he called *Proteo-saurus*. A fifth paper of April 1819 then defended this misnomer, and the last, in 1820, pursued technicalities. Meanwhile, Charles König and other more skillful anatomists had begun to call the fish-like lizard *Ichthyosaurus*.

The identities of both *Ichthyosaurus* and *Plesiosaurus* were not fully resolved until 6 April 1821, when William Conybeare read to the Geological Society his and De la Beche's "Notice of the discovery of a new Fossil Animal, forming a link between the Ichthyosaurus and Crocodile, together with general remarks on the Osteology of the Ichthyosaurus." While collecting the vertebrae of oviparous quadrupeds in exposures of the Lias formation adjacent to Bristol, Conybeare realized that only some of his specimens belonged to species of ichthyosaur. On the basis of a wonderfully complete fossil skeleton found by the Annings and subsequently sold to Colonel Birch, it then became possible to announce the identification of a new prehistoric animal, *Plesiosaurus* ("almost lizard"). Since plesiosaurs and ichthyosaurs were obviously related, however, Conybeare also proposed grouping them together as *Enalio-Sauri* ("marine lizards"). He then went on to describe ichthyosaurian anatomy far more professionally than Home had done, elaborating as well on three species distinctions originally proposed by De la Beche two years earlier. Unable to attend the meeting of 6 April, Mantell remained in Lewes, busily distributing his first prospectus. Yet Conybeare's published paper would be of special significance to him as both Gideon's prospectus and subsequent letter to Conybeare were eventually quoted flatteringly within it.[9]

9. Home's six papers, all of them appearing in the *Philosophical Transactions* of the Royal Society, were "Some Account of the fossil Remains of an Animal more nearly allied to Fishes than to any of the other Classes of Animals," *PT*, 104 (1814), 571–577; "Some further account of the fossil remains of an animal, of which a description was given to the Society in 1814," 106 (1816), 318–321; "Additional facts respecting the fossil remains of an animal on the subject of which two papers have been presented in the *Philosophical Transactions*, showing that the bones of the sternum resemble those of the ornithorhynchus paradoxus," 108 (1818), 24–32; "An account of the fossil skeleton of the Proteo-saurus," 109 (1819), 209–211; "Reasons for giving the name Proteo-Saurus to the fossil skeleton which has been described," 109 (1819), 212–216, Plates XIII–XV; and "On the mode of formation of the canal for containing the spinal marrow, and on the

Imposing as they must have been, ichthyosaurs, plesiosaurs, mosasaurs, and pterodactyls are definitely not "dinosaurs," a term first published only in 1842 and now restricted to the reptilian orders *Saurischia* and *Ornithischia*, neither of which includes any of the four families just named. Like other vertebrate fossils, dinosaur remains had been discovered occasionally for centuries before their true identity was recognized. One such fragment, for example – probably from the femur of a megalosaurus – was described and illustrated so early as 1677 by Robert Plot. Edward Lhwyd, whose book on English fossils (1699) included ichthyosaurian remains, also depicted a megalosaurian tooth from Stonesfield. Other probably megalosaurian remains appeared in a fossil catalog of 1728 and the *Philosophical Transactions* of the Royal Society thirty years later. The earliest probable French dinosaur discovery was reported in 1776, with the first American one (known later to be a hadrosaur) following in 1787. Cuvier published on "fossil crocodiles" in 1800 and 1808. In 1809 a specimen eventually identified as cetiosaurian was found in Oxfordshire. That same year William Smith, the stratigrapher, discovered chunks of what we now know to be *Iguanodon* at Whiteman's Green, near Cuckfield (Gideon's locale also), but could make nothing of them and never published the find. Similarly, Thomas Webster recalled years later that he had found "large saurian bones" near Hastings, Sussex, around 1812; they may have been iguanodontian also. Cuvier's *Recherches* (1812) then appeared, emphasizing vertebrate fossils and reprinting his papers on fossil crocodiles.

John Kidd, the Oxford mineralogist, next affirmed (in *A Geological Essay*, 1815) that the Stonesfield Slate included "remains of one or more large quadrupeds" (38), which he presumed to be mammalian. Visiting Buckland and Kidd at Oxford in 1818, however, Cuvier proclaimed the bones reptilian. Only three years later, in his *Plesiosaurus* paper, Conybeare announced that "An immense saurian animal, approaching to the character of the monitor, but which, from the proportions of many of the specimens, cannot have been less than forty feet long, occurs in the Great

---

form of the fins (if they deserve that name) of the Proteosaurus," 110 (1820), 159–164, Plates XV–XVI.

For "Joseph Pentland – A Forgotten Pioneer in the Osteology of Fossil Marine Reptiles," see Justin B. Delair and William A. S. Sarjeant, *Dorset Natural History and Archaeological Society Proceedings*, 97 (1976), pp. 12–16; together with William A. S. Sarjeant and Justin B. Delair, "An Irish Naturalist in Cuvier's Laboratory: The [sometimes misdated] Letters of Joseph Pentland, 1820–1832," *Bulletin of the British Museum (Natural History)*, 6 (7) (24 April 1980), 245–319. The definitive osteological analysis, however, was not Pentland's but that of W. D. Conybeare and H. De la Beche, *Trans GS*, 5 (1821), 559–594, which is discussed in text. Regarding Gideon, see WDC–GM, 14 Apr 1821 (ATL).

Oolite at Stonesfield, near Oxford." Professors Kidd and Buckland, he added, "have long been engaged in the study of these interesting remains, and it is hoped may soon communicate the result of their observations to the public" (592). In a footnote to this passage, Conybeare then recorded Gideon's own announcements of saurian remains in the Wealden. The dinosaur was shortly to be born.[10]

Megalosaurus

As we know, Gideon Mantell had been significantly exposed to vertebrate paleontology in 1811 and 1812 through books by Parkinson and Ashe. On 17 April 1813, soon after the appearance of Kerr's Cuvier, the Reverend James Douglas advised Mantell that "the great desideratum of geology is now fixed on *strata* in which the bones of quadrupeds are found; they determine the last great revolution of our planet, antecedent to the present order of created life." On 5 August 1817 Douglas wrote Gideon again, enclosing a letter to himself from Elizabeth Cobbold (of Holywells, near Ipswich) that he wanted returned. "You will perceive in my correspondence with this lady," Douglas pointed out clumsily, "that I had enquired particularly respecting animal bones, quadrupeds, etc." Amateur fossil collectors all over Europe were beginning to do the same.

Mantell's own correspondence with Mrs. Cobbold (a poet and naturalist) then began. At Douglas' request she sent him some duplicate fossils and a letter on 19 August. "The quadruped remains found in our Crag stratum," Mrs. Cobbold explained, "are very few." All the fragmentary specimens she went on to enumerate were provisionally attributed to various kinds of mammals. In his reply later that month, Gideon agreed vertebrate fossils were of the first importance but lamented that British examples usually came "so much mutilated and so partially distinct as to require more than Cuvierian sagacity to determine their prototypes." With further letters she sent sketches, lists of fossils, and an occasional poem. He responded with Sussex fossils, stratigraphical inquiries, and news of

10. For what they are and are not, see Alan Charig, *A New Look at the Dinosaurs* (New York, 1979), Chapter 1. Richard Owen's conception of the dinosaur in 1842 is discussed below, in Chapter 9. The fullest previous history of dinosaur discoveries is Justin B. Delair and William A. S. Sarjeant, "The Earliest Discoveries of Dinosaurs," *Isis*, 66 (1975), 5–25, on which I have drawn extensively; for French discoveries, see note 14 below. Regarding Smith, see Charig, p. 50 (the late Dr. Charig kindly showed me the original specimen). John Kidd, *A Geological Essay on the Imperfect Evidence in Support of a Theory of the Earth* (Oxford). Cuvier, *Recherches*, third edition, V, Part II (Paris, 1825), 344. Conybeare and De La Beche (note 9 above), pp. 592–593n.

his forthcoming *South Downs*. But their correspondence lasted only till 1821; Mrs. Cobbold then died three years later, her collected poems appearing posthumously in 1825.

In November 1817 Mantell not only decided to publish a book on his Chalk fossils but sent Etheldred Benett thirty-two specimens, including a fragmentary mandible of "some marine animal" recovered from the Chalk (*South Downs*, Plate XLI); others derived from quite different strata at his new location near Cuckfield. He then wrote similarly that day to G. B. Greenough about his planned book and some remarkable fossils discovered the preceding season. Besides vertebrae and other lacertian relics found in the Chalk, they included "bones and teeth of amphibia" (i.e., crocodiles) from the Weald. By the summer of 1817, therefore, Gideon had already procured a number of saurian fossils from Tilgate Forest, the region he would soon make world famous and his own.

Subsequent letters to Greenough, in November and January, spoke of "alligator" teeth and bones. But on 1 April 1818 Mantell thought his jaw from the Chalk possibly allied to "the fish of the Lias, described by Sir Everard Home" (i.e., *Ichthyosaurus*) and even identified his Cuckfield teeth and bones with some ichthyosaurian ones in the British Museum. When Gideon's first published discussion of Wealden strata appeared in *Provincial Magazine* that fall, however, he was again convinced of having found the teeth and bones of "alligators or of some animal closely allied to that genus" (9). On the evidence, then, Mantell noticed vertebrate remains below the Diluvium in 1817 and began to collect specimens from the quarries at Whiteman's Green, near Cuckfield, that summer. Recognizing the Tilgate beds as unusual, he may have fossilized at Cuckfield initially to ascertain their correct place in the stratigraphical column. By fall, however, Gideon had obtained specimens that were clearly saurian, and from then on was interested in the gigantic remains for their own sake.[11]

With the beginning of 1819, we have Mantell's journal. As of March, for example, he had already hired a quarryman named Leney, who found by far the greater portion of Gideon's Cuckfield specimens for him. Thus, on 30 June, he received a parcel from Leney that included both fossil bones and teeth. Together with his second medical assistant (one Charles Lashmar, dismissed soon afterward), Gideon went to Cuckfield himself on 6 July but was rained out of the quarries. So far as is known, he did not attempt further personal collecting at Cuckfield for the rest of that

11. JD–GM, 17 Apr 1813; JD–GM, 5 Aug 1817; EC–GM, 19 Aug 1817; GM–EC, ca. 24 Aug 1817; GM–EB, 20 Nov 1817 (all ATL). GM–GBG, 20, 25 Nov 1817; 5 Jan, 1 Apr 1818 (all UCL). GM, "Sketch," *Provincial Magazine*, 1 (1818), 8–11, 68–71; incomplete. See also G. Cumberland–GM, 21 Mar 1819 (ATL).

Lign. 45.—Iguanodon Quarry, near Cuckfield, Sussex. 1820.

1. Blue clay, forming the floor of the quarry.
2. Tilgate grit.
3. Soft sandstone.
4. Drift, or diluvium.

Figure 4.2. Four versions of the quarry at Whiteman's Green, near Cuckfield: *Upper left*: Mantell's own two-man version, dated 1820 (from *Petrif*, p. 202; original lost). *Upper right*: first six-man version, GM's frontispiece to *Geol SE England* (1833), original sketch (F. Pollard, prob. 1826; pencil, 145 × 101 mm; Alexander Turnbull Library, Wellington, New Zealand). *Lower left*: second six-man version, GM's frontispiece to some copies of *Illus*. The nine-man version, GM's frontispiece to some copies of *Illus* (1827). *Lower right*: nine-man version, "Strata of Tilgate Forest," commemorates a visit to the quarry by Buckland, Lyell, and Mantell on 7 March 1824. Mantell, the top-hatted figure at far right, stands behind a specimen of *Sphenopteris Mantelli*, a large fossil fern commonly found at this site.

year, remaining content with generous parcels from Leney in August, September, and October. By 16 September, Mantell had so many duplicates from Cuckfield that he could afford to send boxes of them to Miss Benett, Mrs. Cobbold, Colonel Birch, and Greenough. These were, Gideon assured the latter, his best duplicates of bones he had found in the limestone and sandstone of the Weald. "Although from the mutilated state in which the specimens are usually met with, nothing decisive can be concluded respecting the nature of the originals," he advised, "yet it seems probable that they are the remains of some species of *Ichthyosaurus,* for their teeth and vertebrae bear a close resemblance to those of that animal." Miss Benett replied on 2 October that they were certainly new to her.[12]

Mantell recorded nothing more until 25 February 1820, when Leney delivered his next parcel in person. On 9 May Gideon completed his last drawings for *South Downs.* It is therefore useful to ascertain that the plates of his book (which exclude Tilgate fossils) depict no dinosaurian remains. In four instances, however, he had sketched significant reptilia. Thus, Plate XXXIII, Figure 1, is a plesiosaurian tooth, attributed by its discoverer to an "animal of the lizard tribe." Figure 13, on the same plate, depicts mosasaurian vertebrae, which were identified as such. These designations must have been achieved sometime after 2 April 1821, however, as Gideon's first prospectus (which listed all the plates) referred both specimens to "unknown fishes." Plate XLI, Figure 3, also depicts mosasaurian vertebrae, correctly identified even in 1821. Figure 1 on the same plate, a lower jaw from "an animal of the lizard tribe" in 1821, was demoted to a fish's before the book appeared. No ichthyosaurian remains appeared on any of Gideon's plates and no specimens of any other kind were attributed to *Ichthyosaurus.*

Though too late to be depicted, a large bone arrived from Cuckfield on 10 May 1820. Mantell, who had attended Colonel Birch's sale only the month before, then seemed unusually pleased with Leney's exceptional shipment on 16 June, which included "a fine fragment of an enormous bone, several vertebrae, and some teeth of the *Proteo-saurus,*" as Home was still calling his creature. Though Gideon mentioned these supposedly ichthyosaurian teeth (and other saurian remains) in a brief paper of 1 December, he subsequently reidentified them, there being no mention of ichthyosaurian remains from Cuckfield in the *text* of *South Downs* or in any of his later publications. Confronted with numerous fine specimens

---

12. J (19 Mar, 30 June, 6 July, 6 Aug ["Leney of Cuckfield called on me, and I made some further arrangements with him respecting Cuckfield fossils"], 11 Aug, 1, 16 Sept, 5 Oct 1819). GM–GBG, 16 Sept 1819 (UCL); GM–EC, 16 Sept 1819 (ATL); EB–GM, 2 Oct 1819 (ATL); Lashmar indenture (A. Shelley).

of *Ichthyosaurus* in the summer of 1820, however, Mantell dreamed of finding such treasures in his own domain. On 15 August, accordingly, as part of an unusually elaborate family expedition, his brother Thomas drove Mary Ann, her sister Hannah, and himself to Whiteman's Green in his chaise while Gideon rode alongside on horseback. Disappointingly, the latter collected "nothing of consequence" while there. Such distractions as the trial of Queen Caroline and the death of James Moore prevented further excursions for a time. Gideon was therefore gratified to receive some "good bones" from Cuckfield on 9 October and another batch eight days later. He received still more on 12 April 1821 and some unexpectedly fine ones on 7 June, all without knowing what they were.[13]

The matter of positive identification had now become paramount. Beginning in November 1820, therefore, Mantell began to consult books and people regarding a number of his specimens. On the twenty-ninth, for example, he received from Lupton Relfe a book dealing with the Maastricht animal. J. S. Miller (of Bristol) and Charles König (of the British Museum) returned comments on various specimens. Though he rejected the opinions of both that some of his specimens were ichthyosaurian, Gideon accepted other of their identifications in his first prospectus of 2 April 1821. He then sent copies of that prospectus and selected plates to several of his most respected correspondents, including Sowerby, Webster, Greenough, Conybeare, and König. Plate XXXIII, the latter insisted, depicted two ichthyosaurian specimens, while Figure 13 was "Certainly the vertebrae of a species of crocodile." Plate XLI, Figure 1, seemed as unquestionably lacertian to König as Figure 3 was mosasaurian. Only four days after Gideon's prospectus appeared, however, Conybeare announced the discovery of *Plesiosaurus* and defined *Ichthyosaurus* adequately for the first time. This fine paper effectively convinced Mantell that none of his specimens belonged to either animal. Following 6 April 1821, therefore, Gideon regularly alluded to his unknown fossil saurians as crocodiles.

Mantell's increasing commitment to a specific crocodile became evident around September. "In the limestone of the Oak Tree Clay, or Weald Clay of Sussex," he then revealed to Conybeare, "numerous remains occur of an animal of the lizard tribe. Fragments of the ribs, clavicle, radius, pubis,

---

13. For the most part, I have simply compared two printed sources; J. B. Delair, however, kindly identified selected specimens at my request. J (25 Feb, 10 May, 16 June 1820). GM, "Remarks on a Fossil Vegetable," *Annals of Philosophy*, ns 1 (1821), 68: "This specimen . . . derives additional interest from the circumstances of its being associated with the remains of the Ichthyosaurus, and of some species of Lacerta." Though J (15 Aug 1820) does not specify the "ladies" who rode with Thomas, previous entries identify them sufficiently. Hannah Woodhouse, Mary Ann's younger sister, was visiting at Castle Place; she later married Lupton Relfe. J (12 Apr, 7 June 1821).

ilium, femur, tibia, metacarpal bones, vertebrae, and teeth have been discovered; and although the specimens are, for the most part, exceedingly mutilated, yet the structure of the original animal is very clearly indicated." Having by now examined large numbers of these relics, Gideon unhesitatingly identified them as "belonging to the same unknown species of crocodile as the osseous remains discovered at Honfleur and Havre" described in 1812 by Cuvier. But the large reptiles of the past had been far more numerous and varied than either Cuvier or Mantell imagined.[14]

On 26 September, Gideon and George Rollo (his assistant) visited Cuckfield by one-horse chaise to research Tilgate strata for *South Downs.* After dining at the Talbot Inn, they drove to the upper quarries, where thirty men were at work, obtaining "a fine lumbar vertebra of a crocodile and a tooth of the same kind of animal." These were probably among the finds displayed when Charles Lyell visited Castle Place on 4 October and saw at least two kinds of teeth. At the Hunterian Museum later on, Mantell showed a particularly well-preserved tooth – surely the one of 26 September – to William Clift, who identified it as belonging either to a crocodile or a monitor [lizard].[15] If so evidently saurian, this tooth must not have

14. J (29 Nov 1820); B. Faujas de Saint-Fond, *Histoire naturelle de la montagne de Saint-Pierre de Maestricht* (Paris, 1799); JSM–GM, 27 Nov 1820 (ATL); GM–CK, 10 Nov 1820 and CK–GM, undated reply (on same sheet [ATL]); GM–JH, 1 Dec 1820 (WSRO); CK–GM, 12 Dec 1820 (ATL); J (12, 14 Dec 1820); WDC–GM, 14 Apr 1821 (ATL); Conybeare and De la Beche (note 9 above). GM–DG, 9 June 1821 (ESRO); GBG–GM, 24 July 1821 (ATL); GM–GC, 28 July 1821 (Paris); J (30 July 1821); GM–WDC, ca. Sept 1821 (*Trans GS,* 5 [1821], 592–593n); GM–GBG, 21 Sept 1821 (UCL). See also JBP–WB, ca. 22 July 1821 (PL, p. 260). Cuvier announced in 1800 that he had discovered a new type of fossil crocodile at Honfleur. In 1808, he then described two fossil crocodiles (including the one from 1800), comparing them with the modern gavial. Of the two types, one seemed rather like modern crocodiles while the other (his "first") belonged to a type hitherto entirely unknown. (On the basis of Cuvier's drawings and the actual specimens, Philippe Taquet has identified the latter as dinosaurian.) In 1812 Cuvier again attributed both types to fossil crocodiles. Six years later, he visited Oxford and identified *Megalosaurus* bones shown him there as reptilian. In his second edition (1824), which noticed Buckland's paper, Cuvier then remarked that one of his Honfleur "crocodiles" possibly belonged to a similar species. Gideon's identification of his own megalosaurian specimens with Cuvier's was therefore correct. See P. Taquet, "Cuvier, Mantell, Buckland, et les dinosaures,"pp. 475–494 in E. Buffetaut, J. M. Mazin, and E. Salmon, eds., *Actes du Symposium Paléontologique G. Cuvier* (Montbéliard, 1984). I wish to thank Professor Taquet for sending me this paper and additional information.

15. "The Monitors form a genus of Lizards, frequenting marshes and the banks of rivers in hot climates; they have received this name from the prevailing, but absurd, notion that they give warning by a whistling noise, of the approach of Crocodiles and Caymans. One species, the Lacerta nilotica, which devours the eggs of Crocodiles, has been sculpted on the monuments of ancient Egypt" (*G&M,* I, 215n).

shown mastication wear. Clift also commented on a number of Wealden bones. As of 25 October, Gideon's diligent utilization of literature, specimens, and expertise available to him had resulted in at least one success. "The vertebrae of the Maastricht monitor," he wrote Greenough that day, "appear to be completely identified. I have spared no pains to investigate this subject and hope my observations will appear satisfactory." Gideon added that he had made no additions to his collection for some time. We do not know that he either received or collected further Cuckfield specimens prior to the appearance to *The Fossils of the South Downs* in May 1822.[16]

Mantell's description of Tilgate Forest relics in that consecutively written book occupied pages 37 to 60 and was, by all indications, composed early in November 1821, perhaps between the sixth and the eleventh. His chapter could not have been finished before 25 October, when Gideon received the Stonesfield fossils mentioned within it from Lyell, nor prior to 4 November, when he received Volume V, Part 2, of the Geological Society's *Transactions* – referred to by Gideon as "just published" (59n) – containing Conybeare's paper of 6 April on enaliosaurs. There are no entries in Mantell's journal for 6 to 11 November (a sign he may have been writing something else) and his wife and family were gone on holiday to Brighton.

On 4 November, moreover, Mantell also received Lyell's letter of the third, which, regarding Stonesfield and Tilgate, asked provocatively: "What weight of evidence do you require to identify beds?" Huge fossil bones, apparently saurian, had been found in both places but were not obviously from the same animal. Gideon replied indirectly to Lyell (who was now immersed in the study of law and had specifically requested his friend not to distract him with geological matters before Christmas) by appending to the Tilgate chapter his brief essay "On the Analogy Between the Organic Remains of the Tilgate Beds and Those of Stonesfield, near Oxford." In arguing his point about the possible identity of beds, moreover, Lyell had further ventured: "Why should there not be many kinds of monitors in each?" This was the needed spur, for Mantell, who already

16. J (26 Sept 1821). There were two adjacent quarries at Whiteman's Green, the "upper" being near a windmill and the road. Clift: *Foss SD*, p. 50. Regarding *Mosasaurus:* GM–GBG, 1 Nov 1820 (UCL); GM–JH, 25 Oct 1821 (WSRO); GM–GBG, 25 Oct 1821 (UCL); GM–DG, 30 Aug 1822 (ESRO). Parkinson announced the genus *Mosasaurus* ("lizard of the Meuse," named by Conybeare) in his *Outlines of Oryctology* (London, 1822), p. 298. See also J (20 Mar 1820); William Buckland, *Geology and Mineralogy* (1836), Chap. 14; and Adrian J. Desmond, *The Hot-Blooded Dinosaurs* (New York, 1977), Chap. 1. Mosasaurs belonged to the late Cretaceous.

possessed a substantial number of fossilized bones and teeth from Tilgate Forest, was now willing to hazard more than a general identification of some.[17]

"The teeth, vertebrae, bones, and other remains of an animal of the lizard tribe, of enormous magnitude," he wrote in his Tilgate Forest chapter, "are perhaps the most interesting fossils that have been discovered in the county of Sussex" (48). Both teeth and "scales" (as he then believed) were generally well preserved, but the bones proved to be so fragmentary and eroded that positive identification was difficult. Nonetheless, Gideon thought it possible to ascertain at least the genus of the animal and possibly its species. He went on to describe some of the "numerous" specimens in his collection, including teeth, vertebrae, and ribs, all of which he attempted to correlate with Cuvier's papers on recent and fossil crocodiles, as reprinted in the final volume of *Recherches sur les ossemens fossiles de quadrupèdes* (1812).

Mantell's thirty-seventh entry catalogued "Some fragments of a cylindrical bone, probably the femur, indicat[ing] an animal of a gigantic magnitude" – which, in a footnote, he believed as bulky as an elephant and no less than thirty feet long (53). Specimens of its bones measured ten to twenty-seven inches, and from eleven to twenty-five inches in circumference. The remains in question, he had ascertained, belonged to an animal very like the crocodile (or perhaps the monitor lizard) but quite different from any recent species. Gideon also thought his bones "precisely similar" to those of the Honfleur crocodile, which Cuvier regarded as an extinct gavial. Yet his Tilgate species "exceeded in magnitude every animal of the lizard tribe hitherto discovered either in a recent or fossil state" (54). As time would establish, it was genuinely a dinosaur, soon to be known as *Megalosaurus*.[18]

---

17. At this time, nearly all long books were printed in consecutive sections, so that type could be redistributed. Latter portions of the text, certain footnotes, captions for the plates, and the preface and dedication then followed, often in that order. Numerous indications throughout *Foss SD* confirm that its printing was similarly conventional. CL–GM, 3 Nov 1821 (ATL); despite Lyell's argumentation to the contrary, however, the two formations are not identical (as he himself later realized). Gideon's comparison of Stonesfield and Tilgate was further developed by Buckland in his *Megalosaurus* paper (1824), then specifically contradicted two years later by Lyell in *Quarterly Review.*

18. Cuvier, 1812, IV, Part 5, Crocodiles fossiles, pp. 16–29. As Mantell admitted, however, Cuvier's crocodiles had been found in significantly earlier strata (*Foss SD*, p. 52n); they were in fact Upper Jurassic whereas GM's were Lower Cretaceous. Gideon probably had specimens of both *Megalosaurus* and *Cetiosaurus* by late 1817 (*Medals*, II, 727).

Iguanodon

Mantell then went on to enumerate some further specimens as the "Teeth and Bones of Unknown Animals" (pp. 54–55) having no counterparts in Cuvier and remaining too obscure to be identified on present evidence; he included them "not in the hope of being able to elucidate their nature, but to record their existence in the Tilgate beds with a view to future inquiries" (54). Among these, several incisors and molars evidently belonged together. Seemingly different, a very worn incisor crown of cuneiform shape was listed separately, together with several smaller examples "discovered by Mrs. Mantell in the diluvial aggregate" (55). Though Gideon did not know it, both groups of teeth had come from the same animal. By November 1821, then, he already possessed a number of *Iguanodon* teeth, some of which had been found by Mary Ann, ostensibly in strata belonging to the Age of Mammals.

Just when the first *Iguanodon* tooth was discovered and by whom have been obscured by Mantell's diverse recollections of the event, of which he published several. *South Downs,* our earliest and most reliable source, confirms only that Gideon had at least six *Iguanodon* teeth by November 1821, several of them found by his wife. In a journal entry of 28 November 1824, Mantell named himself as discoverer of the teeth. His *Iguanodon* paper of 10 February 1825 (to be reviewed shortly) treats the discovery impersonally, while misdating it as having taken place in the summer of 1822 – after *South Downs* had appeared. In 1827 and 1833 he publicly attributed the find to Mary Ann, and as having been made during the *spring* of 1822. Subsequent publications of 1844, 1850, and 1851, however, credited even the first tooth to himself and omitted any date of discovery. His 1851 account, though fullest, must be approached with particular caution, as Gideon was then in poor health and his memory more than usually unreliable. Arbitrarily endorsing both Mary Ann and the impossible summer of 1822, however, Sidney Spokes conflated this misremembered 1851 version with Gideon's demonstrably erroneous statements of 1825 and 1827, thereby creating the standard legend – which has since been further elaborated with purely fanciful details by Edwin H. Colbert, W. E. Swinton, and many others.[19]

19. Source materials regarding the discovery of *Iguanodon: Foss SD,* pp. 54–55, 299; GM–GC, 9 July 1824 (Paris; nothing specific); J (28 Nov 1824), quoted below; "Notice on the Iguanodon," *PT,* 115 (1825), 179–186, quoted below; *Illus,* p. 71; *Geol SE Engl,* pp. 268–271; *Medals,* II, 739–740; *Pict Atlas,* p. 195; *Petrif,* pp. 228–232. Spokes, pp. 18–24.

    One of Mantell's chief reasons for attempting to keep a journal was that he regularly forgot dates. According to his own conflicting accounts, for example, Gideon discovered his first engraved Norman tiles at Lewes Priory in either 1801 or 1809. Later on, he also

In all likelihood, the real supplier of Mantell's first iguanodon tooth was Leney the quarryman, who in March 1818 provided an allegedly ichthyosaurian specimen that Gideon then mentioned to Greenough on 1 April. While this particular tooth was probably not iguanodontian, it appeared soon after Mantell's "first discovery of bones" at Cuckfield, as his conflated 1851 account requires. Leney surely sold him the tooth without knowing what it was. As related in 1851, therefore, Gideon instructed Leney and other quarrymen to be especially alert for teeth, which (as Mantell knew from Cuvier) constituted the best evidence available for the identification of fossil vertebrates. From 1818 on, he often received teeth from Cuckfield.

In June 1820 (just after the drawings for *South Downs* had been completed), Leney sent Mantell a fine shipment, including an enormous bone, several vertebrae, and some teeth. Though Gideon had at first glance called these teeth proteosaurian (attributing them to ichthyosaurs), none were subsequently published as such. Some of these bones, therefore, and perhaps at least one tooth may have been iguanodontian. The unusual promise of this shipment, surely, drew Mantell to Cuckfield, on that unusual family expedition of 15 August. While there, Mary Ann probably found the several smaller iguanodon teeth that *South Downs* attributes to her. (It does *not* state they were the first.) But Gideon's journal entry for that date fails to mention any teeth at all. Less striking than others already in his possession, they may not have come to notice for a day or two. He then spent almost two years, when not distracted by more immediate concerns, collecting further examples and puzzling over their significance.[20]

"remembered" composing sad boyhood verses to commemorate the graves of two fine young women at Penshurst. But Fanny (1796–1820) and Susannah (1799–1821) Allnutt did not die of tuberculosis – the younger having nursed the elder – till Mantell was over thirty. In his first journal entries, Gideon misdated his barrow opening of 1818 by approximately four months. Similarly, he misdated Ellen Maria's birth by a day. (In 1851 and 1852, he would misdate his own birth by a year.) Mantell also thought Buckland's visit to Lewes in 1824 had followed the GS meeting by three weeks rather than two. As I have learned, a number of Gideon's dated letters do not accord with journal entries noting he had written them. Several of his publications are also misleadingly dated. His paper of "1" June 1822, for example, responded to Lyell's letter of the sixth. The *Iguanodon* essay itself is similarly misdated, 1 Jan being most improbable.

If, as I suspect, Mantell originally intended "the summer of 1820" for his discovery of *Iguanodon*, he may simply have written a o that looked like a 2 and then failed to correct the error in proof. Alternatively, he may have conflated events in retrospect (as happened with both the Norman tiles and the Allnutt girls). His substitution of "spring 1822" for "summer" was surely in response to someone's having pointed out the impossibility of the later date. In his last major lecture on *Iguanodon*, however, Gideon finally attested to the proper year (*Lit Gaz*, 17 Apr 1852, p. 355).

20. Since Mantell had no patients at Cuckfield (pronounced Cookfield), it is reasonable to ask how he came to be associated with the area. In the latter eighteenth century, and

Taking the originally discovered tooth and other Tilgate specimens with him to the Geological Society meeting of 21 June 1822, Mantell finally revealed his long-hidden treasures (and a defiantly heterodox interpretation of them) to some of the more important members, including Buckland, Conybeare, and Clift – only to find that none of the three thought the teeth particularly interesting. Buckland, probably, remarked that they belonged either to some large variety of wolf-fish or were mammalian and from the Diluvium. Only William Hyde Wollaston (who was not a geologist) agreed that the remains were those of an unknown, plant-eating reptile; he generously encouraged Gideon to ignore his critics and persevere. By June 1822, then, Mantell had weighed the seemingly mammalian character of those teeth against their stratigraphic occurrence below the Chalk and invented the first herbivorous dinosaur.[21]

We do not have another specific date in the history of *Iguanodon* until 11 June 1823, when Lyell wrote to announce that he would be leaving for Paris toward month's end, to improve his French, and offered to deliver books or other presents as desired; Lyell would use them as introductions to the savants. Mantell availed himself of this opportunity to

throughout the Regency, Cuckfield was a major posting stage on coach runs between Brighton and London (a route Gideon must have traversed frequently between 1811 and 1816). While horses were being changed, passengers either relaxed at one of the inns or wandered about. Beginning in 1817, Whiteman's Green began supplying stone to rebuild the Regent's Pavilion at Brighton. This important industry employed about thirty men, who swiftly enlarged the existing quarry. By so doing they probably exposed saurian bones that came to Gideon's attention the same year ("This Is Cuckfield," anon. booklet, 1967; *Petrif*, pp. 204–206).

Mary Ann's geological interests and collecting activities are noted in J; for example, 31 Mar, 20 Apr, 20 July 1819; 29 Sept 1820; 11 July 1823; also *Illus*, p. 64. For the importance of 1820 to Gideon, see *Petrif*, pp. 155, 197, 202, 204. CK–GM, 12 Dec 1820 (ATL), identifying specimens, strongly suggests that GM possessed several *Iguanodon* teeth and a fine vertebra by then.

21. GM submitted two short papers to the GS on 21 June (OM): one on "Some specimens of the Blue Chalk Marl of Bletchingley," *Trans GS*, ns 1 (1824), 421; and the other "On the Iron-Sand Formation of Sussex," ns 2 (1826), 131–134; they were read 17 Jan 1823 and 21 June 1822, respectively. In the latter, GM realized (with help from Lyell) that his Tilgate beds were part of the Iron Sand formation; he also presented "Teeth and bones of crocodiles and other saurian animals, of an enormous magnitude"; "Teeth of an unknown herbivorous reptile, differing from any hitherto discovered either in a recent or fossil state"; and "Teeth of an animal of the lacertian tribe, resembling those found at Stonesfield near Oxford, and figured by Lhwyd" (p. 134). Mantell, therefore, had remarked publicly on *Megalosaurus* several times before either Parkinson or Buckland had done so once, while mentioning *Iguanodon* as well. The names, of course, came later. Though Gideon never dated the meeting at which his new animal was derided (*Petrif*, p. 229), 21 June 1822 is the obvious choice. OM, moreover, show that Wollaston was present.

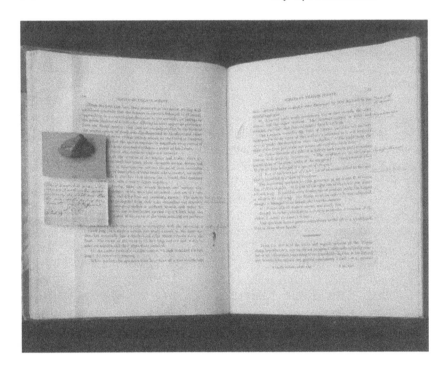

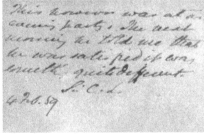

Figure 4.3. Historic teeth composite. Gideon Mantell's own copy of *The Fossils of the South Downs,* open at pages 54 and 55, "Teeth and Bones of Unknown Animals," with his later annotations identifying "Tooth of the Iguanodon" and "Megalosaurus" (Mantell Collection, Alexander Turnbull Library, Wellington, New Zealand). *On page 54*: The first-found *Iguanodon* tooth, mounted on a card identifying it as such, with inscriptions on the obverse by Mantell and on the reverse by Lyell (latter 4 February 1859; collection of the Museum of New Zealand, Te Papa Tongarewa). Close-up photographs by Michael Hall, Museum of New Zealand, Te Papa Tongarewa, Wellington, New Zealand, no. F.003325.

send specimens of his temporarily discredited reptilian herbivore. Arriving in Paris on 25 June, Lyell was graciously invited to attend Cuvier's next weekly soiree, held as usual in his drawing room at the Jardin des Plantes. There, on Saturday, the twenty-eighth, Lyell showed Gideon's worn, original specimen to the world's greatest comparative anatomist – who promptly dismissed it as the upper incisor of a fossil rhinoceros and went on with his party. "When the first-discovered teeth were shown to Baron Cuvier," Gideon later recalled at second hand, "he pronounced them to be the incisors of a rhinoceros; the metatarsals, those of a hippopotamus; the fragment of a femur, with a medullary canal, that of some large mammalian." But these identifications were not immediately challenged.[22]

"A geological friend of mine [i.e., Lyell], who has been at Paris for many months," Mantell reported to Davies Gilbert on 18 November,

> has just returned me a box of specimens which he undertook to submit to the notice of Cuvier, who has favored me with his opinion respecting them. Without going into details, you will I am sure, my dear Sir, be pleased to learn that our beds of limestone, etc. in Tilgate Forest, in addition to the interesting assemblage of organic remains described in my work, contain the bones and teeth of the rhinoceros; and the teeth of a quadruped which Cuvier declares is distinct from any now known either in a recent or fossil state.

Actually, both the "rhinoceros" and the unknown quadruped were *Iguanodon*.

Lyell himself then wrote on 4 December, sending what few specimens he had been able to procure for Mantell in Paris (all the quarrymen there being in the pay of Cuvier or other French fossil collectors). "I paid a visit to Professor Buckland at Oxford the other day," he continued,

22. CL–GM, 11 June 1823 (ATL); Wilson, pp. 115–117. The national Museum of New Zealand, Wellington, preserves (among other GM specimens) a worn *Iguanodon* tooth mounted on cardboard. The top, annotated in GM's hand, has "Vide *Oss. Foss.* Tome V" in the upper righthand corner beside the tooth. Below it he wrote: "Left lower abraded molar. This was the first tooth of the *Iguanodon,* sent to Baron Cuvier, who pronounced it to be incisor of Rhinoceros." On the back, in Lyell's hand, there is a further note: "This however was at an evening party. The next morning he told me that he was satisfied it was something quite different. Sir C. L. / 4 Feb 59" (I wish to thank Garry J. Tee for calling this specimen to my attention). See also John C. Yaldwyn, Garry J. Tee, and Alan P. Mason, "The status of Gideon Mantell's 'first' *Iguanodon* tooth in the Museum of New Zealand Te Papa Tongarewa," *Archives of Natural History*, 24 (1997), 397–421. GM, *Pict Atlas*, p. 195 (quoted). See also GC, *Recherches*, second and third editions, V, Part 2, 350–351; and *Petrif*, p. 228. According to J (21 July 1823) and *Petrif*, p. 230, the metacarpal bones were sent separately.

and mentioned your rhinoceros tooth of the Tilgate beds. He seemed as much inclined to believe it as if we had asserted that a child had been discovered there, and he made a remark that well deserves your attention; viz., that Diluvium debris, with its accompanying fossils, is sometimes mixed up with and as it were introduced into the upper and exposed surface of older strata . . . and thus the remains might be found intermingled in the same rock with fossils of an ancient date.

Buckland's point was well taken, for if Gideon's mysterious tooth could not be firmly placed within older strata, its seemingly mammalian character was more easily explained by assuming it to be no older than the Diluvium.[23]

Annus Mirabilis

During the year beginning 1 February 1824, study of prehistoric reptiles – dinosaurs especially – came of age in Britain, with Gideon Mantell's surely the most important role. Always liking to complete (or at least date) projects on the first of months, he finished a substantial essay on the natural history of Lewes at the beginning of February. Intended primarily for Thomas Horsfield's history of the town, it also appeared independently. In his essay Gideon briefly surveyed his organic remains from the limestone of Tilgate Forest. They now included

the bones, scales, and teeth of a gigantic crocodile or alligator. . . . Also the teeth of the rhinoceros; bones of the elephant, and of some large unknown quadruped; and the teeth, and probably bones, of an herbivorous animal, which M. le Baron Cuvier (who did me the honor to examine them) assures me are perfectly distinct from any previously known, either in a recent or fossil state. (xiv–xv)

The accompanying plate reveals Mantell's "gigantic crocodile" to be *Megalosaurus* and his "herbivorous animal," *Iguanodon*. There could not have been rhinos and elephants in the same stratum with these, but the layers were subjacent at Cuckfield, Quaternary deposits lying directly atop Cretaceous ones, with no intervening Tertiary. Confused by the pro-

23. GM–DG, 18 Nov 1823 (ESRO); CL–GM, 4 Dec 1823 (ATL). The argument had become confused at this point because Lyell, Buckland, and Mantell all accepted Cuvier's identification of the worn *Iguanodon* tooth as rhinoceran. Gideon, therefore, was supposing that a large mammal had existed during the early Cretaceous. Very probably, Lyell and Buckland helped him see how anomalous that idea was. Through further study and collecting, Mantell then realized that the "rhinoceros" and unknown quadruped were identical. (The same kind of logic underlay his *Alcyonia* paper of 1814, in which supposedly distinct fossils were shown to be differing aspects of one.)

longed opposition of his most respected colleagues and the unusual strata, Gideon became temporarily unclear regarding Tilgate fauna and the succession of life generally. Though probably still convinced in his own mind that *Iguanodon* must have been reptilian, he was not yet ready to say so in print.[24]

This prudent conviction may have changed within two weeks. On Friday, 20 February 1824, Mantell attended a regular evening meeting of the Geological Society, at No. 20 Bedford Street, Covent Garden, to hear a pair of spectacular papers. William Conybeare's "On the discovery of an almost perfect skeleton of the Plesiosaurus," his third on that animal, described one of Mary Anning's most stunning finds (which was displayed in the vestibule). Even more interesting to Gideon, however, was William Buckland's "Notice on the Megalosaurus or great Fossil Lizard of Stonesfield," in which the president of the Society, as Buckland now was, officially ascribed huge bones preserved in the Ashmolean Museum at Oxford to an extinct saurian. The idea was not his alone and had obviously been latent since 1818, when Cuvier visited Buckland's museum to identify the Stonesfield relics (which John Kidd had noticed three years before) as reptilian. Gideon then announced related discoveries from Tilgate in his *South Downs* prospectus of 1821 and subsequent letter to Conybeare. After noticing "an immense saurian animal" from Stonesfield in his *Plesiosaurus* paper of 6 April, Conybeare published Mantell's related description (of *Megalosaurus* as Honfleur crocodile) together with his own. On 14 May, Gideon wrote Buckland of the "bones and teeth of an unknown species of lacerta" he had discovered at Cuckfield. By September 1821 Cuvier's assistant, Joseph Pentland, was inquiring about Buckland's "Stonesfield reptile," which Mantell and Lyell discussed on 4 or 5 October. As Cuvier then assured Buckland, "the great Stonesfield beast was a monitor forty feet long and as big as an elephant."

Conybeare's paper of 6 April, with Mantell's remarks appended, circulated in November, while Gideon was further describing megalosaurian specimens in the manuscript of his forthcoming book. When published in May 1822, *South Downs* then became the first English book ever to anatomize dinosaurian remains as such. Gideon also wrote both Buckland and Adolphe Brongniart in June to discuss vertebrate remains from Cuckfield,

24. GM, "Outlines of the Natural History of the Environs of Lewes," pp. iii–xvi (citing geology only) in Thomas W. Horsfield, *History and Antiquities of Lewes and Its Vicinity,* I (Lewes, 1824; also as a separate). Plate XXIX, Fig. 14, is the earliest depiction of an iguanodon tooth ever published. "You say that the tooth – plate XXIX, fig. 14 – is of an herbivorous animal," Lyell responded importantly; "how do you know this? Cuvier, as I understand him, when searching for an analogy to it, looked among the *reptiles*" (CL–GM, 17 Feb 1824 [ATL]).

once more emphasizing how closely they resembled Cuvier's fossil crocodiles from Havre and Honfleur.

A few weeks later, James Parkinson noticed Stonesfield material in his *Outlines of Oryctology* (July 1822), under the heading "Megalosaurus." Crudely but significantly combining teeth and bones, Parkinson envisioned "an animal, apparently approaching the monitor in its mode of dentition, etc., not yet described" but as much as forty feet long and eight high (298). Though sketchy and derivative at best (it was almost pure Cuvier), this was the first reasonably accurate reconstruction of a specifically named dinosaur to be published. Gideon surely read Parkinson's book as soon as it became available to him, and had certainly done so by 20 February 1824, when Buckland gave his belated paper – the first ever concerning a whole dinosaur.[25]

What Buckland actually said to the Geological Society that evening is slightly conjectural in that he rewrote extensively (as was sometimes allowed) before publication. The Oxford professor no doubt began verbosely, as his paper does, by saying he was prompted to lay before the Geological Society some drawings of fossil bones from the skeleton of an enormous animal "in the hope that, imperfect as are the present materials, their communication to the public may induce those who possess other parts of the same reptile to transmit to the Society such further information as may lead to a more complete elucidation of its osteology." Though only a few parts of its skeleton were presently known, these were sufficient to ascertain its zoological affinities. "While the vertebral column and extremities much resemble those of quadrupeds," Buckland observed, "the teeth show the creature to have been oviparous, and to have belonged to the order of saurians or lizards" (390). By comparing its thighbone (which, in one specimen at Oxford, was two feet nine inches long) with that of an ordinary lizard, Cuvier had deduced that the Stonesfield reptile must have been more than forty feet long and as bulky as an elephant.

25. GS (OM), 20 Feb 1824; *Petrif,* p. 343. WDC, "On the discovery of an almost perfect skeleton of the Plesiosaurus," *Trans GS,* ns 1 (1824), 381–389. WB, "Notice on the Megalosaurus or great Fossil Lizard of Stonesfield," *Trans GS,* ns 1 (1824), 390–396.
Cuvier, Kidd, GM, and WDC are all cited in note 10 above. GM–WB, 14 May 1821 (ATL); GM–GC, 28 July 1821 (Paris) is more cautious. JBP–WB, 20 Sept 1821 (PL, pp. 261–262). CL–GM, 3 Nov 1821 (ATL), quoting Cuvier via Buckland and Greenough. Captain James Vetch (of the Royal Engineers) presented some megalosaurian specimens of his own to the Geological Society in March 1822: GS (OM), 15 Mar and 17 May 1822; also *DNB.* GM–WB, 2 June 1822 (ATL); GM–Ad. Brongniart, 22 June 1822 (ATL). JP, *Outlines of Oryctology* (London, 1822); in the same book, Parkinson renamed GM's *Ventriculites* as *Mantellia* (pp. 53–54). WB–GC, 9 Feb 1823 (Paris) mentions WB's forthcoming *Megalosaurus* paper, together with many details relevant to WB's *Reliquiae Diluvianae* (1823).

Buckland then proceeded to further discoveries at Stonesfield (the fuller historical import of which would not be realized for years). Among these were two minute portions of mammalian jaw identified as such by Cuvier. Without "the highest sanction," Buckland confessed, he would have hesitated to acknowledge these specimens, which formed "a case hitherto unique in the discoveries of geology; viz., that of the remains of a land quadruped being found in a formation subjacent to Chalk" (391; for him, *Megalosaurus* was not primarily a land animal but "probably . . . amphibious" [390], like the crocodile and tortoise). The real significance of Buckland's tiny mandibles was that they were, by millions of years, the oldest mammalian relics yet discovered and, together with further evidence, would prove dinosaurs and mammals to have been contemporaries. Among other discoveries, Buckland had also recovered "bones of large cetaceous [whale-like] animals" (392). They actually belonged to *Cetiosaurus,* a brontosaur-type creature, but the Oxford professor remained thoroughly unaware that he was announcing two dinosaurs rather than one.

Following the papers of Conybeare and Buckland, Mantell provided a third, less formal presentation, having brought with him to the meeting enough megalosaurian specimens to convince Buckland that his researches had been based on insufficient evidence. Two weeks later, Gideon recorded in Lewes, "Professor Buckland came express from Oxford, with my friend Mr. Lyell, to inspect my Tilgate fossils. I had met the Professor at a meeting of the Geological Society about three weeks since and shown him some specimens of bones and vertebrae of the *Megalosaurus* from Tilgate Forest." The next day Gideon escorted Buckland and Lyell to Cuckfield, but we do not know what else happened that rainy afternoon – because the rest of this journal entry and all others for the next eight months have been torn out.[26]

26. Re Buckland, GS (OM) reads in full:

"A notice was read on the Megalosaurus or great Fossil Lizard of Stonesfield, near Oxford, by the Revd Wm Buckland, F.R.S., President of the Geological Society of London & Professor of Mineralogy & Geology in the University of Oxford, etc., etc.

The author observes that he has been induced to lay before the Society the accompanying representations of various portions of the skeleton of the fossil animal discovered at Stonesfield, in the hope that such persons as possess other parts of this extraordinary reptile may also transmit to the Society such further information as may lead to a more complete restoration of its osteology. No two bones have been discovered in actual contact with one another, excepting a series of the Vertebrae. From the analogies of the teeth they may be referred to the order of the Saurians or Lizards. From the proportions of the largest specimen of a fossil thigh bone, as compared with the ordinary standard of the Lacerta, it has been inferred that the length of the animal exceeded forty feet & its height seven feet. Professor Buckland has therefore assigned to it the name of Megalosaurus. The various organic remains which are found associated with this gigantic

The impact of Mantell's collection on Buckland is adequately attested by the latter's published paper. *Megalosaurus* aside, however, Gideon remained quite unable to convince Buckland that the mysterious teeth from Tilgate Forest were saurian. Mantell next sent a further selection of them to Cuvier, who on 20 March confessed his bewilderment to Buckland. "Cuvier is much puzzled about your large brown teeth," Buckland then advised Gideon in June. "I think they must have come from some fish like the *Diadon* or *Tetradon*," he added. Well along with his new edition, Cuvier hoped Mantell would send him sheets and plates of his forthcoming book on Tilgate fossils in time for them to be included. On 28 February, Cuvier's English assistant (J. B. Pentland) had written Buckland at Cuvier's request to find out whether or not the *Megalosaurus* paper had yet appeared. As of early June it just had, for Buckland sent Gideon a copy of it with his letter. After having relevant portions translated into French, Cuvier included them in the second (1824) and third (1825) editions of *Recherches sur les ossemens fossiles.*[27]

In these almost identical editions, Mantell was treated extremely well. An entire section discussing fossil crocodiles, for example, now bore his name. *The Fossils of the South Downs,* copies of which Gideon probably sent to Paris with Lyell in 1823, emerged as "un ouvrage très-intéressant," the chapter on Tilgate especially. After its publication, moreover, the author has continued to make even more discoveries, some of them megalosaurian; others included extremely singular teeth apparently from a reptile, but worn like those of an herbivore. Cuvier depicted several *Iguanodon* teeth in his plates. Gideon appeared also in Cuvier's sections on fossil tortoises, on *Megalosaurus,* and again separately regarding his unknown, puzzling saurian. Originally, Cuvier admitted, he had thought the worn

Lizard form a very interesting and remarkable assemblage. After enumerating these the author concluded with a description of the plates and observations on the anatomical structure of such parts of the Megalosaurus as have hitherto been discovered." J (6 and 7 Mar 1824). As subsequent chapters will establish, the two small mammalian fossils from Stonesfield, identified by Cuvier in 1818, would soon prove to be of immense theoretical importance.

27. Regarding WB's *Megalosaurus* paper, it is obvious that page 391, lines 4–22, must have been added later; Mantell subsequently reassigned the femur in question to *Iguanodon*. Pages 393–394, the parts dealing with GM, were also added. Omitting these sections probably allows us to reconstruct the original paper – and demonstrates that fully half of what WB eventually published on *Megalosaurus* actually came from Gideon.

Cuvier probably received a handwritten version of Buckland's paper soon after it was given. Just one month later he wrote Conybeare in praise of Buckland's work (GC–WDC, 20 Mar 1824; ATL Stutchbury Papers). GC–WB, 20 Mar 1824 (RS); WB–GC, 2 June 1824 (Oxford); WB–GM, 2 June 1824 (ATL). JBP–WB, 28 Feb 1824 (PL, p. 304); Cuvier's discussion of *Megalosaurus* appears in *Recherches,* second and third editions, V, Part 2, 343–350. See also pp. 161–163; 232; and, for unnamed *Iguanodon,* pp. 350–352 with plate.

teeth shown him by Lyell to be mammalian, like those of a rhinoceros; only after Mantell had sent him a graded series of deeply worn to nearly unworn teeth did the author become convinced of his error.

On 20 June 1824, a heroic day in the annals of paleontology, Cuvier wrote to Mantell in French:

> I have waited to give you my opinion [he began, regarding Gideon's unusual specimens] until I had time to examine them. Today, since I have done just that, I hasten to express my gratitude, and to offer a few ideas inspired by the curious teeth which are part of your package, and the drawing from the memoir you are going to publish concerning them.
>
> These teeth are certainly unknown to me. They are not from a carnivore; nonetheless, I believe they belong . . . to the order of reptiles. From their exterior one could also take them for fish teeth, similar to tetradons or diadons – but their interior structure is very different. Might we not have here a new animal, an herbivorous reptile?

Many of Mantell's large bones, Cuvier suggested, probably belonged to the same creature. Above all, he admonished, seek for a fossil jawbone containing a number of teeth, for such a specimen would unlock the real structure of this enigmatic animal. He asked to be informed of any further discoveries.

Mantell had at last obtained from Cuvier that validation of his own conjectures which at the time seemed necessary. Gideon's reply of 9 July was naturally enthusiastic. "Nothing could be more gratifying to me," he wrote, "than your confirming me in the opinion that the curious teeth of Tilgate Forest belong to an unknown animal, and are not those of the *Diadon,* as Professor Buckland and others of my friends would insist upon." Further passages in the same letter concur that the "unknown reptile" must have been "herbivorous."[28]

Gideon next attempted to find out what kind of reptile *Iguanodon* might have been. In August or September, probably, he went to London for a day, ransacking collections at the Hunterian Museum (of the Royal College of Surgeons) with William Clift, the curator, for whatever jaws and teeth of living reptiles they could find.[29]

---

28. The date of Mantell's parcel is unknown; it must have been within the six months following November 1823. GC–GM, 20 June 1824, is known only from GM's transcriptions of it in his publications (*Petrif,* p. 231, e.g.) and a version by Richard Owen at BMNH erroneously dated 26 June. All GC–GM letters (at least three) have been lost. GM–GC, 9 July 1824 (Paris). Gideon later claimed (correctly, I believe) to have declared *Iguanodon* an herbivorous reptile on 21 June 1822 (note 21 above). Cuvier had long since generalized that herbivores were larger than carnivores.

29. Though based on assumptions about the history of life that differ from our own, this comparative method of Cuvier's was utilized to some considerable degree by everyone

The Hunterian Museum, Royal College of Surgeons

Figure 4.4. The Hunterian Museum, Royal College of Surgeons, from *Illustrated London News,* 1844.

who seriously attempted to interpret the bones of unknown animals. As an approach it seemed reasonable enough, but Cuvier had developed his procedures while dealing with mammalian fossils of the recent past – *earlier* horses, elephants, cats, and so on. The method worked less well with animals more geologically remote, and while indispensable to the discovery of dinosaurs, possibly retarded their interpretation. (These remarks on Cuvier's methodology endorse others made earlier by William Coleman and Adrian J. Desmond.)

Nothing that either Clift or Mantell could turn up seemed promising –
until Samuel Stutchbury (1798–1859), Clift's assistant since 1820, directed Gideon's attention to the three-and-a-half-foot skeleton of an
iguana he had recently prepared from an old specimen long preserved in
alcohol. Osteologically, the large tropical American lizards (whose Spanish name derived from a native Haitian one) were clearly distinct from
both monitors and crocodiles, as Cuvier had established prior to 1812.
But Stutchbury was not the first to suppose them analogous with reptiles
of the past; Parkinson (1811) and Pentland (1821) were earlier. Gideon
himself mentioned "inguanas" in *The Fossils of the South Downs* (1822),
with so little familiarity as consistently to misspell their name. Despite
these predecessors, therefore, Stutchbury's became the decisive suggestion, and the coincidence of his curatorial labors with Cuvier's change of
mind helped to give the dinosaur a form it has ever since retained.

To his great delight, Mantell found that the iguana's teeth, though small,
closely resembled the much larger ones from Tilgate. On 28 September, he
wrote Clift from Castle Place.

> My dear Sir,
>
> Since I had the pleasure of seeing you in Town, I have endeavored
> to obtain a specimen of the *Iguana tuberculata,* that I might intro-
> duce a sketch of the jaw to illustrate the history of my fossil herbiv-
> orous reptile, but I can neither borrow nor purchase one. Under these
> circumstances may I again intrude on your indulgence and beg per-
> mission to make a sketch of the lower jaw of the iguana in the Mu-
> seum? Mr. Stutchbury, who is an old acquaintance of mine [son of
> the London dealer], would I am sure execute it for me. I only want
> an outline of part of the jaw, with the teeth magnified so as to show
> the mode of dentition.

Clift replied on 26 October, enclosing a beautiful drawing of the iguana's
jaw and teeth that Gideon used in his sequel to *The Fossils of the South
Downs* three years later. The specimen, Clift added, belonged to Stutchbury himself, and was "the common edible iguana of the West-India Islands," but how it had come to London from Barbados was unknown.[30]

---

30. For Stutchbury, hitherto little known, see Michael D. Crane, "Samuel Stutchbury
(1798–1859), Naturalist and Geologist," *Notes and Records of the Royal Society of
London,* 37 (1983), 189–200; and David F. Branagan, "Samuel Stutchbury and His
Manuscripts," pp. 7–15 in M. E. Hoare and L. G. Bell, eds., *In Search of New Zealand's
Scientific Heritage* (Wellington: RSNZ Bulletin 21, 1984). *Petrif,* p. 230. Parkinson,
Org Rem, III, 287, 293, 294–295, 299; *Pentland letters,* p. 274 (2 July 1821). GM–WC,
28 Sept 1824 (BMNH); WC–GM, 26 Oct 1824 (ATL; note *Petrif,* p. 230n); GM–GC,
12 Nov 1824 (Paris) includes *Iguanosaurus* as herbivorous reptile.

Mantell probably failed to attend the Geological Society's first meeting of the season on 5 November – they met on the first and third Friday of every month, November to June. He went on the nineteenth in any case, evidently with Cuvier's precious letter of 20 June in hand. What a pleasure it must then have been for him to confront his former critics, who, more than two years earlier, had sneered at his epoch-making fossils and, rather openly, at himself – the nonuniversity son of an unlettered shoemaker. Now all that was changed, for Gideon had done more than win Cuvier's sanction or approval. Almost singlehandedly, and despite the virtually unanimous disapproval of his most esteemed colleagues, he had persuaded the French authority to change his mind, yet so graciously that Cuvier had willingly recanted in public. It was an astonishing accomplishment for a relatively obscure provincial surgeon still only thirty-four. Even before the appearance of his now-awaited paper on the creature he was calling *Iguanosaurus,* Gideon Mantell had entered the inner circle of the Geological Society, attaining from Cuvier's letter alone a status that his expensive and much-labored book, good as it was, had been unable to procure.

The ensuing year would be one of recognition and successive honors. As if knowing that, Mantell was soon in an exultant mood; he celebrated it repeatedly by aping another of his heroes, the poet Byron, who had died nobly in Greece that April. Thus, when a severe hurricane ravaged the coast of Sussex on 23 November (an event remembered in several local histories) Gideon chose to relish the destructive energies of nature, which, as the poet had written of them, reduced all that man built to insignificance. He therefore drove to Brighton, arriving between 1 and 2 P.M. when the sea was raging so violently that the recently completed Chain Pier was seriously damaged, its railings washed away and much of its platform destroyed. Nevertheless, Gideon ventured to its farthest end; though drenched thoroughly by the waves, he felt amply compensated by the grandeur of the scene.[31]

Meanwhile, Cuvier's endorsement of Mantell's remarkable saurian discovery was creating a storm of its own, for on that same day Buckland wrote obligingly: "I readily agree to the disposition Mr. Clift has made of your iguana-like teeth, and see myself no objection to your name of 'Iguanosaurus' for this animal." He was also willing to give up to *Iguanosaurus* some bones previously claimed for his own *Megalosaurus.* Buckland urged

---

31. A partisan on behalf of Greek independence from the Turks, Byron died of fever at Missolonghi on 19 Apr 1824; all of Europe was shaken by the news. For the next decade at least, he remained a cult figure in England, with GM among his most ardent devotees. J (23 Nov).

Gideon to acquire skeletons of iguana and crocodile for comparisons, hoped to see him and some specimens at the next meeting of the Geological, and invited him to dine beforehand with the exclusive Geological Club at the Thatched House Tavern in St. James Street.

Conybeare, writing probably the same day, urged Mantell to rename his creature slightly: "Your discovery of the analogy between the iguana and the fossil teeth," he wrote, "is very interesting, but the name you propose, 'Iguano Saurus,' will hardly do because it is equally applicable to the recent iguana. 'Iguanoides' (like an iguana) or 'Iguanodon' (having the teeth of an iguana) would be better." Gideon, who received both this letter and Buckland's on the twenty-fourth, accepted Conybeare's second coinage almost immediately, for he noted in his journal on Sunday, the twenty-eighth, that he had had numerous applications during the past week "respecting the new animal whose teeth I have discovered in the sandstone of Tilgate Forest, and which I have named the *Iguanodon*."[32]

32. WB–GM, 23 Nov 1824 (ATL); WDC–GM, ca. 23 Nov 1824 (ATL). J (28 Nov). See also *New Monthly Magazine*, 12 (Dec 1824), 575.

5

The Geology of Sussex

On New Year's Day 1825, when otherwise occupied, Gideon Mantell supposedly completed and signed his most famous publication, "Notice on the Iguanodon, a newly discovered fossil reptile, from the sandstone of Tilgate forest in Sussex," which he cast in the form of a letter to Davies Gilbert, who then read it before the Royal Society of London on 10 February 1825.

The sandstone of Tilgate Forest, Mantell explained, "is a portion of that extensive series of arenaceous strata which constitutes the Iron Sand formation" (179). It contains fossil plants, shells, fishes, birds, turtles, and especially saurians, including *Crocodile, Megalosaurus, Plesiosaurus,* and (we now know) *Iguanodon.* Teeth of the first three are easily distinguished, he wrote glibly, "but in the summer of 1822 others were discovered in the same strata which, although evidently referable to some herbivorous reptile, possessed characters so remarkable that the most superficial observer would have been struck with their appearance" (180). Gideon had then transmitted specimens of the teeth to naturalists at home and abroad – but among them all only the illustrious Baron Cuvier offered a helpful opinion (and here Mantell quoted, in the original French, most of Cuvier's letter to him of 20 June 1824).

Though led by Cuvier's remarks to further research, Gideon still had only teeth to work with. "Among the specimens lately collected," he went on,

> some however were so perfect that I resolved to avail myself of the obliging offer of Mr. Clift (to whose kindness and liberality I hold myself particularly indebted) to assist me in comparing the fossil teeth with those of the recent lacertae in the museum of the Royal College of Surgeons. The result of this examination proved highly satisfactory, for in an iguana which Mr. Stutchbury had prepared to present to the College, we discovered teeth possessing the form and structure of the fossil specimens. (181–182)

His accompanying plate displayed part of the upper jaw of Stutchbury's iguana and seven iguanodon teeth, including an especially large one; the

first-found tooth, "much worn by mastication"; and the small but unusually distinct tooth of a young iguanodon. While the precise nature
of the new animal was still uncertain, Mantell believed it probably amphibious though not marine, inhabiting rivers or freshwater lakes instead.
In either case, "the term *Iguanodon,* derived from the form of the teeth
(and which I have adopted at the suggestion of the Rev. William Conybeare), will not, it is presumed, be deemed objectionable" (184). *Iguanodon* was the second dinosaur to be named.

The skeletons of both *Megalosaurus* and *Iguanodon* were so little known,
Gideon continued, that it was impossible to assign bones to one animal
or the other. Since iguanodon teeth had not been found at Stonesfield, however, those bones found in Tilgate Forest which resembled others found
at Stonesfield may be megalosaurian, while those unique to Tilgate may
be iguanodontian. On this basis, the large femur in Mantell's collection that
Buckland, in his paper of 1824, had assigned to *Megalosaurus* probably
belonged to *Iguanodon* instead. If so, *Iguanodon* was clearly the larger of
the two animals and may have attained a length of more than sixty feet.
Daring as this conjecture was, it seems to have been respectfully accepted.
With Gilbert's presentation of Gideon's paper, then, on 10 February 1825,
the herbivorous, land-dwelling dinosaur became a scientific reality.[1]

Aftermath

On 4 February 1825, a day after his thirty-fifth birthday, Mantell ranked
among the elite few named to seats on the prestigious governing body of
the Geological Society, its Council. That same month, the final volume of
Cuvier's new *Recherches* (1821–1824 edition) arrived in England, to be
much appreciated there for its generous commendation of recent British

---

1. GM, "Notice on the Iguanodon, a newly discovered fossil reptile, from the sandstone of
   Tilgate forest in Sussex," *PT,* 115 (1825), 179–186 and Plate XIV. Summarized in *New
   Monthly Magazine,* 15 (Oct 1825), 444–445. ("If I had not found the teeth of the iguan-
   odon, neither the vertebrae nor femurs in my possession would have led me to hazard the
   conjecture that the bones of Tilgate were not those of the megalosaurus of Stonesfield"
   [GM–RIM, 14 May 1826 (GS)]). As Peter Bowler (*Fossils and Progress,* p. 20) has noted,
   except for the pterodactyl, all the earlier saurian discoveries had been of marine genera:
   Mosasaurus, Goniopholis, Ichthyosaurus, and *Plesiosaurus.* Though Buckland and
   Mantell clearly distinguished *Megalosaurus* and *Iguanodon* from the enaliosaurs (at
   least), both creatures were initially regarded as amphibian. GM's *Illus* (1827) is the first
   book anywhere to place dinosaurs firmly on land (p. 83). His "Age of Reptiles" and the
   poem derived from it (both 1831) include unequivocal statements. Decisive evidence then
   followed with Gideon's discovery of the armored dinosaur *Hylaeosaurus,* which was evidently not a swimmer.

work on fossil saurians, a field in which Gideon's name stood preeminent. As he noted proudly, "Cuvier has made handsome mention of me in the second part of his fifth volume." To Davies Gilbert, Mantell then wrote:

> You will I am certain be pleased to learn that in Cuvier's last volume, which has just appeared, he has paid most particular attention to my discoveries in Sussex, and has mentioned in the most flattering manner my intention of publishing the fossils of Tilgate Forest. He has figured and described several of the teeth of my new animal, but his remarks do not in the slightest degree affect the views I have taken in my paper. Nor is there indeed a word in that memoir which, since the perusal of Cuvier's work, I could wish to alter. This is truly gratifying.

Cuvier himself received Gideon's "most respectful acknowledgments" by letter, for the handsome manner in which his researches had been noticed. The name of Gideon Mantell had now become a familiar one within the scientific circles of both London and Paris.[2]

Once his *Iguanodon* paper had so clearly succeeded, Gideon resumed some earlier maneuvering, through Gilbert and others, for election to England's most prestigious scientific organization, the Royal Society. These efforts blossomed so early as 24 March, when a petition on his behalf circulated easily among the membership. "Gideon Mantell Esq. of Lewes in Sussex," it read, "a gentleman well skilled in general science and particularly in geology, and known to this Society as the author of a paper on the fossil genus *Iguanodon,* being desirous of becoming a fellow," was duly nominated by William Buckland, Adam Sedgwick, William Conybeare, William Fitton, Davies Gilbert, Charles König, G. B. Greenough, John Abernethy, William Clift, and others. As was customary, Gideon's official election would not take place until fall (24 November), but with such powerful nominators supporting him, its pro forma outcome was already certain. In a short ceremony on 22 December, with Sir Everard Home presiding, Charles Babbage and John Herschel then formally introduced Mantell to the Society. Afterward, he wrote, "It was with no small degree of pleasure that I placed my name in the Charter Book, which contained that of Sir Isaac Newton and so many eminent characters." As his friend Robert Bakewell would assure him later on, Gideon had ridden on the back of his *Iguanodon* into the Temple of Immortality.[3]

2. He served for only one year. After moving to Clapham Common and then to London itself, however, Mantell was reelected to the Council in 1841–1844 and 1847–1852; he also served as vice-president from 1848 to 1850. J (Mar 1825). GM–DG, 1 Feb 1825 (ESRO). GM–GC, 21 Mar 1825 (Paris, Inst). *Petrif,* p. 171.

3. GM–DG, 12 Nov 1823 ("I have long been anxious to become a fellow of the Royal Society"); 1, 18 Feb 1825 (all ESRO); nomination/election petition (RS). RS [J. F. W. Her-

Illustrations

Mantell, we recall, had already begun to plan an additional work devoted exclusively to his Tilgate Forest discoveries even before *The Fossils of the South Downs* appeared. Various hints within its text manifested themselves more concretely when the author announced to G. B. Greenough so early as 21 June 1822 that he intended to publish an "appendix" to *South Downs* dealing with the extraordinary Tilgate fossils. As Gideon proceeded, his enthusiasm increased. "The strata at Tilgate Forest," he wrote Davies Gilbert on 30 August 1822, "are a mine of interest. Every week brings me some new acquisition from the laborers I have there in my pay. . . . In fact, I believe those beds will vie with the far-famed quarries of Stonesfield." He then reported further to Gilbert in November 1823 regarding Cuvier's various identifications of Tilgate fossils. They made him "more than ever desirous of publishing sketches of the organic remains of Tilgate Forest." Though such an opportunity did not immediately materialize, solicitations must have gone out in November 1824 because Conybeare responded by subscribing; that same month, Lyell similarly alluded to Gideon's forthcoming book. A fortnight in advance of his *Iguanodon* paper, the *Sussex Advertiser* announced Mantell's projected "Fossils of Tilgate Forest," with nearly thirty plates. When the final volume of *Recherches* arrived in February 1825, Cuvier's voice then augmented those already urging Gideon to publish specifically on Tilgate Forest.

Late in July, a formal prospectus appeared. Cuvier's insistence (it said), and several exciting discoveries made since *South Downs* had appeared, seemingly necessitated a further tome. The promised book, slated for publication later that year, would be "A Supplement to the Geology of Sussex, comprising The Fossils of Tilgate Forest." All twenty-two plates were then listed, together with the names of twenty-four subscribers. This last was, of course, a disappointing number, and Gideon, who had incurred serious financial losses with his first book, saw no good reason to repeat the experience. Thus, *Illustrations* languished once again.

On 14 May 1826 Mantell wrote Roderick Murchison (1792–1871), whose outstanding geological career had only recently begun, to discuss some reptilian bones the latter had found. "I hope you will oblige me with a copy of your paper as soon as possible," Gideon urged, "for I am just

schel]–GM, 25 Nov 1825 (ATL); GM–JFWH, 27 Nov 1825 (RS); also GM–DG, 21, 26 Nov 1825 (ESRO); DG–GM, 2 Dec 1825 (ATL); GM–DG, 5 Dec 1825 (ESRO); DG–GM, 8 Dec 1825 (ATL). J (Nov–Dec entries, esp. 22 Dec); RB–GM, 30 Nov 1829 (ATL). Gideon also celebrated his election to the RS by considering expensive ichthyosaur specimens offered by Mary Anning (–GM, 24 Nov 1825 [ATL]). J (17 Sept 1829, re RB, iguanodon, and GM's immortality).

now writing the description of my saurians." Mantell had also surveyed
the coast from Beachy Head to Winchelsea, so that his remarks concerning the Hastings Sands formation would be accurate. Next day, he drove
to Eastbourne to make sections of the cliffs and afterward went overnight
to Hastings for the same purpose. On 30 May, Mantell was at Cuckfield,
drawing the quarries (his frontispiece). At Godstone on 5 June he then
sketched a section of Tilvester Hill. These researches are clearly reflected
in Gideon's text, in his geological map, and in colored sections accompanying the latter. By mid-June his drawings were done.

In a public letter of 2 August to Robert Jameson, the Edinburgh professor and editor, Mantell once more promised that his work on the Tilgate fossils would soon appear. According to his journal, he actually began its relatively brief text nine days later. The final stages of production
were then easily accomplished. During the first week in December, Gideon
sent Lupton Relfe all the completed plates, wrote the dedication and preface, and on Friday the eighth received the first book off the press. Two days
later he sent Murchison, as secretary of the Geological Society, a copy, together with some illustrative specimens. Davies Gilbert, once more the
dedicatee, was sent his special copy on the fourteenth, while those to other
subscribers were delayed until Christmas Day. By then, a first review (possibly by the author himself) had already appeared, for the *Sussex Advertiser* was full of praise on 18 December.[4]

Despite its lengthy title, *Illustrations of the Geology of Sussex,* "Containing a general view of the geological relations of the south-eastern part
of England; with figures and descriptions of the fossils of Tilgate Forest,"
consists of only ninety-two pages and twenty-two plates. Though dated
1827, it was published a month early, in an edition of but 150 copies.
Gideon's attractive frontispiece appropriately commemorates the only
joint visit to his quarry at Whiteman's Green by Lyell, Buckland, and him-

4. *Illustrations:* GM–GBG, 1 June 1822 (UCL); GM–DG, 30 Aug 1822, 12 Nov 1823 (both
ESRO); GM–GC, 12 Nov 1824 (Paris; re Plate XVII. Plates XLV and XLVI had already
appeared in GM's fossil vegetable paper of 1824); WDC–GM, ca. 23 Nov 1824 (ATL);
CL–GM, 24 Nov 1824 (ATL); *SA,* 31 Jan 1825, p. 3 (as *The Fossils of Tilgate Forest*);
prospectus (W. Lee, Lewes, 1825) to GBG, cancelled 30 July 1825 (GS). GM–RIM, 14
May 1826 (GS); RIM, "Geological Sketch of the North-western Extremity of Sussex,"
Trans GS, ns 2 (1826), 97–107, with numerous mentions of GM and an important
GC–RIM excerpt; J (15, 17–18, 30 May, 5 June 1826); GM, "Remarks on the Geological Position of the Strata of Tilgate Forest in Sussex," *ENPJ,* 1 (1826), 262–265 (p. 262);
J (11 Aug 1826); W. C. Trevelyan–GM, 26 Aug 1826 (ATL); *SA,* 2 Oct 1826, p. 3; J (29,
30 Nov, 8 Dec 1826); *SA,* 18 Dec 1826, p. 2. Two years later, John Phillips published
Illustrations of the Geology of Yorkshire (York, 1829; London, 1836) with GM as a subscriber. What might have been a series of such titles effectively terminated in 1835 with
the foundation of the Geological Survey.

self; it took place, as we saw, on 7 March 1824, two weeks after Buckland had presented his paper on *Megalosaurus*.

The text then began with a synopsis of the geology of Sussex and a comprehensive table of its strata, starting with the uppermost and newest. Alluvial Deposits included Alluvium ("The effect of causes still in operation") and Diluvium ("The effect of causes no longer in action"). Tertiary Formations, partly marine and partly freshwater, included the Plastic Clay and London Clay. Secondary Formations included the Chalk and Shanklin Sand, both wholly marine; and the Weald Clay and Hastings Sands and Clays, both of which were freshwater. Gideon now regarded the strata of Tilgate Forest, in which his saurian relics had been found, as part of the Hastings Sands rather than the Iron Sand. Lowermost in Sussex were the Ashburnham beds, at the base of the Hastings Sands. Each division passed in review, beginning with Alluvium, but almost half this section concerned the Hastings formations, which he had traced in a number of new localities.

Throughout his stratigraphical review, Mantell's geology was more stridently catastrophist than formerly. Thus, in discussing the Alluvium, he remarked that geological changes now in operation, "even if carried on upon an extended scale, are manifestly so unimportant, and so inadequate to produce any of the grand revolutions which constitute the principal objects of geological inquiry" that he thought it unnecessary to devote much time to their investigation (4). Subterranean forests, for example, had "evidently been torn up by some sudden eruption of the sea" (6), and the Chalk itself had suffered extensively from successive catastrophes. Lyell excepted, Buckland, Conybeare, Sedgwick, and most of Gideon's geological friends held roughly similar beliefs, but Mantell himself owed more than any of them to the generosity of Baron Cuvier and responded in kind by endorsing the Parisian's views almost totally.[5]

Despite its more advanced stratigraphy and evident catastrophism, the most important part of Gideon's *Illustrations* was his section on "Saurian Animals, or Lizards" beginning on page 63. Remains of at least four gigantic genera – *Crocodile, Megalosaurus, Iguanodon,* and *Plesiosaurus* – had been discovered in the strata of Tilgate Forest, but never as connected skeletons. Though teeth of the four genera proved readily distinguishable, isolated bones were not. Mantell therefore discussed in turn remains

---

5. Lyell's *Quarterly Review* essay of 1826 (34: 507–540), in contrast, had been deeply ambivalent toward Cuvier. GM's rather one-sided correspondence with GC, including fossils and drawings, began in 1821. They met briefly in 1830 (see below) and GM was later invited to visit Paris (CL–GM, 10 Oct 1830 [ATL]). Four GM–GC letters (1821–1824) are preserved at the Museum National d'Histoire Naturelle, Paris. Cuvier's replies were sold at Gideon's estate auction in 1853 and are now lost.

assignable to each of the four types and then noticed others that could not be so positively identified.

Crocodiles' remains had been found in all the Secondary formations of England, from the Oolite to the Chalk, and in Tertiary strata also. Gideon himself (in *South Downs*) first pointed out their presence in the Hastings Sands. Since then, Cuvier had affirmed Mantell's tentative identification of these specimens with gavial remains found in France. It could now be shown that there were at least two species of crocodile in the strata of Tilgate Forest, and perhaps so many as four.

Gideon's discussion of *Megalosaurus* too generously credited the first discovery of its remains to Buckland. Stonesfield aside, however, its teeth, ribs, vertebrae, and other bones had now been found in Tilgate Forest. Though Buckland also attributed a number of other unconnected bones to *Megalosaurus,* Mantell demurred, believing many of them more probably iguanodontian. From an examination of their teeth, Gideon concluded that megalosaurs resembled both the modern-day crocodile and monitor in certain respects but were more nearly related to the latter. The fact that *Megalosaurus* and other ancient saurians had vertebrae with concave faces (resembling those of a fish) led him to speculate that they probably lived in marshes; "there cannot be a doubt," he argued, "from the immense number of aquatic animals enclosed in the Secondary formations, that water once covered a much more considerable portion of the surface of the globe than it does at present" (69n). He cautioned, however, that the osteology of *Megalosaurus* was still very little known.

As for "a nondescript herbivorous reptile" called *Iguanodon,* the "first" specimens of its teeth had been "found by Mrs. Mantell in the coarse conglomerate of the Forest, in the spring of 1822" (71); since then, Gideon had assembled a full series of them, from young and perfect to old and worn. When the first, worn tooth was discovered, Mantell related, he thought it so like the molar of a grazing animal as to wonder if the Tilgate strata might not belong to the Age of Mammals. Subsequent discovery of a graded series proved that the teeth belonged to an unknown herbivorous *reptile,* but their structure then became so extraordinary that Gideon determined to consult Cuvier, which Lyell helped him to do. In his letter of 20 June 1824 (quoted at length) Cuvier replied that the teeth might well be those of a new animal, a plant-eating saurian; the second edition of his *Recherches* subsequently depicted several of the teeth and described their structure. Gideon himself, meanwhile, had named the new animal and analyzed its teeth in a paper read before the Royal Society.

Technical discussion then followed, combining Cuvier's observations on the teeth with Mantell's own. Among all recent lizards, only those of the iguana were similar. But Gideon now emphasized some important

Figure 5.1. Iguanodon teeth composite. Plate IV of Mantell's *Illustrations of the Geology of Sussex* (London: Lupton Relfe, 1827), as drawn by F. Pollard, with four of the depicted specimens (BMNH).

differences. In particular, no existing reptile (including the iguana) masticates; herbivorous amphibia exist, but they gnaw rather than chew. Since, as Cuvier had decreed, "every organic individual forms an entire system of its own," it follows from the peculiarity of *Iguanodon*'s teeth that its muscles and jaws must have differed greatly from those of any existing reptile. No portion of the iguanodon would be more welcome, therefore, than some significant fragment of its jaw.

Of iguanodon bones less could be said. In February 1825, probably, Mantell had gone through his entire collection of vertebrae from Tilgate Forest with Conybeare, sorting out those attributable to crocodiles, plesiosaurs, and megalosaurs. Several enormous vertebrae remained, presumably belonging to *Iguanodon*. Similarly, one portion of a femur was twenty-three inches in circumference. "Were it clothed with muscles and integuments of suitable proportions," Gideon marvelled, "where is the living animal with a thigh that could rival this extremity of a lizard of the primitive ages of the world?" Some metatarsals, too, were "so large that they appear more like the bones of mammoths or elephants than of reptiles" (71; we forget how anomalous the very idea of a large reptile was). Mantell had shown that the teeth and thighbone of his creature were at least twenty times larger than those of the iguana. (But some of these bones were actually from an as yet unrecognized and much larger dinosaur named *Pelorosaurus*.)

At this time he also rendered to posterity the notorious "horn" of the iguanodon, a mysterious spiky bone probably discovered by Mary Ann in 1824 but tentatively identified as rhinoceran till Joseph Pentland deduced from a cast (prepared by Francis Chantrey for Buckland) that it must have belonged to a reptile. One species of iguana had such a horn; Mary Ann's fossil therefore established (in Gideon's mind) "another remarkable analogy between the iguanodon and the animal from which its name is derived." Whatever disputes remained concerning its nature, "*Iguanodon* was one of the most gigantic reptiles of the ancient world, and a colossus in comparison to the pigmy alligators and crocodiles that now inhabit the globe" (78).[6]

---

6. The "horn," discovered by MAM in a quarry near Crawley, is actually a thumb bone, but this wasn't known till much later. For the original discovery, see CL–GM, 23 May 1825 (ATL); GM–WB, 16 June 1825 (RS); WB–GM, 17 July 1825 (ATL); Dr. Thomas Hodgkin–GM, 18 July 1825 (not 1828) re JBP's opinion (ATL); GM–GC, 26 Aug 1825 (Paris, Inst.); ms of lecture, 3 Oct 1836 (ATL); and *Petrif*, p. 198. Francis Chantry prepared casts of GM's best iguanodon specimens in 1825 and 1826; they were then sent to collectors and museums in England and France. GM's contemptuous comparison of *Iguanodon* with today's "pigmy" reptiles echoes Byron (*Cain* [1821], Act II, Scene ii; *Don Juan*, Canto II [1823]), who in turn reflects the influence of Cuvier.

Plesiosaurus, the final saurian, had been fully and ably described by Conybeare in a series of papers; teeth, vertebrae, and other bones from it, Mantell could now report, occur in the strata of Tilgate Forest. Though not yet in a position to identify species within the genus more particularly, Gideon discussed what specimens he had and summarized portions of Conybeare's work. He thought the long-necked variety of plesiosaur, which Conybeare had announced so dramatically on 20 February 1824 (the same evening on which Buckland announced *Megalosaurus*), a "most wonderful animal" (79n).

Among several quite different bones – apparently saurian, but of indeterminate character – Mantell noticed one of considerable size that Buckland (in 1824) had assigned to a whale. Though it differed from the humerus of any known plesiosaur, Gideon supposed this one to belong to a new species within the same group. (It was actually from the as yet unrecognized dinosaur *Cetiosaurus,* but Mantell typically attempted at first to accommodate new material within genera already known to exist.) Other bones seemed to be those of birds. (They actually belonged to pterosaurs and were the first examples to be discovered in England.)

In conclusion, Gideon affirmed that the Hastings beds had formed very differently from those immediately above and below them. They were primarily freshwater deposits (though including occasional marine exuviae) and probably represented the delta or estuary of a great river. He then described, for the first of several times, what this tropical region must have looked like, with its lush fern-covered valleys and plains, its great tree ferns or palms, and its sandy riverbanks or lakeshores, with turtles of various kinds, and "groups of enormous crocodiles basking in the fens and shallows." *Plesiosaurus* lurked within the ocean depths, while "gigantic *Megalosaurus* and yet more gigantic *Iguanodon*" towered over all on land (83).

Historical Importance

Gideon Mantell often referred to *The Fossils of the South Downs* by its subtitle, "Illustrations of the Geology of Sussex." Consistently, but confusingly, he also referred to *Illustrations of the Geology of Sussex* by *its* subtitle, "The Fossils of Tilgate Forest." *Illustrations,* moreover, cites *South Downs* as "Volume I," clearly implying that the two are parts of a single work; deliberately of a size, they were often bound together. Whether one takes them separately or not, both are of foremost importance in the history of dinosaur discoveries. *South Downs* was the first book anywhere to describe a systematic collection of dinosaur remains, to include remains from more than one kind of dinosaur (*Megalosaurus* and *Iguanodon*), and to which dinosaur remains are more than incidental. Neither Parkinson nor Buckland significantly improved on Gideon's attempted reconstruction of *Megalosaurus*. As its plates attest, *Illustrations* was the first book in English to be *primarily* concerned with dinosaurs and other fossil reptiles. It was also the first in English to include more than a single *named* dinosaur and the first anywhere to speculate in detail about the environment in which dinosaurs may have lived. We must, of course, remember that the term "dinosaur" did not yet exist, that only two dinosaurs (as opposed to other saurians) were known, and that Cuvier had published on them both in 1824 and 1825. *Illustrations,* however, depicts bones from at least four different kinds of dinosaur.[7]

Whether bound with *South Downs* or by itself, *Illustrations* is the rarest and most historic dinosaur book in English. While at last reaffirming Mantell's belief in the freshwater origin of the Hastings Sands, however, it also revealed his inadequate attention to geological causation and the origin of landforms. Though a book, moreover, *Illustrations* was ineptly merchandized, consequently attracting less attention than two or three important papers would have. After dawdling for years, Gideon wrote and published on his saurians abruptly, when his mind was too often distracted with other concerns and some of his most significant ideas were changing. With the advent of more efficient printing methods, his book's attractive but costly pages had become old-fashioned. Subscriptions aside, fewer than fifty copies were ever sold.[8]

Gideon's Museum

On 25 March 1829, following negotiations the previous October, Mantell agreed to a medical partnership of fourteen years with one George Rickword, formerly of Horsham. Under its terms, Gideon would receive

7. GM's personal copy of *Foss SD* at ATL includes emendations later appearing in *Illus;* similarly, his personal copy of *Illus,* also at ATL, became a repository for still further improvements. GM showed specimens depicted on plates XIV, XV, and XVI to Cuvier in August 1830, changing his identifications of several in consequence. Plate XV, Fig. 1, eventually became *Cetiosaurus* and XVI, Figs. 1–4, *Pelorosaurus.* (For some modern researchers the two names are synonymous.)

8. Gideon admitted to the poor sale of his *Illustrations* in *Petrif,* p. 226n. Only sixty-nine copies had been subscribed. For the most remarkable of GM's interruptions, see my "Hannah Russell" (biographical essay) in DRD, *Gideon Algernon Mantell: A Bibliography with Supplementary Essays* (Delmar, N.Y.: 1998); Sidney Spokes, "A Case of Circumstantial Evidence," *Sussex County Magazine,* 2 (1937), 118–122; and GM's Hannah Russell file (ATL). His medical arguments and pro bono detective work saved an innocent woman from the gallows.

two-thirds of the profits for the first seven years and, on 25 March 1836, an additional five hundred pounds from Rickword, who would also pay nine hundred pounds to initiate the partnership. After seven years, profits would be split evenly between them. Mantell probably sought this partnership for several reasons: his last parent had recently died; his fourth (and last) surviving child, Reginald Neville, had been born; he was enjoying considerable prosperity; he wished to minimize the irregular hours inherent in his profession; he was experiencing poor health; and he desired more time for both his family and his avocations, fossil collecting in particular. More fundamentally, Gideon was nearing forty and thought it time to claim whatever dreams now lay within his reach.[9]

He therefore spent part of the Rickword money on a more suitable memorial to his parents in St. John's churchyard and much of the rest to build a new room atop Castle Place – like a cupola on the roof – which included uniquely designed cabinets and was specifically intended as his museum. "I am hard at work removing my collection into a larger room," he then informed Buckland on 15 June. In August, when his new showroom was done, Gideon even compiled a printed guide of thirty-two pages, published by Lupton Relfe the same month. Mantell's collection, it stated, could be seen gratis on the first and third Tuesday of every month, from 1 till 3 P.M., upon previous application by letter. (But this generous and orderly procedure did not work out, for Gideon's friends dropped by whenever they chose and naturally expected him or other members of his family to show them the collection in accord with their convenience.) After some diligent preparations, Gideon officially opened his free museum on Thursday, 17 September, by inviting selected guests for tea, supper, and a dance.

The museum consisted of six cabinets, five against the walls and one in the center of the room. Case I contained minerals (including native gold and silver), recent marine fossils, and some polished sections of ammonite. Case II was devoted to the strata of Tilgate Forest and other subdivisions of the Hastings formation, which Mantell supposed to have originated in "the bed of a river or estuary." His printed guide invited visitors to "imagine a river flowing through a country inhabited by lizards, turtles, etc., and clothed with forests of plants allied to the palms and arborescent ferns" (11). Remains of four enormous reptiles had also been identified: *Iguanodon, Megalosaurus, Crocodile,* and *Plesiosaurus.* Of *Iguanodon,* an "herbivorous reptile, related to the iguana," Gideon possessed "bones of the head, teeth, vertebrae, clavicles, coracoid bone, ribs, chevron bone, femur,

9. GM's Lady Day agreement with George Rickword, 25 March 1829, is at ESRO; its terms were reviewed fully in the further partnership between both of them and Andrew Doyle on 29 Oct 1831 (ATL).

leg bones (tibia and fibula), metatarsal bones, phalanges, ungueal bone, and *horn*" (12). Some of these had been recently acquired, and a few were casts. He had other bones also; long and slender, they were probably remnants of the flying reptile, *Pterodactylus.*

Further Tilgate specimens in Case II included turtles, fishes, freshwater shells, and plants. Among the latter were two species recently named for Mantell by Adolphe Brongniart, as well as *Clathraria Lyellii,* a large tree fern Gideon had named. On the evidence, he wrote boldly, the following inferences concerning Tilgate Forest seemed justified:

> 1st. The reptiles and plants must have been inhabitants of a country enjoying a much higher temperature than any part of Europe, and the former, from their enormous magnitude and osteological characters, clearly belong to an order of things of which the present state of the earth affords no example; the epoch of their existence may indeed be termed the *Age of Reptiles.*
>
> 2dly. The broken and rolled state of the greater part of the bones, the pebbles, and the conglomeritic character of many of the deposits prove that the strata were formed in the bed of a river or an estuary.
>
> 3dly. It is equally obvious that the Hastings, or Tilgate, strata must have been formed and consolidated before the Chalk (which rests upon and once covered them) was deposited. It follows that after the Hastings beds were formed, they must have been submerged beneath the ocean which deposited the Chalk formations. . . .
>
> 4thly. The ocean of the Chalk, in its turn, must have passed away, and the consolidated Chalk have been covered by the waters which deposited the Tertiary strata, for the latter contains fossils entirely distinct from those of the Chalk.
>
> Lastly, the Tertiary, in common with the Chalk and Hastings beds, must have been subsequently broken up, probably by volcanic energy, and the wealds of Kent and Sussex formed, and the Chalk dislocated and separated, by the upheaving of the central strata of the Hastings formation; the lateral fissures in the Chalk now constitute the valleys through which the existing rivers flow, and effect the drainage of the country. To this epoch may probably also be referred the formation of the beds of diluvium. (17–18)

Compared with its predecessors in Mantell's writings, this analysis was remarkably forthright (though not strikingly original). At a good time, Gideon had found a new, more feasible vehicle for the expression of his views, which continued to develop – and being answerable neither to a learned society nor to the vagaries of public support, he spoke out as he pleased. This was Mantell's earliest significant acknowledgment of volcanic forces underlying landscape and a fairly definite rejection of the Deluge. The revolutions and catastrophes of which he had been so sure only two years before were now unmentioned, though still implicit.

Case III was filled with donated fossils from locations mostly beyond Sussex. Shelves one and two had remains of elephants, aurochs, and Irish elk, presented in part by G. B. Greenough; three and four offered mammoth and other remains from Burma, as recently collected there by John Crawfurd; five included ichthyosauria from Lyme Regis, the skull of a cave bear from Germany, bones in limestone from Gibraltar, and a mammoth tooth from Siberia. Six featured an assemblage of fossil fishes and insects from Aix, in Provence, collected the previous year by Lyell and Murchison. On the remaining shelf a series of almost seventy models, presented by Cuvier at Lyell's instigation, recalled specimens from the museum of natural history in Paris; they had arrived only on the fifteenth of April.

Drawers in Case III and all of cases IV and V – the latter, a mahogany cabinet – stored various British (mostly local) fossils, some of them on hand since 1809; several zoophytes and shells were gifts from Etheldred Benett. Under the glass top of Case VI, in the center of the room, Mantell displayed crustacea, his most striking fossil fishes, and some *Mosasaurus* vertebrae. Its drawers housed extensive collections of reptilian teeth from Tilgate Forest; Stonesfield fossils, presented by Lyell in 1821; Suffolk fossils, from Mrs. Cobbold; Wiltshire fossils, from Miss Benett; shells from the London Clay and Isle of Sheppey that were among Gideon's earliest acquisitions; Scottish and Isle of Wight specimens from Lyell; Dutch specimens from Maastricht; French and Italian shells from Cuvier and Brongniart; teeth and bones from Kirkdale Cave, Yorkshire (a site immortalized by Buckland in 1822); hippopotamus and rhinoceros bones from the Vale of Arno; and miscellaneous others. Finally, on one wall a pair of antlers from the extinct Irish "elk" spread nearly eleven feet. England had no other private fossil collection of equivalent scope.[10]

Colleagues

On 29 September 1829, in Hampstead, Robert Bakewell wrote "A Visit to the Mantellian Museum at Lewes," which his literary associate John

---

10. J (30 Apr, 16 June, 5, 20 Aug, 31 Dec 1829); I wish to thank the late Stephen Moore for hospitality at Castle Place, the interior of which has since been gutted. Gideon's "new room" can still be seen from atop Lewes Castle. GM–WB, 15 June 1829 (RS); GM, "An Abridged Catalogue of Mr. Mantell's Collection of the Organic Remains of Sussex" (London: L. Relfe, 1829). For GBG's contributions, see *Wonders*, sixth edition (1848), I, 224n. The cave bear and some other specimens had come from the sale of James Parkinson's collection (Apr 1827).

C. Loudon then published in the year-old *Magazine of Natural History* on 1 January 1830. Anxious to please a newly discovered friend, Bakewell relied heavily on Mantell's published and private comments. Among the latter was a new insistence on *Megalosaurus* and *Iguanodon* as *land* dwellers "evidently formed for walking on solid ground." The strata in which their remains were found, moreover, "must have been deposited in a freshwater lake or estuary, or in the bed of a mighty river, on the sides of which lived and flourished plants and animals analogous to those of tropical climates" (10). Bakewell had also heard Gideon say that the relative failure of *South Downs* to attract as much scientific attention as it deserved undoubtedly reflected metropolitan prejudice against provincials.

Enthusiastically praising Mantell's collection of Chalk fossils as "the finest in the kingdom" (11), Bakewell extolled its splendidly preserved fossil fishes, which (he thought) must have been petrified suddenly by submarine hot springs. Gideon's ventriculites were also outstanding, and some of his ammonites remarkable. But the most interesting specimens were undoubtedly those from freshwater Wealden strata below the Chalk, including abundant remains of terrestrial plants and large animals. While many bones and teeth remained unidentified, those of four enormous reptiles had been ascertained: *Crocodile* (at least two species), *Plesiosaurus, Megalosaurus,* and especially *Iguanodon.* The latter's peculiar teeth (evidently adapted to chew rough, thick plant stems) had been "first discovered by Mrs. Mantell in the coarse conglomerate of Tilgate Forest in the year 1822" (13). On the whole, Bakewell's knowledgeable evaluation of the collection mirrored Gideon's own, and – understandably enough – favored his saurian discoveries.[11]

Robert Bakewell (1768–1843) was the outspoken author of a delightful *Introduction to Geology* (editions in 1813, 1815, 1827, 1833, and 1838), which being nearly the first of its kind interested many in the science, including Charles Lyell. Though over sixty now, Bakewell had lost none of his witty cantankerousness and cynical wisdom; in his tough, old-fashioned outlook he resembled G. B. Greenough but was fully ten years older and went well beyond Greenough in disliking "organized" science, which Bakewell tended to equate with the stifling orthodoxy of official opinion. Bakewell therefore never joined the Geological Society and, while respected by its members, was scarcely on equal terms with most of them. Despite

11. Robert Bakewell, "A Visit to the Mantellian Museum at Lewes," *The Magazine of Natural History,* 3 (1829), 9–17. In RB–GM, 1 Oct 1829 (ATL), Bakewell explained that he had coined the adjective "Mantellian" by analogy with "Hunterian" and "Ashmolean." J (1 Jan 1830). See also "Museum of Gideon Mantell," *AJSA,* 23 (Oct 1832), 162–179.

their difference in age, however, he and Gideon had much in common – both were highly intelligent, articulate provincials – and Bakewell had the further attraction of having written a scientifically acceptable, yet popular exposition of what had initially been current geology.

Having somehow heard of him (through Lyell, I imagine), Mantell wrote Bakewell around 6 September, inviting him to preview the museum before its official opening. Accepting gladly, Bakewell and his wife spent the evenings of the eleventh, twelfth, and thirteenth with Mantell, who found the elderly curmudgeon "a most intelligent and agreeable man." Similarly impressed with Gideon, Bakewell wrote Loudon from Hampstead, proposing a review of Mantell's museum – and to Gideon himself, for advice on what to emphasize. Unexpectedly, he received not only facts and guidance but also copies of *South Downs* and *Illustrations*. Bakewell then reciprocated with equal generosity, sending an extensive collection of basic rocks and the two volumes of his *Travels* (1823). On 3 November, he wrote again to report the steady sales of his new third edition. Its publisher, Thomas Longman, now considered *Introduction to Geology* a standard work, to be updated and reissued every few years. So, Bakewell promised, his fourth edition would include a full account of Gideon's discoveries – and, in 1833, it did.[12]

In the same letter, Bakewell announced his having written the previous Saturday to an American friend, Benjamin Silliman (1779–1864), of New Haven, Connecticut, in the United States. Destined to be Mantell's most intimate correspondent, as well as the most discreet, Silliman had abandoned a possible career in law to become in 1804 the first lecturer on chemistry at Yale College (not University till 1887). Fortified the next year by a visit to Europe (of which he wrote a popular account), Silliman developed particular interests in mineralogy and geology. He then established, in 1813, the first geology course at any American institution of higher learning, and, five years later, founded the *American Journal of Science and Arts,* remaining editor-in-chief till 1856. When the American Geological Society formed in 1819, Silliman was likewise among its organizers. In 1829, he edited Bakewell's *Introduction* for American students, adding to the Yankee edition of it a summary of his own lectures on geology. Anxious to expand Yale's pathetically small mineral and fossil collections, he now looked forward to obtaining specimens from an accomplished British collector.

---

12. "It is noteworthy," H. B. Woodward observed, "that the three prominent authorities on practical or applied geology, William Smith, John Farey, and Robert Bakewell, were not members of the Society" (*Hist GS,* 1907, p. 53). RB–GM, 8 Sept, 19 Sept, 1 Oct, 16 Oct, 3 Nov 1829 (all ATL; further RB–GM letters to 1843).

Addressing Silliman on 31 October, Bakewell had promised that Mantell would gladly exchange Sussex fossils for American ones, but Yale heard nothing from Lewes for almost six months. Gideon belatedly wrote Silliman for the first time on 20 April 1830, sending him copies of *Illustrations* and his museum catalog. "In common with my geological friends in this country," he continued graciously, "I cannot but feel the highest respect and gratitude for your exertions in the science which it is our object to promote." Bakewell, described as one of Mantell's best friends, had suggested while visiting Lewes last September that Silliman might appreciate receiving duplicates. He would have complied earlier, Gideon apologized lamely, but those on hand had all been sent to Silliman's fellow American, G. W. Featherstonhaugh, thus necessitating the delay.[13]

Silliman, meanwhile, had written an initial letter to Mantell on 3 July 1830, only to receive his of 20 April on the twelfth. Reading through Gideon's *Illustrations* the next day, he immediately lectured on it to his geology class of two hundred at Yale. Nothing done in geology within the last twenty years, Silliman then declared, "has enriched it more than the observations made in the English and French Secondary, Tertiary, and Diluvial." In accordance with Bakewell's suggestion, he invited Mantell to trade Sussex material for American and wondered if Gideon would be willing to receive a gift set of the *American Journal of Science and Arts,* eighteen volumes of which had thus far appeared. Mantell, of course, accepted this unexpectedly noble offer and would exchange not only specimens but letters, news, opinions, ideas, publications, and mementos with Silliman for the rest of his life.[14]

On 4 December 1829, Buckland read a paper to the Geological Society "On the Discovery of Fossil Bones of the Iguanodon in the Iron Sand of the Wealden Formation in the Isle of Wight and in the Isle of Purbeck." He began by acknowledging that "We are indebted to the researches of Mr. Mantell for our knowledge of the existence of that curious and most gigantic herbivorous reptile," which Gideon had thought seventy feet long (425). The femur reported by Murchison in a paper of December 1825 was probably from *Iguanodon* also – as Gideon had earlier proposed – and Buckland was now prepared to add three more localities, two in the Isle

13. RB–GM, 3 Nov 1829 (ATL). George P. Fisher, *Life of Benjamin Silliman* (2 vols., New York, 1866) reprints letters from both Bakewell (II, 51–65) and Mantell (II, 182–232). RB–BS, 31 Oct 1829 (Yale). GM–BS, 20 Apr 1830 (The Historical Society of Pennsylvania; Philadelphia); Joan M. Eyles, "G. W. Featherstonhaugh (1780–1866)," *JSBNH,* 8 (1978), 381–395, reprints all six GWF–GM letters (ATL) fully, together with further information; J (16, 21 July; 6, 18, 23–25 Oct 1827).

14. BS–GM, 3, 28 July 1830 (Yale). Except for 20 Apr 1830, all known GM–BS correspondence is at Yale and ATL.

of Wight and one in the Isle of Purbeck. As Lyell noted in a letter the next day, Mantell and his creature were treated with due honors, though Buckland – evidently incredulous about its size – had jokingly compared *Iguanodon* to "the small, *genteel* lizards of our days." In his relatively brief account, Buckland managed to cite Gideon by name nine separate times.

In the same letter Lyell announced that he would be going to press on Monday, the seventh, with his two-volume *Principles of Geology,* "an attempt to explain the former changes of the Earth's surface by reference to causes now in operation." The first volume, Lyell guessed, would be quite done by the end of the month. Having undertaken some extensive additions, however, Lyell did not finish it until June; Volume 1 was then published on 24 July. Gideon probably received his complimentary copy about three days later. On 14 August, having read it carefully, he noted the previous day's "West Indian shower, a deluge which in a few minutes washed down twenty tons of chalk and flints from Cliffe Hill into the street below," adding that "My friend Lyell would have gloried in witnessing the 'effects of modern causes'!" Clearly, he had found Lyell's *Principles* even more persuasive than expected.[15]

Next morning, however, Mantell received a most disappointing letter from Baron Cuvier in London, whose brief trip to England had been abruptly truncated by political upheaval in France. Compelled now to leave for Paris on the seventeenth, he could not visit Lewes as had been planned. Seizing the only opportunity he had, Gideon immediately determined to go see him. After hastily packing a box of fossils, he drove himself to Brighton, then took the 3 P.M. coach for town. Reaching Charing Cross at half past eight, Mantell directed a hackney coach to Sablionier's Hotel in Leicester Square, only to find that the Baron was out. Leaving his fossils there, Gideon next drove to Tottenham Court Road, just in time for the Kentish Town stage, which delivered him to Mrs. Woodhouse's, where his wife and older son were staying.

On Monday, 16 August, Mantell rose at seven and returned with Walter in a cabriolet to Charing Cross, from which Gideon sent a letter to his sister Jemima in Lewes (she was occupying the former house of their deceased mother), advising her to have his carriage waiting for him that night in Brighton. At nine he and Walter then called on Baron Cuvier, who received them both with open arms. "He is a stout, square-built man,

---

15. WB, "On the Discovery of Fossil Bones," *Trans GS,* ns 3 (1829), 425–431; Plate XLI. But this version includes an appendix dated May 1835, effectively demonstrating how laggard the *Transactions* had become (and why papers circulated much earlier as authors' separates than as part of the completed volume). Peter John Martin, *A Geological Memoir on a Part of Western Sussex* (London, 1828) introduced the term "Wealden formation." CL–GM, 5 Dec 1829 (ATL); Wilson, Chapter 9. J (14 Aug 1830).

about the middle size," Mantell recorded, "his face large and not so expressive as I had anticipated; a noble expanse of forehead; his eyes bright and penetrating." Cuvier expressed his delight at seeing Gideon, was kind to ten-year-old Walter, regretted that he could not speak English, and proceeded to examine Mantell's specimens. "I laid before him a considerable number of our Sussex fossils," Gideon told Davies Gilbert a month later, "and had the satisfaction of finding his opinions coincide with mine in most instances."

According to Mantell's journal, Cuvier pointed out the extremities of a reptilian jaw in one of Gideon's two Chalk fossils but otherwise provided no new information. A related memorandum, however, belies this. "Called on Baron Cuvier at Sablionier's Hotel, Leicester Square," Mantell scrawled in a geological notebook:

> my son Walter accompanied me. Among the fossils which I had brought up to show him, the Baron remarked on several. Os quadratum of a reptile. . . . Bones of birds rather than of *Pterodactyl,* probably of some wader. Conical teeth from the Chalk, probably crocodile – most certainly reptile. One of the vertebrae was the axis [second cervical] of a reptile. The larger vertebra, concavo-convex and thin, is reptile – not cetaceous, as Buckland etc. supposed. The large, flat, slightly curved rib from Tilgate is reptile. The sternum he considered very curious; would not determine it. I still think it is of a young iguanodon. Teeth of reptiles. . . . The Baron kept a few of the fossils to draw.

Gideon also preserved a number of Cuvier's identifications as marginalia in his own copies of *South Downs* and *Illustrations,* utilizing them in subsequent publications.

In parting, Cuvier assured Mantell he would return to visit him in the spring, a promise that was not kept. "I left this distinguished man with great regret," Gideon concluded. "We had never met before, and he was the idol of my scientific idolatry. In truth, the whole time that I was with him I was in a state of feverish excitement which I cannot describe. Some twenty years ago [when he first visited James Parkinson], such feelings were not unusual with me." Together with Walter, Gideon then met Mary Ann at the Panorama and all three of them went to the Geological Society to see fossils. He left town alone a little before four, and after a dull ride reached Brighton at nine, in the midst of an illumination honoring the first visit of England's new king, William IV. From there, he returned in his own coach to Lewes.[16]

16. GM and GC in London: J (15, 16 Aug 1830); GM–DG, 13 Sept 1830 (ESRO); GM, "Geological Expeditions" notebook (ATL), 16 Aug 1830; *Foss SD* and *Illus,* author's copies (ATL); J (16 Aug 1830); *Petrif,* pp. 269n, 489. Cuvier became a baron in 1819.

Despite the timing of this unique interview with Cuvier, Mantell continued to be impressed with Lyell's defense of modern causes and began to revise his own thinking accordingly. Thus, he had written in 1822 (*South Downs*) that "the present effects of the ocean appear to be wholly inadequate to produce changes like those which have formerly taken place." Five years later, in *Illustrations,* he dismissed modern causes as insignificant. But as a result of Lyell's book such facile conclusions were no longer admissible. Consequently, Gideon wrote beside the passage in *South Downs:* "The effects of volcanoes and earthquakes may, however, be perfectly adequate to produce dislocations of the strata equal to any that have been noticed in this volume." Though undated, this addition was almost certainly made between July 1830 and March 1831. By the latter month, Mantell had gone over to Lyell's camp entirely. "My friend Mr. Lyell's book, *The Principles of Geology,*" he wrote Silliman then, "is the most important contribution to our favorite science that has appeared for a long while." Eight months later Gideon was even more emphatic: "Lyell's book," he then attested, "is considered the most philosophical geological work that has appeared in our time," adding that a second volume would soon be forthcoming and that the first pages of each had been written at Castle Place. By then Lyell himself had jokingly referred to Gideon as his apostle.[17]

17. *Foss SD,* p. 304; author's copy (ATL), with annotations by GM (quoted); *Illus,* p. 4. GM–BS, 29 Mar, 24 Nov 1831 (Yale); CL–GM, 30 Aug 1831 (ATL).

6

Hylaeosaurus

The dissolute and thoroughly unloved George IV died on 26 June 1830, an event Gideon Mantell did not think sufficiently important to record. A necessary proclamation bestowed the crown upon William the Fourth, who had been his predecessor's younger brother. Already sixty-five at coronation, William wisely devoted the seven years of his important but relatively colorless reign to conciliatory reforms and the enhancement of public confidence in hereditary monarchy. He is most often remembered today for the Reform Bill of 1832 that, after much debate and heated opposition, extended the franchise a bit and achieved some preliminary degree of parliamentary reform.[1]

Reptiles and Reform

Early in June 1831, Mantell published "The Age of Reptiles," his first popular account of saurian discoveries. Appearing originally in the *Sussex Advertiser* on the thirteenth, it was later reprinted in Scotland and America. "Among the numerous interesting facts which the researches of modern geologists have brought to light," it began, "there is none more extraordinary and imposing than the discovery that there was a period when *the earth was peopled by oviparous quadrupeds of a most appalling magnitude,* and that reptiles were the *Lords of the Creation* before the existence of the human race!" Though the concept had originated more vaguely with Cuvier, Gideon himself was responsible for establishing a comprehensive picture of that era in the public mind. He fully appreciated how numerous saurians must have been and how rarely any of them would be fossilized; the climate at that time was wholly different from our own, but there were still correlations with present species, as "in the pigmy monitor and iguana of modern times we perceive striking resem-

---

1. Death of George IV and accession of William IV. J (15, 31 July, 30 Aug, 3, 7 Sept, 2 Oct, 1 Nov 1830); *Town Book,* pp. 262–279; GM, *Narrative* (1831).

blances to the colossal *Megalosaurus* and *Iguanodon* of the ancient world."

The rest of this brief paper assigned known saurians to their appropriate strata. Thus, ichthyosaurs, plesiosaurs, and pterodactyls all belonged to the lower reaches of the Oolite formation, the Lias. Oxfordshire's Stonesfield beds, of approximately the same age, included numerous crocodiles, *Megalosaurus,* and some mammalian remains that had not been satisfactorily explained. In freshwater formations between the Oolite and the Chalk, turtle remains became frequent, together with continued evidence of *Megalosaurus, Plesiosaurus, Pterodactylus,* and an "enormous herbivorous reptile" called *Iguanodon,* which may have been nine or ten feet high and sixty to one hundred feet long. Reptiles were less common in the Chalk, but crocodiles, turtles, and the extraordinary *Mosasaurus* could be found. By the end of this era, however, the great saurians had all become extinct, lesser ones only surviving, and with the Tertiary formations a new order of things began (later to be called the Age of Mammals). This popular description was the first to emphasize the succession of reptilian genera as well as their abrupt extinction, important concepts with which Gideon should be credited.[2]

On a different level of achievement, he also became a dinosaur poet. "My good friend," Bakewell had written him on 27 May, "how comes it to pass that you say not one word about Reform? Have Sir John and Lady Shelley transformed you into a Tory? Let me recommend you to get the jaw of a Tory into your museum, for they will soon be extinct animals." Responding to this clever idea with gusto, Gideon versified it on 22 June, smuggling his result into the next issue of the *Sussex Advertiser* as an anonymous contribution:

THE AGE OF REPTILES

"The earth was peopled by oviparous quadrupeds of a most appalling magnitude, and reptiles were the 'Lords of the Creation,' before the existence of the human race" – *Mantell's Age of Reptiles.*

---

2. GM, "The Age of Reptiles," *SA,* 13 June 1831, p. 4 (also as a separate); "The Geological Age of Reptiles," *ENPJ,* 11 (1831), 181–185; *AJSA,* 21 (1832), 359–363. On 3 Nov 1829 GM "Wrote a little paper, 'The Age of Reptiles,' for the *Scientific Annual* edited by Mr. [William] Higgins" (J), but no such periodical ever appeared. Cuvier had earlier postulated a four-stage sequence of life on earth, the age of reptiles being first (*Essay on the Theory of the Earth,* fifth edition [Edinburgh and London, 1827], 253–264, 295–296. GM, rather than Cuvier, was the target of William Kirby's objections to the age of reptiles idea in his Bridgewater Treatise of 1835, *On the Power, Wisdom, and Goodness of God As Manifested in the Creation of Animals* (2 vols., London), I, 38–40. See J (16 June 1829); Kirby–GM, 25 July 1835 (ATL, with draft reply to a magazine editor by GM); and *Medals,* II, 686n.

There once was a time, as geologists say,
That reptiles alone o'er our planet had sway,
And the "Lords of Creation" were all creeping things,
Some crawling on earth and some soaring with wings!
These monsters so greatly polluted the air
That nothing but reptiles inhabited there,
Till a grand revolution destroyed the whole race,
And refined earth and air, and then men took their place –
So Cuvier hath taught, and by Mantell more lately
It seems (if he do not exaggerate greatly)
That their size was appalling and far surpassed all
The monsters who've lived since the time of the Fall!
Now I cannot but think, and perhaps so will you,
That these sages so shrewd had a moral in view,
And that by the reptiles they meant to sketch out
The Anti-Reformers now put to the rout.
For have not those seat-mongers, evil betide 'em,
Like reptiles corrupted whate'er came beside 'em?
They were "Lords of Creation" and all but their order
Were slaves –

.   .   .   .   .

By the grand revolution, Reform is intended
And the empire of reptiles forever is ended;
For although toads and vipers will still much abound,
The chief monsters no longer encumber the ground.
Our political atmosphere healthy will be
And England once more be the Isle of the Free!

The omitted lines referred to specific Tories, all of whom were opponents of Reform. A few days later, on the thirtieth, Gideon drafted another protest poem, this time about the starving Irish.[3]

"My time," Mantell wrote testily the same month, "is fleeing away and sadly, sadly unimproved – a dull round of visiting; scenes of misery constantly before me, and which affect me as acutely as ever; scribbling verses, letters, and geological scraps and a hundred other nothingnesses." And thus his all-too-few remaining days were running down. "Almighty Being," he prayed, "oh, grant that in another state of existence my intellectual powers may have full exercise and development. Oh, this clay!"

3. RB–GM, 27 May 1831 (ATL). Anon. but GM, "The Age of Reptiles" (poem), *SA*, 27 June 1831, p. 3; J (30 June 1831). Despite Gideon's seemingly cavalier attitude toward Reform, he had been at the center of local agitation concerning it since Sept 1830 and was much gratified by its successful outcome in June 1832. Beginning in Mar 1831, he repeatedly visited the poorhouse of St. John's parish to treat cases of typhus. The famed cholera epidemic then began in Apr, with a more virulent form arriving in late Oct.

Like Byron, he found "society where none intrudes," taking long walks by himself to lighten the irritations thrust upon him daily.

One such irritation was the "proud priest" Peter Crofts, churchwarden of St. John's, who obliged Mantell to dig up the cypress he had planted on his mother's grave. Another was Robert Trotter of Borde Hill, near Cuckfield, who had begun to usurp specimens from the quarries of Tilgate Forest. When Gideon visited Whiteman's Green on 12 December 1830 quarrymen he had dealt with for years refused to sell him further specimens, having found a more lucrative customer nearby. So much for Mantell's hopes of discovering the jaw of the iguanodon! Though Trotter, a decent sort, came calling in January, Gideon refused to see him. After some further preparation Trotter called again in July, this time presenting the model of a magnificent tibia he had recently obtained ("and which," Mantell growled, "ought to have been mine"). That September, Gideon visited Cuckfield with John Phillips and got a few fossils but found all the best had been "poached" by Mr. Trotter. He felt no more amiable at month's end after seeing the magnificent bones in Trotter's collection. Later on, however, Mantell regularly stopped at Borde Hill whenever he was near Cuckfield, obtaining from the generous Trotter a number of excellent specimens.[4]

In January 1832 Gideon must have received a copy of Lyell's second volume, which began with a summary of Lamarck's evolutionary theories (probably written at Castle Place), then proceeded to refute those theories at length. The entire volume was partially Mantell's in that he had called Lamarck to Lyell's attention and was also cited three times within it, once for an archeological discovery. Some of Lyell's discussion, moreover, accorded with what Gideon had suggested about the succession and extinction of species in his "Age of Reptiles" paper the previous year. But the analysis was entirely Lyell's, Mantell being largely unprepared at this time even to contemplate such bold ideas as that one species "passes by insensible shades" into another or that there is a "universal struggle for existence." Though Lyell himself accepted the second idea and would eventually concede the first, Gideon (like Lyell at this time) remained thoroughly unsympathetic to any transformist view of nature. For him, as for Cuvier, each species would always represent a new creation.[5]

4. J (15, 18 June 1831). The latter quotes "society where none intrudes" from Byron's *Childe Harold's Pilgrimage;* GM's "clay" of 15 June echoes Byron's *Manfred;* there had been other quotations from the same poet on 9 and 13 May. J (15 June 1831, 12 Dec 1830, 18 Jan, 1 July, 13 Sept, 2 Oct [re 30 Sept] 1831). Geol Exp, 30 Sept 1831; *Geol SE Engl,* pp. 332–333n.
5. Charles Lyell, *Principles of Geology,* Volume II (London, 1832). The seventeen-page first chapter is wholly concerned with Lamarck's theory of the transmutation of species (parts

Following passage of the Reform Bill in June 1832, Mantell attended the second annual meeting of the British Association for the Advancement of Science at Oxford. Created the year before by a meeting at York (to which John Phillips was important), the Association developed, in broad terms, as an outgrowth of the Industrial Revolution in England. Reflecting social changes that revolution engendered, it was a response within the world of learning to stresses like those in the political world of Reform. London's Royal Society, especially, had become for many little more than an effete gentleman's club. The Association, contrarily, emphasized the social, cultural, and technological *power* of science, casting scientists (a word to be invented in 1834) in far more central and dynamic roles.

On Thursday, 21 June, after breakfasting with Mantell, Buckland led the company ("all Oxford was in attendance") on one of his famous geological field trips to Shotover Hill, with frequent stops along the way for marvelously engaging lectures. Riding with Lord Northampton in his stately carriage, Gideon was jocularly appointed surgeon to the expedition. After a splendid party that night and a meeting of the Geological Section, the surgeon chatted with his friend Fitton until two in the morning. Following another breakfast with Buckland on Friday, he addressed the Geological Section on Wealden strata and their saurian remains, emphasizing freshwater origins and *Iguanodon*. His remarks, like others presented there, were widely reported.[6]

of which now strike us as surprisingly modern); quoting pp. 4, 56 in Lyell, *Principles,* II (his translations from Lamarck).

6. BAAS. For general background, see Jack Morrell and Arnold Thackray, *Gentlemen of Science: Early Years of the British Association for the Advancement of Science* (Oxford, 1981). Re GM specifically, J (17–30 June 1832); Curwen, pp. 102–109; WDC–S. Stutchbury, 26 June 1832 (ATL); and GM, *BAAS Report,* 1 (1833), 587. GM–Northampton, 21 July 1832 (Castle Ashby), adds details. John Phillips (1800–1874; *DNB*), nephew of William Smith the stratigrapher, resided in York from 1825 to 1840 and was a secretary of the BAAS from 1832 to 1859; he served also as professor of geology at King's College, London, from 1834 to 1840 and lectured annually. Phillips' letters to GM (1831–1848) are at ATL and there are about twenty mentions of him in J. See also *Hist GS, Gent Sci,* and Cecil J. Schneer, ed., *Toward a History of Geology* (Cambridge, Mass., and London, 1969), p. 146 (J. Eyles on W. Smith). William Henry Fitton (1780–1861; *DNB*) practiced medicine until he married wealth. As president of the GS (1827–1829), he inaugurated their *Proceedings* (1827ff.), which allowed preliminary reports of important papers to circulate far more rapidly than did the hopelessly overloaded *Transactions.* Sharing his interest in the Cretaceous, Fitton exchanged letters (ATL), visits, and mutual esteem with GM from 1822 onward. See also *DSB* (J. Eyles), *Hist GS,* and PL.

A New Dinosaur

In his letter to Silliman on 20 July 1832, Mantell included a number of predictable topics. Cholera was on the march again, as every day brought news of some further death. Amidst it all "the Reform bugbear" went on as usual and would "amuse John Bull till some new phantasm supplies its place." The meeting of the British Association at Oxford (attended by himself) had passed off admirably. Gideon had there enjoyed special attention from Buckland and visits with many colleagues. After "a week of the highest possible intellectual enjoyment," he had returned to Lewes via Bristol, Bath, and Lyme Regis, seeing W. D. Conybeare, Samuel Stutchbury, and the famous Mary Anning en route.

"I have made a grand discovery in the grit of Tilgate Forest," Mantell then continued unexpectedly.

> A mass of stone blown into fifty pieces or more by the quarrymen was found full of bones. They reserved it for me, and with much difficulty and great labor I have succeeded in uniting and clearing a slab 4 1/2 feet by 2 1/2 feet exhibiting twelve vertebrae, *eight in place,* with many ribs, coracoid bones, omoplates, chevron bones, etc. and several of those curious dermal bones which support the scales.

He planned to exhibit this "grand morceau (the finest yet found in the south of England)" to the Royal Society that fall, and promised Silliman he would hear more of it in due course.

On 6 September Mantell wrote Buckland, to thank him and his wife Mary for their hospitality at Oxford. "I have had little leisure since my return," he continued,

> and that little has been devoted to the labor of chiselling out a magnificent specimen which by good luck I obtained (in spite of *Trotter & Co.*) from Tilgate Forest after my return from Dorsetshire. It was so unpromising a mass that the quarrymen thought no-one would look at it but me, and therefore wrote me word that they had a 'great consarn of bites and boanes' if I would come up and see them.

With some reluctance Gideon did so, finding a pile of fragments in which bones were discernible but so imperfect and poorly defined that he resolved only after great hesitation to send up a cart and horse. Having gotten the bones home, Mantell had them washed, then successfully pieced together a block four-and-a-half feet long by two-and-a-half wide, though this left him with more than a wheelbarrowful of additional fragments too broken to fit in anywhere.

Initially, Mantell assumed his new bones belonged either to *Megalo-*

Figure 6.1. Remains of the *Hylaeosaurus,* discovered in Tilgate Forest. The original, 4 1/2 feet long, is in BMNH, but not on public display. According to Mantell's caption (in *Wonders*): "This plate represents the highly interesting specimen
discovered in Tilgate Forest in the summer of 1832; it consists of the anterior portion of a skeleton of the *Hylaeosaurus,* or Fossil Lizard of the Weald, lying on the
back." He then identified: (1) part of the cranium; (2) neck vertebrae (i.e., cervical);
(3) back vertebrae (i.e., dorsal); (4) ribs; (5) dermal processes, exterior to the skeleton; (6) very large dermal spines, each fifteen inches long; (7) coracoid bones
(which, as found in reptiles, extend from the scapula toward the breastbone [ster-
num]); (8) scapulae (shoulder blades, adjacent to the neck); (9) the glenoid cavity,
or socket for the head of the arm bone, formed by the union of the coracoid and
scapula; (10) detached bones.

saurus or to *Iguanodon.* Yet in some important ways these remains were
clearly different. He planned, therefore, to examine reptilian skeletons in
London and eventually to write a paper on his present specimen. When
this trip had to be postponed, William Clift and an associate came to Lewes
instead. "Since I had the pleasure of addressing you," Gideon then wrote
Buckland on 5 October, "Mr. Clift and [John Edward] Gray have been here
to examine my new fossil, but neither has thrown any light upon it." Clift
suggested that the large processes (bony spines) evident in Mantell's specimen might belong to a dermal fringe, like those adorning some species of
iguana. The sternum was also very curious, and while vertebrae, omoplate,
and ribs were all vaguely crocodilian, the coracoid bone distinguished this

genus from all known recent or fossil crocodiles. It could not be mega-losaurian either. Meanwhile, having found similar bones in his own col-lection, Gideon thought his specimen might be *Iguanodon,* but as no teeth had been found in the block, it seemed premature to say so. He could make nothing of the bony spines – unless they *were* some sort of dermal fringe.

By mid-October, Mantell knew he had discovered a second unknown reptile, though its appearance remained vague. In response to something of his, Bakewell replied on the seventeenth: "To prevent anyone giving a name to your new reptile, I would, till its precise character among sauri-ans be made out, call it *S. paradoxicus.*" On 2 November he further advised caution when publishing on the beast, as later discoveries would proba-bly clear up many of its ambiguities. Ignoring both of these good coun-sels, however, Gideon arranged to present his paper and unique saurian to the Geological Society (not the Royal, as originally planned) the next month. He had decided to call the new creature *Hylaeosaurus,* meaning "Wealden [or forest] lizard."[7]

On 30 November, Mantell sent his already celebrated fossil to London, together with some large iguanodon bones and a painting by Warren Lee (his boyhood friend) of *Iguanodon*'s back leg – as restored by Gideon, who had calculated its musculature. Mantell then left for London Monday morning, 3 December, taking the Lewes coach at nine, which delivered him at four. After dressing, Gideon went immediately to the Geological Society; finding his treasures safe, he hurried on to Lyell, who had mar-ried charming Mary Horner in Bonn on 12 July and was just returned. He gossiped with the Lyells till midnight, came back next morning for break-fast, then unpacked and arranged his fossils at Somerset House. Dining that night with Murchison, Gideon met Buckland and accompanied him to his hotel; he stayed till 3 A.M. looking over, and helping to revise, parts of a new book on geology, mineralogy, and natural theology for which Buckland was to receive a thousand pounds.

After his customary few hours' sleep, Mantell went busily about on Wednesday, calling on Lyell and visiting the College of Surgeons, then re-turning once more to Somerset House to prepare everything for that

7. *Hylaeosaurus.* GM–BS, 20 July 1832 (Yale); GM–Northampton, 21 July (Castle Ashby); GM–SGM, 24 July (APS). The late Dr. Charig kindly showed me the original specimen at BMNH; it is not on public display. GM–WB, 6 Sept, 5 Oct (RS); *Medals,* II, 734–739. Robert Trotter of Borde Hill, near Cuckfield, had begun to purchase and collect Tilgate fossils by Dec 1830; a most generous "poacher," he would donate several fine specimens to GM's collection (e.g., *Geol SE Engl,* pp. 332–333n). Gray (*DNB*) had been König's as-sistant at the BM since 1824. RB–GM, 17 Oct, 2 Nov 1832 (ATL). GM–WB, 12, 23, 28 Oct, 24 Dec 1832 (RS); GM–SGM, 19 Dec 1832 (APS). Re the name "Hylaeosaurus," see *Geol SE Engl,* p. 328&n.

evening. When Buckland came in, Gideon read major sections of his paper and explained parts of his fossils, including the puzzling spines. The meeting began at eight, with Mantell pleased to see how many of his friends had come – even Bakewell, ill though he was. While allowed to read his paper, Gideon was required to abridge it by one-third, an unanticipated limitation which annoyed him greatly. But all went well, and when Lee's painting of the restored iguanodon leg was at last let down, the audience seemed much impressed. The huge size of ancient saurians was becoming common knowledge.[8]

Following a day of miscellaneous activity and packing up, Mantell spent most of Friday, the seventh, with twenty-two-year-old Thomas Hawkins (1810–1889; *DNB*), a recklessly extravagant lover of fossil marine reptiles who was already well-known for his remarkable collection of ichthyosaurs and plesiosaurs from Lyme Regis. (Bought there, his specimens were then prepared superbly by an Italian sculptor.) On first seeing examples the previous month, Gideon had assumed Hawkins chiseled them himself. Now visiting the young man at home, he found more splendid fossils to admire, including an *Ichthyosaurus platydon* some twenty-five feet long. But Hawkins restored too freely, not hesitating to supply missing portions wherever they might be needed; though he never amended without reasonable analogies, Gideon still thought so much artifice unwarranted. The two men became good friends nonetheless and would meet again the following March, by which time Gideon had gotten over most of his objections.[9]

Publishers

That Saturday, 8 December, in Albemarle Street, Mantell waited on the famous Scottish publisher John Murray, whose authors included Byron in literature and Lyell in science. Together, they looked through *Illustrations* (1827), which Gideon hoped to expand into a book devoted exclusively

---

8. J (31 Dec, covering 30 Nov onward). Charles Lyell, 35, wed Mary Horner, 24, in Bonn on 12 July 1832 (Wilson, pp. 361–364). Buckland's long-awaited *G&M* did not appear until 1836; like Kirby, its author had been selected for the task (an appeal to nature on behalf of Anglican theology) by Davies Gilbert, in his role as president of the Royal Society. Unlike Kirby, however, Buckland collaborated openly with other members of the scientific community, including Gideon, who would later recommend *G&M* to his own readers (*Medals*, I, 10–11).
9. Hawkins: J (Oct, 6, 7 Dec 1832; 9 Feb, 15 Mar 1833); *Geol SE Engl*, pp. 322–323n; GM–BS, 18 Jan 1834 (for the Italian sculptor), 25 Sept 1835 (Yale). Many of Hawkins' specimens are now at BMNH; see Chapter 7 below.

to giant saurians, such as Bakewell had been urging him to write. Revising that plan somewhat, Murray preferred a thick octavo volume of regional geology; he therefore encouraged Mantell to prepare its text and illustrations, upon which judgment would then be made. Ebullient at having garnered so prestigious a publisher, Gideon stood beneath Murray's famous portrait of Byron, looked up into its haughty face, and perhaps struck a revealingly similar pose.

Mantell spent the rest of December scribbling furiously. Despite considerable professional activity, he wrote 280 pages of text (much of it compiled from his previous books) in less than three weeks and drew 130 sketches for woodcuts, besides maintaining his scientific correspondence and sending off packages of fossils to Europe and America. On Lyell's advice, he planned now to withdraw his paper on *Hylaeosaurus* from the Geological Society, publishing it in his book instead. Naming the new animal fully became of particular importance because it had emerged that only those who designated species, as opposed to genera, were commemorated in scientific nomenclature. Thus, Henry De la Beche (1796–1855), a recent geological acquaintance, wrote Gideon on the seventeenth to advise him that the German fossilist Hermann von Meyer had specified Iguanodon *Mantelli*. "Don't forget to give your new animal a specific as well as generic name," De la Beche therefore cautioned Gideon. "See how the name will even now stand with your own *Iguanodon*. It will be Iguanodon Mantelli *Von Meyer* – the said Von Meyer very probably never having seen a bone of it in his life, his whole information being derived from your works." As of 19 December, then, Gideon's new animal became *Hylaeosaurus armatus* and his forthcoming book, "The Geology of the South of England."[10]

Mantell finished his manuscript on New Year's Eve, after some all-night sessions. He immediately sent Murray a nearly complete draft, dozens of drawings, a copy of *South Downs* (to aid the lithographer), and proofs of a copperplate map and sections, which were necessarily to be colored. On receiving Murray's note of 2 January that the box had been received, Gideon emphasized the next day that his *Hylaeosaurus* paper would appear nowhere else. When a whole week dragged by with no acceptance

10. H. De la Beche [pronounced "Beach"]–GM, 8, 17 Dec 1832 (ATL). With his letter of the eighth, De la Beche sent a copy of Hermann von Meyer, *Palaeologica, zur Geschichte der Erde und ihrer Geschöpfe* (Frankfurt, 1832), in which GM is repeatedly cited (pp. 50, 110–117, 345–352); *Iguanodon Mantelli* is at p. 110. De la Beche, founder of the Geological Survey (1835), and GM corresponded between 1832 and 1851 (ATL). See De la Beche, *Researches in Theoretical Geology* (London, 1834); GM–BS, 18 June 1834 (Yale); *DSB* (V. A. Eyles); and Paul J. McCartney, *Henry De la Beche* (Cardiff, 1977). J (8, 31 Dec [collective entry] 1832); GM–SGM, 12, 19 Dec 1832 (APS).

forthcoming, Mantell became distinctly nervous. On the tenth, therefore, he unwisely complained to Lyell that Murray had kept his manuscript for nearly a fortnight. Murray then responded almost immediately, saying that such a book would never repay its costs. Greatly dismayed, Gideon offered on the thirteenth to reduce expenses and provide subscribers, recalling disingenuously how well *South Downs* had sold ten years before, when geology was far less popular. By no means taken in, Murray declined the book for a second time, sending its manuscript and drawings to Lupton Relfe as Mantell had instructed. On receiving Murray's definite rejection, Gideon wrote without delay to Relfe, who dutifully visited Lewes on 22 January but was on the verge of bankruptcy and could offer no acceptable terms.[11]

Desperate now, Mantell wrote Bakewell, sending him bad news and the manuscript. True friend that he was, Bakewell immediately notified his own publisher, Thomas Longman, explaining the salient facts. He had just received a manuscript from Gideon Mantell, who wished him to recommend a publisher. More than half the text had already appeared, in two expensive folios out of print for several years. Mr. Mantell proposed to publish the present work in octavo, comprising about seventeen sheets, together with woodcuts (some of them from his former works), copper engravings or lithographs, and a map. He would expect an edition of one thousand copies at joint risk, to sell at fifteen or eighteen shillings per copy. Mantell's former publisher, Bakewell said, had failed. But Mantell himself was "a clear writer, an excellent draftsman, and I think beyond doubt the most scientific anatomist in England . . . , the British Cuvier." One copy of these remarks went to Longman (who, like Bakewell, lived in Hampstead); the other was sent to Gideon, accompanied by some sage advice concerning the proposed terms.

Longman called on Bakewell the next morning to see Mantell's "Geology of the South of England" and seemed very well disposed to publish it (as Bakewell reported to Gideon the same day) but would need to look at the drawings and map as well, so that he could calculate expenses. As to the firm itself, Bakewell confided, "I have much greater dependence on their honesty and straightforwardness than I could have on Murray, who is a pompous, self-sufficient, arrogant Scot but humble to those whom he thinks he can make a good profit by." Longman, however, would soon read a notice of Gideon and his *Hylaeosaurus* in *The Penny Magazine*

11. GM and Murray. GM–JM, 1 Jan 1833 (Murray archives); JM Jr.–GM, 2 Jan 1833 (ATL); GM–JM, 3 Jan (Murray), and GM–JH, 3 Jan (WSRO), both misdated 1832; GM–CL, 10 Jan (Murray); JM–GM, 12 Jan (ATL), GM–JM, 13 Jan (Murray); JM–GM, 15 Jan (ATL); J (12, 21, 22 Jan 1833).

(where Bakewell had placed it) and think well of the already reputable author as a result. In some concluding remarks, Bakewell stressed the importance of publishers and periodicals to authorship.

On the twenty-sixth, as it happened – the same day that Bakewell's notice appeared – Longman wrote Mantell directly to offer terms: He would cover *all* the expenses of publishing and divide the profits evenly. When Gideon, of course, happily accepted this arrangement the next day, Longman immediately sent his manuscript to the printer. A prospectus for the book appeared on 8 February, and Gideon received his first proof sheet on the ninth. "You will be pleased to learn," he then wrote Silliman, "that I have a new work in the press – *The Geology of the South-East of England.* . . . It consists of an abstract of my two former volumes on the geology of Sussex, the memoir read before the Geological Society, and notices of all the discoveries that have been made in this part of England since my *Fossils of Tilgate Forest.*" Probably appearing in April, the octavo volume would include sixty or more wood engravings, "which in England are all the fashion," and several lithographs.

Gideon's book received some unexpected publicity from Roderick Murchison in his presidential address to the Geological Society on 15 February. "Mr. Mantell," he observed,

> whose energies seem to expand in each succeeding year, notwithstanding the limited field to which his researches are necessarily confined, has presented us with an account of an undescribed and singular species of saurian, to which he assigns the name of *Hylaeosaurus*. This fortunate exhumation has, I am happy to say, encouraged the enterprising ranger of Tilgate Forest to make it the nucleus of a new and comprehensive work, in which he will not only describe all the vertebrate animals in his rich domain, the Weald of Sussex, but will embrace in it a geological description of his own and of the adjoining counties.

Murchison's widely reported commendation then spread throughout England, interesting no one in the realm more obviously than its king.

To his complete surprise, five days later Mantell received a letter from Sir Herbert Taylor, the royal secretary, announcing that William IV desired to have the forthcoming book dedicated to himself. In consequence, it was. But Gideon regretted that the original dedicatee, Sir Henry Halford (William's physician, and president of the Board of Health) had then to be informed. A new prospectus, out by the twenty-fifth, listed both King William and one of his sons among Gideon's subscribers. Longman, meanwhile, had announced Mantell's book as a future publication. By 14 March Gideon could assure correspondents like Samuel Woodward of Norwich (to whom he wrote on the back of a prospectus) that his book was in the press and would be out in three or four weeks. In London that day, Mantell had breakfasted with Bakewell at Hampstead, going over sheets of the latter's new fourth edition. At noon he then visited Longman to pick up proofs and arrange plates. After breakfast with Thomas Hawkins on the fifteenth (to see his wonderful *Ichthyosaurus platydon* again), Gideon read some of his own proof sheets to Lyell and looked over chapters soon to appear in the third and final volume of Lyell's *Principles*. On the sixteenth he devoted all morning to augmenting and correcting his proofs, so as to incorporate Lyell's suggestions. Later in the day, he returned the galleys to Longman and formally signed their publishing agreement.

That night, Mantell accompanied Buckland to a conversazione at Kensington Palace given by the Duke of Sussex (Augustus Frederick, 1773–1843; *DNB*), who was the sixth son of George III and, from 1830 to 1839, president of the Royal Society. Arriving at nine, he was soon conducted to an anteroom and shortly thereafter presented to the Duke, who received him very graciously. Among that distinguished assemblage of more than four hundred, Gideon met such luminaries as Michael Faraday, Prince Tallyrand, and Henry Brougham. Thomas Moore, one of his favorite poets, then appeared with old Samuel Rogers, author of *The Pleasures of Memory* (a very popular poem of 1792) and *Italy* (1822, 1828). Murchison and Rogers were soon talking with him, and when Francis Chantrey, the sculptor, joined in, Mantell showed some plates from his new book. Sir John Herschel next engaged him in a long, enthusiastic conversation on geology and astronomy; Gideon may have recast his "Introductory Observations" in proof because of it. Among many other scientific friends, he saw Lyell, Clift, Greenough, König, and Lambert. When the party broke up at midnight, Murchison gave him a ride to his hotel. Gideon had never before entered into such distinguished company as an equal, and his doing so characterized a new age, which, in the spirit of the Reform Bill, had begun to supplement an aristocracy of birth with a meritocracy of accomplishment.

After correcting further proofs on Sunday, the seventeenth, Mantell left for Lewes that evening. Back in London to visit a dying patient on 8 April, he continued to emend the press. His dedication and preface were written late in the month. An advance copy of *The Geology of the South-East of England* reached Gideon at Lewes on 2 May, its style pleasing him very much. Nine days later, he received ten additional copies from Longman, distributing them all. Bakewell then assured Mantell his work was "everything that your best friends could wish it to be." Mrs. Bakewell, too, found the book "most pleasant reading." In a subsequent letter Bakewell added that "the style is so excellent and the facts so clearly stated, it cannot fail to become generally known and approved. Nothing but the inconvenient size and the expense of your former works impeded their speedy sale." Though originally advertised for eighteen shillings, Gideon's book was offered for general purchase at twenty-one, the plates and map having proved more expensive than anticipated. As a commercial venture, it was no more than marginally successful.[12]

The New Book

The Geology of the South-East of England, an octavo volume of 416 pages, five plates, and a colored map (with geological sections), consisted of some introductory observations, eleven misnumbered chapters, and an appendix – Mantell's "Tabular Arrangement of the Organic Remains of Sussex." The "Introductory Observations" proposed to adumbrate "leading principles of modern geology," derived in general from Bakewell, Lyell, and Herschel, for this work aspired to be more than a regional study. Though clearly willing to broaden any potential readership, Gideon did not regard *South-East England* as an introduction to geology. Rather, he saw it as an opportunity to synthesize and correct his previous discoveries within a newly accepted framework of Lyellian beliefs.

Thus, in his stratigraphic terminology and throughout the book, Mantell frequently revealed a pervasive indebtedness to the *Principles of Geology,* which would have been even more explicit if some late changes attempted in proof had been accepted. Chapter 3, for example, on Diluvium, was to have been extensively revised, including Lyell's new term "Pliocene," which Gideon knew only because he had seen his friend's as yet unpublished third volume in manuscript. Its persuasive arguments on behalf of gradual, rather than catastrophic, geological change influenced him significantly. Thus, in Chapter 5, on the Chalk formation, Gideon accepted from his *Illustrations* a remark that the chalk with flints stratum had "suffered most extensively from the effects of . . . catastrophes" (73).

12. GM and Longman. RB–Thomas Longman [GM's copy], 23 Jan 1833; RB–GM, 23 Jan 1833 (both ATL). RB, "The Fossil Iguanodon," *The Penny Magazine,* 26 Jan 1833, pp. 27–28 (incl. *Hylaeosaurus*). TL–GM, 26 Jan, 2 Feb 1833 (Longman archives); TL–GM, 26, 30 Jan 1837 (ATL); J (8, 9 Feb 1833); CL–GM, 9 Feb 1833 (ATL; supports title change to that adopted); GM–BS, 10 Feb 1833 (Yale). Murchison: GS (OM), VI, 143–144. J (20 Feb 1833). Prospectus, attached to GM–WC, 25 Feb 1833 (RCS); and GM–SW, 14 Mar 1833 (Norwich); GM–JSM, 1 Apr 1833; J (14–17 Mar, 8 Apr, 2, 11 May [correcting Curwen]); GM–SW, 12 May (re price increase; Norwich); RB–GM, 17 May (ATL); also GM–A. Boué, 21 May (GS, Paris); A. Boué–GM, 25 June 1833 (ATL). RB–GM, 1 June 1833. Longman archives (Univ. of Reading) reveal printing costs totaling £328/8/8. Of the 1,000 copies, 492 remained in 1834, 413 in 1835, and 277 in June 1836; half of the unsold copies were then dispatched to GM in batches (J, 4 Nov 1836).

In proof, however, he attempted to substitute "operations" for "catastrophes," which would have moved him from Cuvier toward Lyell. Understandably, Longman allowed none of these important but expensive resettings. Parts of *South-East England*, consequently, did not reflect Gideon's final intentions.[13]

Among other noteworthy additions to his previous material, Mantell supplemented Chapter 3, Diluvium, with new information about the Brighton cliffs, large portions of which comprised what he now called the Elephant Bed, because of their abundant proboscidean remains. He had also discovered within them some embedded blocks (later to be recognized as glacial erratics) of a sandstone found nowhere else in England. Chapter 4, on Tertiary formations, introduced some American material and included the Isle of Wight. Discussing fossils in the Chalk (page 146), Gideon recalled his own discovery of *Mosasaurus* in England (as noted in *South Downs*), then credited S. G. Morton's report of a similar find in the Cretaceous of North America. Finally, he once more foresaw that the remains Buckland had incorrectly called cetacean would eventually be reclassified; though they actually belonged to *Cetiosaurus*, a not yet identified dinosaur, he still thought them plesiosaurian.[14]

Chapter 10, "Observations on the Fossil Remains of the Hylaeosaurus and other Saurian Reptiles Discovered in the Strata of Tilgate Forest in Sussex," comprised Mantell's lengthy paper of 5 December 1832. The five genera of oviparous quadrupeds currently known included *Plesiosaurus, Cylindricodon, Crocodile, Megalosaurus,* and *Iguanodon,* each of which he would discuss in turn. Of these, the first required only a paragraph, as Conybeare had done it all. *Cylindricodon,* an herbivorous saurian discovered by Dr. Georg Friedrich Jaeger of Stuttgart, was actually a large Triassic amphibian. Gideon, however, whose correspondence with Jaeger began in 1828, thought the latter's drawings resembled specimens he had collected from Tilgate Forest and was then convinced of their identity by the French geologist Ami Boué, who had seen both the German originals and Mantell's. In his recent listing of reptilian specimens, John Edward Gray (of the British Museum) had identified the two Wealden crocodiles

13. Gideon's prospectus could muster only 24 supporters. A final subscribers' list (loosely inserted, usually missing) had 159 names, for 173 copies. GM's proof copy of *Geol SE Engl* is at ATL; Lyell probably suggested a number of changes on 15 March. Besides Lyell, another important influence on the book was Dr. Samuel G. Morton (1799–1851; *DAB*) of Philadelphia (*Medals*, I, 293, 430, 441; *Petrif*, p. 200). SGM–GM letters (1831–1837) are at ATL; GM–SGM, at APS; see also GM–RNM, 11 July 1851 (ATL). The Herschel epigraph is from *A Preliminary Discourse on the Study of Natural History* (London, 1831).

14. *Geol SE Engl*, Chapters 4–6.

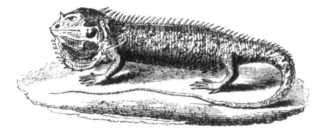

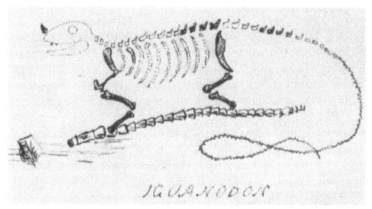

Figure 6.2. *Top: Iguana cornuta,* as sketched by Mantell (*Geol SE Engl,* 1833, p. 325). *Bottom: Iguanodon,* as reconstructed by Mantell (ca. 1833), a drawing not published in his lifetime (BMNH). Gideon's specimen of *Hylaeosaurus* (above) was the most nearly intact skeleton of any dinosaur that had so far been discovered. Of *Iguanodon,* contrarily, he had loose bones only. Encouraged by the relative completeness of *Hylaeosaurus,* he now attempted to assemble *Iguanodon,* taking the iguana as his model. The sketch at lower left, if anything, may be a tooth. The lower drawing is that of *Iguanodon*'s spine, as seen from above. Following these two preliminaries, he outlined the missing skull, placed the "horn" upon it, and then assembled what vertebrae and limb bones he had. (But the scapula he included belongs to *Hylaeosaurus,* not *Iguanodon.*) This sketch of Mantell's is the earliest known scientific attempt to piece a dinosaur together.

as *Crocodilus Mantelli* and *Gavialis Lamourouxii,* respectively. But Gideon reasonably believed that still further species (if not genera) were represented among his specimens, which he then described. A two-page review of *Megalosaurus* added little.

The section on *Iguanodon* was, of course, much longer. Though various specimens purporting to be iguanodontian had been discovered elsewhere

in the Weald, in the Isle of Wight, and even in Germany (all legitimate localities), Mantell did not yet accept such identifications as conclusive. His
own collection now included several head bones, especially the os tympani
(which in reptiles unites the lower jaw with the skull). Others, however,
were still difficult to ascertain, including vertebrae, ribs, sternum, clavicles, humerus, radius, ulna, femur, tibia, fibula, and various bones of the
feet and toes. Comparing his specimens with bones of the modern, diminutive iguana, Gideon tentatively assigned some fossil ones to *Iguanodon,* but
could be certain only of one femur. From it and other evidence, however,
he reiterated that *Iguanodon* had been larger than *Megalosaurus.* Having
assembled parts of a hypothetical skeleton for his beast, Mantell went on
to speculate what kind of animal it may have been. Even on present evidence, he could not entirely decide between a crocodilian model and a lacertian one. Assuming that *Iguanodon* probably resembled the iguana more
closely, however, Gideon proceeded to map out a creature seventy feet long
(most of that, tail) and nine feet high at the head.

In the next section of his paper, Mantell recalled how *Hylaeosaurus* had
been discovered and what the four-and-a-half-foot fossil contained: ten
vertebrae, two coracoid bones, two omoplates, and a series of bony spines,
together with fragmentary ribs and other bones, several fossil plants, and
casts of freshwater shells. Apologizing for the possible tedium of his approach, Gideon proceeded to compare and describe each bone. Among
them, the bony spines were clearly most anomalous. Once more taking a
modern iguana as his model, however, Mantell argued that the spines were
probably analogous to its cartilaginous dermal fringe. He also plausibly
reconstructed a posture for the dying animal that would have placed its
spines as found. (Alternatively, he admitted, the spines may have been on
its tail.) In any case, these relics were unquestionably those of a new, *armored* saurian, for which the name *Hylaeosaurus armatus* ("armored lizard
of the Weald") would not be inappropriate. Gideon then concluded by reconstructing portions of the environment in which *Hylaeosaurus* must
have lived and speculating that this particular carcass had probably been
transported a long distance by the river whose delta comprised the
Wealden strata of Sussex. He closed with a deeply felt eulogy of Cuvier,
who had died from cholera in May 1832.[15]

Chapter 11, "Results of the Geological Investigation of the South-East
of England," began with the Isle of Portland, which Mantell had visited

---

15. *Geol SE Engl,* pp. 289–333. Boué (1794–1881) visited GM at Lewes on 9 Aug 1831
(J; Geol Exp; *Geol SE Engl,* p. 293). For Georg Friedrich Jaeger (1785–1866), director
of the State Natural History collection in Stuttgart, see *Geol SE Engl,* pp. 292–294,
304n; *Medals,* II, 784; letters to GM, 1828–1851 (ATL); and ten entries in J.

Figure 6.3. "Reptiles restored, the remains of which are to be found in a fossil state in Tilgate Forest, Sussex," in watercolor, pencil, and ink (225 × 320 mm) by George Scharf, 1833, from conceptions by Gideon Mantell (Alexander Turnbull Library, Wellington, New Zealand). "This served as a Sketch for a Picture 3 Yards long." Scharf's painting, previously unpublished, is the earliest known attempt to reconstruct the external appearance of dinosaurs and their environment. The freshwater origin of the Wealden formation is clearly indicated. *Iguanodon,* said in a penciled note alongside to have been one hundred feet long, is at left. There are then, counterclockwise, two turtles, a crocodile, a short-necked plesiosaur, *Hylaeosaurus,* two wading birds, *Megalosaurus* (as monitor lizard), and a long-necked plesiosaur, together with invertebrates and plants. The penciled notes (which misidentify *Hylaeosaurus* as *Megalosaurus*) are not by Gideon. Scharf's painting may have been intended as a frontispiece for Mantell's never-completed book on saurians, the idea rejected by John Murray in 1833. The three-yard version (no longer extant) probably adorned Mantell's museum or his lectures.

in July 1832 (while touring after the BAAS meeting at Oxford), though only for a few hours. The tropical forest he saw there, now petrified, must have "gradually and tranquilly subsided . . . beneath a body of fresh water" (338–339) – the bed of a vast lake or estuary, into which a river flowed, forming a delta. In another geological change, the Weald sank into that vast ocean which had deposited the Chalk formation, but how quickly or when was still unknown. Gideon next discussed the flora and fauna of the "Cretaceous epoch," incidentally referring his readers to the third volume of Lyell's *Principles,* then in press. He also described changes in the geology of Sussex during the "Tertiary era." Further discussion announced the "Elephant epoch" (our Pleistocene), after which Tertiary strata were uplifted to their present position, the English Channel created, and transverse valleys of the North and South Downs produced or enlarged. Following that great upheaval, however, no further changes of equivalent magnitude occurred. Nothing but the most ordinary effects of air, sea, and rivers have been detected within the modern era. Despite Lyell's evident influence upon him, Gideon remained convinced – with Cuvier, Buckland, Bakewell, and Silliman, among others – that the past had been geologically more active than the present (how else could such large creatures have been produced?) and was not quite so assured a disciple as Lyell might have wished.

During the next two years, Mantell's new book attracted predictable attention from correspondents and publications. Among these notices, that by Robert Bakewell in Loudon's *Magazine of Natural History* was both laudatory and shrewd. "We regard this volume as one of the most valuable works on local geology that has yet appeared," it concluded, "and as it relates to a country that is more frequented than any other in England for its watering places, it cannot fail to become the travelling companion of the numerous intelligent visitors of the southeastern coast" (360–361). *Monthly Review* thought the scientific world greatly indebted to Mantell, his book being filled with "materials for solemn and lasting meditation, conveyed in language worthy the great subject" (345). Though nominally limited to southeastern England, it was "substantially a key of the most easy application to the general doctrines of geology" (346). *The Lancet,* in May 1834, believed Mantell's work an "admirable essay . . . distinguished by a correct arrangement of the subjects, an accuracy of description, an acuteness of deduction, a comprehensiveness of theory, a perspicuity of style, and an excellence of graphic illustration which reflect infinite credit upon the zeal and industry of the author" (988), who was then highly praised for combining his medical career with such devotion to science. In his *American Journal of Science and Arts* for July 1834, Benjamin Silliman treated *South-East England* more generally, utilizing subsequent personal information from Gideon's letters. Praise of another kind infused two further publications. One, translating Cuvier's survey of the animal kingdom (1834), included numerous genuflections toward Mantell by an anonymous English editor, Edward Griffith (1790–1858; *DNB*). The other, Horsfield's history and topography of Sussex (1835), featured long quotations from *South-East England*. Gideon also contributed a routine essay on geology and mineralogy to the same work; written quickly, it was nothing more than a facile précis of his book.[16]

Despite its favorable reception, *South-East England* was alive with faults, as Mantell admitted candidly to Silliman in a letter of 3 October. "The only parts I am satisfied with," he wrote then, "are the introductory remarks, the chapter on *Hylaeosaurus*, and the last, on the results of geological researches in the southeast of England." From our point of view, one of the book's chief failings was Gideon's still distorted view of *Iguanodon*, resulting from his obsession with size. Nevertheless, despite its many small lapses, *South-East England* succeeded for a time, as a remarkably adequate regional survey – weak on geological processes perhaps, but strong on available chronology. Besides including several previously undescribed species, it announced the discovery of an important new dinosaur (the most accurately reconstructed thus far) and, with *Mosasaurus*, established the first intercontinental saurian.[17]

16. Reviews of *Geol SE Engl.* RB, *Magazine of Natural History,* 6, (1833), 359–361; *Monthly Review,* 8, (1833), 337–347 (345–347); *The Lancet,* 1 (1834), 937–939; *AJSA,* 26 (July 1834), 216–217. Georges Cuvier, *The Animal Kingdom* (an augmented translation of Cuvier and P. A. Latreille, *Le Règne animal distribué d'apres son organisation pour servir de base a l'histoire naturelle des animaux et d'introduction a l'anatomie comparée* [5 vols., Paris, 1829–1830]), 4 vols., London, 1834; see esp. II, 10n, 15n, 32–33n, 44–45n; Thomas W. Horsfield, *The History, Antiquities, and Topography of the County of Sussex* (Lewes, 1835); see esp. pp. 63–64, 184. GM's "Geology and Mineralogy" is at pp. 8–24. GM–BS, 3 Oct 1833 (Yale).

17. GM's scientific influence was now at its height. Among the current publications in which he is mentioned prominently (besides those of Cuvier, Lyell, Meyer, Kirby, Buckland, and Horsfield) are WHF, *Geological Sketch of the Vicinity of Hastings* (London, 1833); SW, *A Synoptical Table of British Organic Remains* (London, 1830) and *Outline of the Geology of Norfolk* (London, 1833); RB's *Introduction to Geology,* fourth edition (London, 1833) and second American (New Haven, 1833); and SGM, *Synopsis of the Organic Remains of the Cretaceous Group of the United States* (Philadelphia, 1834), which was dedicated to Mantell. The transatlantic collaboration of SGM and GM (letters at ATL and APS) helped to establish the Cretaceous Period worldwide. Regarding later fossils, GM also mentions Timothy A. Conrad, *Fossil Shells of the Tertiary Formations of North America* (Philadelphia, 1832–1837; GM quoted, p. 11); see TAC–GM, 10 June 1835 (BMNH).

7

Old Steine

On Monday, 24 June 1833, Gideon Mantell left Lewes for London by the early coach. Arriving shortly after noon, he visited both Longman and Relfe on business, saw his sister Kezia at Bromley the next day, and then proceeded at 7 P.M. by the South Mail to Cambridge, arriving at 2 A.M. on the twenty-sixth. Come to attend the third annual meeting of the British Association for the Advancement of Science, Mantell soon arose to meet Davies Gilbert, who accompanied him to the Philosophical Society, where he registered, paid his fees, and obtained tickets of admission to the various sections. After returning to his hotel for breakfast, Gideon went immediately to Trinity College; there he found Lord Northampton, Buckland, Murchison, Sedgwick, Fitton, and Whewell. The latter gave him a ticket for a grand dinner in Trinity College Hall that night, at which Mantell sat next to Dr. Richard Harlan of Philadelphia. After renewing many other scientific friendships, he attended the Geological Section that evening. During its second meeting, on the twenty-seventh, Gideon was permitted to display and discuss a large iguanodon femur he had recently acquired.[1]

On Friday the twenty-eighth Mantell breakfasted with William Clift at St. John's College, attended the Association's meeting in the Senate House, dined again at Trinity College, and took tea with Lonsdale, Phillips, Clift, and old William Smith, the venerable stratigrapher. After attending the evening Geological Section once more, he went to Lord Northampton's rooms at eleven for trip planning and did not get to bed until two. Up four hours later, he breakfasted with Northampton in Buckland's rooms and then joined an excursion to Ely, where Edward Blore (William IV's architect and the builder of Scott's Abbotsford) elucidated the cathedral. Back

---

1. J (24–27 June 1833); *Gent Sci*, pp. 165–175; 411–416; *DAB*. Richard Harlan (1796–1843) of Philadelphia, also interested in fossil saurians, met Gideon again at the GS in London on 6 July, and then visited him at Lewes on the twenty-second and twenty-third (J); they subsequently exchanged specimens and letters into 1836 (ATL). The femur on which Gideon lectured may have come from a summer of 1832 find by workmen at Tetham, Kent (noted in GM's copy [ATL] of *Illus*, opposite page 30).

at Cambridge by two, Gideon packed up his things and ducked in on a party of Sedgwick's before departing with Northampton in his Lordship's elegant carriage.

Mantell had first met the Marquess on 28 November 1831, when (following a letter of introduction from Buckland) Northampton spent three hours seeing only a part of Gideon's museum; "a very agreeable, intelligent man" of the proprietor's own age, he returned to peruse the rest on 6 December. As the Northamptons were currently at Brighton, a series of visits and letters followed. Mantell and his noble friend then devoted considerable time to each other at the Oxford meeting of the BAAS in June – going on Buckland's geological excursion together, for example. They also went fossil collecting in Sussex that November. Ensuing months and the further BAAS session just ended had only strengthened their warm relationship. As a result, Gideon now savored his long-anticipated journey to Castle Ashby, the Marquess' nobly proportioned Elizabethan/Jacobean mansion in Northamptonshire, and was gratified to arrive there at 9 P.M., along a magnificent tree-lined avenue.

After rising early next morning to walk the grounds by himself, Mantell shared a busy day with his host, including late breakfast, some charming gossip, church, noble conservatories, greenhouses, gardens, luncheon – and an afternoon reviewing cabinets filled with fossils and shells. At six the Bucklands and Murchisons arrived for dinner and the evening. Less accustomed to hospitality from aristocrats than they, Gideon felt himself living in a fairy tale and when he finally retired at 1 A.M. was so overcome by the romantic splendor of a brilliant moon that he could hardly sleep. Mantell spent Monday visiting nearby quarries and looking over further specimens with Northampton. Dinner and fireworks that night honored his Lordship's daughter Lady Ann, who was celebrating her birthday. Gideon revisited the quarry town of Whiston next morning to inspect its highly ornamented church. Afterward, he discussed the geology of western England with Murchison, viewed surrounding countryside from atop the Castle, and joined an innocuous excursion by carriage to nearby Yardley Chase, seeing the oak made famous by William Cowper. On Wednesday, 3 July, Gideon regretfully bid adieu to Castle Ashby, where he had spent three days of uninterrupted happiness.[2]

---

2. J (28 June–3 July 1833). For Spencer Joshua Alwyne Compton, Lord Northampton (1790–1851), see *DNB*; *Gent Sci*, pp. 434–439; and the many references to him in J. ATL preserves more than sixty letters from him; Gideon's replies are at Castle Ashby (private). Later president of the BAAS, and then of the RS (he had been president of the GS earlier), Northampton remained one of Gideon's closest and most supportive scientific friends.

The Lure of Brighton

Having secured reliable friends in London, Oxford, Cambridge, and now among the aristocracy, Mantell began to look askance at Lewes. He had long thought of relocating to Brighton, which, though only eight miles away, was vastly more cosmopolitan and had enjoyed frequent royal visits ever since the Regency. The *one* royal visit to Lewes in October 1830 helped to impress Mantell with his town's inferiority to Brighton, especially after his *Narrative* of that visit appeared. "At this season of the year," Gideon then advised Silliman in November 1831, "the Court resides at Brighton, and the greater part of the nobility follow in its train." Already well-known there, he hoped, after somehow acquiring an honorary M.D. diploma, to establish a medical practice in Brighton, with its wealthy aristocrats as his clientele.

By July 1833, following his return from Castle Ashby, Mantell began to discuss his possible relocation quite seriously with Bakewell and some other trusted friends. Playing their advisory roles in all good faith, they told him what he wanted to hear. "In confidence," he then wrote Silliman on 3 October, "I have almost resolved to move to Brighton, where there is a larger field for my professional engagements, and a chance of being able to turn my successes to some account." But, he continued pragmatically, "when a man is turned forty, and has four children and but a very moderate fortune, a removal is a subject of no little anxiety, particularly to a mind like mine." How could he wangle some money?[3]

Mantell had a particularly useful friend in Brighton, the elderly, powerful, and very rich third Earl of Egremont (1751–1837). A noted patron of artists and civic betterment, he had been lord lieutenant of Sussex since 1819 (when Gideon's journal first mentioned his name). Two years later, he became one of the most prestigious subscribers to Gideon's *Fossils of the South Downs*. In December 1825 Egremont called at Castle Place to congratulate Mantell on his election to the Royal Society, then dropped by occasionally thereafter. They reciprocated visits in May 1833 when Egremont inspected the Castle Place museum and Gideon spent three hours at Petworth, the Earl's magnificent estate in western Sussex, looking through his outstanding collection of paintings and statuary. Having now realized on 3 October that moving to Brighton was a real possibility, Gideon

---

3. The move to Brighton. J (31 Dec 1822, 1 Jan 1823, 29 Jan 1831). GM, *A Narrative of the Visit of Their Most Gracious Majesties William IV and Queen Adelaide to the Ancient Borough of Lewes on the 22nd of October 1830* (London: Lupton Relfe, 1831), esp. p. 14. GM–BS, 24 Nov 1831 (Yale); Lyell, *LLJ,* I, 377 (with important observations on GM); RB–GM, 3 Aug 1833 (ATL); GM–BS, 3 Oct 1833 (Yale).

promptly informed Egremont, who responded the next day with an invitation and a gift of game. When Mantell expectantly waited upon his Lordship in Brighton on the fifth, the octogenarian gentleman not only approved of Gideon's relocating there but, in his characteristically blunt, unceremonious manner, offered him one thousand pounds to do so! A few days later, after conferring with Lyell and other friends, Gideon decided to accept the Earl's extraordinary offer.[4]

Having then worked out the numerous details, Mantell was as of 4 November "in all the bustle of a removal to Brighton," where he expected to be domiciled by year's end. He had just agreed to lease house No. 20 on the Steine from a Mr. Budd. In December, Gideon trundled all his fossils over from Lewes, reinstating them in the drawing room of his new home. On the twenty-first, his wife, children, and servants moved in, so it was "Farewell forever to Castle Place!" This transition marked the beginning of a new phase in Mantell's life and would be far more drastic in its outcome than his temporary euphoria allowed him to foresee.[5]

The New Museum

No. 20 The Steine, an imposingly tall but narrow structure three stories high, with an ominously swollen facade, would cost more than £350 per year in rent and taxes. When he took it, Gideon deliberately selected a house larger than his needs and more expensive than his means. These two discrepancies were to have been mutually resolved by devoting his drawing room floor to a commercial geological museum. By the time he had installed his collection, however (in some new, expensive cases), his scientific and professional friends had persuaded him that such obvious proprietorship would damage his reputation and profits as a surgeon. With considerable misgivings, therefore, Gideon acceded to their opinions and on 14 January 1834 opened his museum to the public every Tuesday, gratis upon prior application by letter.

4. Sir George O'Brien Wyndham, third earl of Egremont (1751–1837; *DNB*), has been remembered for his wealth, civic generosity, artistic patronage (Turner, Leslie, Constable, Carew, Flaxman, and others), horse breeding, and illegitimate children. Hating ceremony and narrow-mindedness, he was a blunt, plainspoken man. Gideon first mentioned him in J (3 Aug 1819); also 25 Apr 1821, 10 Dec 1825. Several of GM's books, including a unique colored copy of *Illus*, are at Petworth (private). GM's own copy of *Illus* at ATL includes notes to him from Egremont; fourteen others have also been preserved (ATL). Re the move to Brighton: J (4 May, 19 June, 4, 5, 6–13 Oct 1833); Egremont–GM, 23 Oct 1833 (ATL). The name "Brighton" became official in 1810; Gideon's "Steyne" soon became the Old Steine (pronounced "Steen").

5. GM–WC, 4 Nov 1833 (BMNH); J (ca. 4 Nov, 21 Dec 1833).

Located one flight up from his entrance hall, Mantell's new museum overlooked the trees, plantings, and gravel roundabout of the Steine. At left, where one entered, the nearest exhibit was a long case filled mostly with iguanodon, turtle, fish, bird, and plant relics from Tilgate Forest, the remainder being shells. Cases along the facade wall displayed fishes and other Chalk fossils, an iguanodon tibia, and a mammoth femur. Against the right-hand wall, on both sides of a fireplace, local fossils could be seen. In a squarish recess at right, parallel cases housed skeletons of recent animals (including iguana), donated specimens from all over the world, and various individual fossils of note, including *Iguanodon, Hylaeosaurus,* ammonites, crustacea, and plants. Miscellaneous antiquities, diagrams, and a bust of the Earl of Egremont appeared above the cases; below them, drawers not open to the public held still further specimens. Before long, Mantell's collection became one of the most famous in Europe.[6]

Lectures

Besides establishing his museum, Gideon worked diligently in other ways to popularize geology at Brighton. Thus, on Saturday, 8 February 1834, at 2 P.M., he lectured nearby in the card room of the Old Ship (Brighton's best-known hotel) on the geology and organic remains of Sussex. By donating all proceeds to Sussex County Hospital, Mantell tacitly emphasized the compatibility of his medical and scientific vocations. Visitors to his museum, he said, often wanted to know more about geology than he had time to tell them, so he was taking this opportunity to inform them further, being pleased to benefit the hospital as well. Throughout this initial lecture, which he read too fast, Gideon utilized a variety of specimens and diagrams, including fossil shells, bones, and tusks; stratigraphical sections; and even some attempted restorations (neither specified nor preserved) of the extinct plants and animals he described. Though the audience had dif-

6. Though No. 20 The Steine is now in commercial use, its original exterior (drawn by EMM) has been preserved; there is a memorial tablet to GM (8 May 1930). GM–BS, 18 Jan, 18 June 1834 (Yale); GM–SGM, 18 Jan 1834 (APS); GM–SW, 19 Jan 1834 (Norwich). GM, "A Descriptive Catalogue of the Collection Illustrative of Geology and Fossil Comparative Anatomy in the Museum of Gideon Mantell, Esq. F. R. S." (London: Relfe & Fletcher, 1834). Mantellian Museum Visitors' Book, 1834–1838 (ATL). J (1 May 1834 [collective entry]; 18, 20 Apr; 5 & other May, 13 Oct, 26 Nov, 8 Dec 1835; 16 Aug 1836). Dr. John Forbes of Chichester wrote GM on 13 Mar 1834 to warn of prejudice against "scientific doctors" and advised Gideon to reestablish his museum as a public institution. On 3 Apr Forbes wrote again, assuring Gideon once more that his museum would injure rather than promote his professional interests (both ATL).

ficulty following him at times, everyone was impressed with the man and remained eager to hear from him again.

On 15 February, therefore, Mantell repeated his lecture, this time at the National Schools in Church Street, on behalf of the All Souls Chapel building fund. His crowd, estimated at nearly four hundred, was even larger than before. Having been so well accepted previously, Gideon extended his remarks to include additional facts and illustrations. As for *Iguanodon*, assuredly, he had rather understated than exaggerated its size, for "(using a felicitous illustration) like Frankenstein, he was actually appalled at the being which rose beneath his meditations." Besides having much more to say about the geology of Sussex, Mantell also introduced an important new diagram, of the Temple of Jupiter Serapis near Naples, a striking example of significant geological changes within historic times that Lyell was using as frontispiece to volume one of the *Principles*. Gideon also described the geological changes of the Weald more elaborately and dramatically than before.[7]

Following his initial success, the *Brighton Gazette* had suggested that Mantell offer a course of public lectures on geology. After the even more successful second appearance, however, he graciously declined, alleging professional obligations. In reality, Gideon had none, for having arrived in Brighton without a practice he became part of a medical glut in that popular watering place and soon discovered how unneeded he was.

When the passage of some further weeks afforded no relief, Mantell unburdened his troubled mind to Silliman. "Soon after I had settled my family in our new abode," he related,

> I gave two lectures on geology for the benefit of two public institutions, a hospital and a church. Both lectures were well attended and passed off very agreeably. They were the means of introducing me to all the first people in the town, and my society was courted by the fashionables: in fact, I was the *lion* of the season. Hundreds of the nobility and gentry flocked to my museum every Tuesday.

But the "season" at Brighton ended with the court's departure in February; since then the town had been comparatively empty. None of Gideon's social or scientific triumphs brought him any money, and his expected practice had failed to materialize. "So here I am," he lamented, "confessedly one of the most successful practitioners in the country, . . . with more reputation as a man of science than I deserve, and yet without a patient."[8]

7. First and second lectures. Broadsheet and newspaper clippings (ATL); *BGaz*, 13 Feb 1834, p. 3; *BGaz*, 20 Feb 1834, p. 3; *BGrd*, 9 July 1834, p. 2.

8. According to J. D. Parry (*An Historical and Descriptive Account of the Coast of Sussex*)

He was not, however, without friends, and soon assembled a distinguished coterie of them, including Moses Ricardo (brother of the economist David) and the novelist Horace Smith. On 29 May, Ricardo, Smith, and other members of the circle joined a group of about forty persons (including several clergy, as many doctors, and almost a dozen ladies) on a public geological excursion "à la Buckland," which had been organized by Gideon. Mantell led his participants along the shore to Rottingdean, then on by carriages to Castle Hill, near Newhaven, where he lectured on the strata, helped everyone to collect some fossils, and had arranged a light meal in a tent. Despite his triumph on this occasion, Gideon felt obliged to demur from organizing any others. "I dare not," he confided to Silliman, "for would you believe it, certain detractors have attempted to ruin my professional prospects by asserting that I devote all my attention to geology!"[9]

The Maidstone Iguanodon

A more important episode, in which his coterie was vital to him, began on 6 May, when Mantell received drawings and descriptions of some recently found fossil bones from W. H. Bensted of Maidstone, Kent. Gideon was unable to pursue the matter until 3 June (shortly after his triumphant excursion), when he went to London by the afternoon stage and the next morning met William Devonshire Saull, a wine and brandy merchant whose geological collection of twenty thousand specimens – about the size of Gideon's – at 15 Aldersgate Street was open to the public every Thursday. (He and Mantell had begun to exchange fossils in November 1830, a year before Saull was elected to the Geological Society.) They left London together shortly after noon, taking outside seats on the Maidstone coach so that Gideon could make notes on the landscapes en route.

Arriving at 5 P.M., Mantell and Saull put up at the Mitre; after suitable

---

in 1833, Brighton, with its population of about forty thousand, had ten physicians and thirty-two surgeons (p. 171). GM–BS, 18 June 1834 (Yale).

9. Moses Ricardo (1776–1866) was among the seventeen children of Abraham Ricardo, an Amsterdam Jew who moved to England and became prominent on the London Stock Exchange. Unlike several of his brothers, Moses became a surgeon rather than a stockbroker, but retired early from ill health. His more prominent brother David (*DNB*), an extremely successful financier himself, helped to found both the GS and modern economic theory. See GM–BS, 4 Oct 1834 (Yale). For Horatio ("Horace") Smith (1779–1849), see *DNB*; Arthur H. Beavan, *James and Horace Smith* (London, 1900); Gordon Hake, *Memoirs of Eighty Years* (London, 1892); and biographies of Shelley and Keats. A dozen letters to GM and some relevant poems by HS about him have been preserved (ATL). J (29 May 1834); GM–RIM, 6 June 1834 (GS); GM–BS, 18 June 1834 (Yale).

Figure 7.1. *Top:* The Maidstone Iguanodon, as found (W. H. Bensted, 1834; watercolor, 355 × 230 mm; Maidstone Museum, Kent); *bottom:* and as prepared (G. A. Mantell, *The Wonders of Geology*, third edition, 1839). Size of the original: eight feet by six. In his caption, Mantell identified: (1) and (2) right and left thigh bones (femurs); (3) a shin bone (tibia); (4) toe bones (metatarsial and phalangeal) of the hind feet; (5 - upper) a claw bone (ungual); (5 - lower) a claw bone (later, vertebra); (6) foot bones (but possibly lower arm bones instead); (7) arm bone (radius; later, humerus); (8) eight vertebrae in a consecutive series; (9) ribs; (10) two collar bones (clavicles); (11) pelvic bones (iliar); (12) a chevron bone, or haemal arch.

Figure 7.2. The quarry of Kentish Rag, near Maidstone, in which Mr. Bensted discovered bones of the Iguanodon (W. H. Bensted original drawing, previously unpublished; ATL).

refreshments, they went to see Bensted and some unusual fossil bones found in his quarry. Gideon then recorded his historic observations fully:

> At Mr. Binsted's [*sic*], the proprietor of a large quarry about half a mile southwest of the town, I saw a fine specimen of fossil bones imbedded in a mass of Kentish Rag. They belong to the iguanodon and consist of:
>
> Two femurs, one so perfect as to show both extremities . . . they are 33 inches in length. One tibia very like that which I have figured, except that the distal extremity is more even [sketch] than mine [sketch]. A fibula: a fragment lying near the tibia is probably of this bone. On the sinestral aspect of the femur is a clavicle in every respect similar to the one I have from Tilgate Forest; it is 2 *8 inches long* and a fragment of another lies near it.
>
> Portions of large flattish bones, which may be either of the pelvis or sternum.
>
> Vertebrae: there are several caudal which resemble the fossil crocodilian type, having the faces almost flat, and the visceral aspect channelled and therefore belong to the posterior third of the tail. There are others which are probably lumbar or last dorsal: these are pressed so very flat that their original shape can be but obscurely defined.

Ungual bones: There are two of a very peculiar form, being somewhat flattened, and grooved like the claw-bone of a land tortoise. The
largest is 4 inches long and 2-1/8 inches wide at the base – it lies in
contact with a large bone, the precise nature of which I could not determine. There are fragments of many ribs; one shows an obscure
double termination, like the fifth rib of the hylaeosaurus. There are
the remains of a tooth and the impression of another, decidedly of the
iguanodon [sketches of both].

Though without a jaw, this Maidstone find was the first in which bones
and teeth of *Iguanodon* were unquestionably associated. While confirming many of Gideon's conjectures, it altered some others and has often
been regarded as the type specimen.

After dinner Bensted showed Mantell the quarry in which his iguanodon
had been found. While sketching geological sections there, Gideon realized
that the embedded fossil shells (including ammonites and terebratulae)
were not freshwater species, like those associated with iguanodon remains
in Tilgate Forest, but marine – an unexpected discovery that needed to be
explained. Despite its immense scientific worth, however, the Maidstone
Iguanodon never became a paper for either the Royal or the Geological
societies.[10]

On 16 July, Bensted visited Mantell at Brighton, bringing him a fragment
of the Maidstone Iguanodon so that its restoration could be discussed.
Probably at this time Gideon offered him ten pounds for the entire specimen and was declined; Gideon's friends then intervened on his behalf, as
his journal entry of 14 August recorded:

Mr. Bensted, the owner of the Maidstone Iguanodon, refused my offer of £10 for it; he had been offered £20 and required £25. My *very,*

10. The Maidstone Iguanodon. J (6 May, 3–5 June 1834); Geol Exp, 4–5 June (quoted in
part); GM–RIM, 6 June 1834 (GS); *Petrif,* pp. 301–307; WHB–GM, 24 Mar, 9, 30 Apr
1834 (ATL). William Harding Bensted (1802–1873) owned several quarries in the Kentish Rag near Maidstone. Between 1834 and 1852, he wrote about seventy letters and
notes to GM (ATL), who mentioned him in *Wonders, Medals,* and *Petrif.* William Devonshire Saull (1784–1855) appears in the same books, but I have found no letters.
Other contemporary sources regarding the discovery of the Maidstone Iguanodon include [George Fairholme], "Recent Discovery of Bones of the Iguanodon," *Philosophical
Magazine,* ns 5 (1834), 77–78; GM, "Discovery of the Bones of the Iguanodon," *ENPJ,*
17 (1834), 200–201; GM–SGM, 18 June 1834 (APS); GM–BS, 18 June 1834 (Yale); *AJSA,*
27 (1835), 355–360. [GM], "News in Science," *BGaz,* 24 Apr, p. 4; and 12 June, p. 4;
GM, "A Descriptive Catalogue," third edition (Nov 1834); RB, "On the Maidstone Fossil Skeleton in the Museum of Gideon Mantell," *Magazine of Natural History,* 8 (1835),
99–102; *BGaz,* 13 Aug 1836, p. 3 (punning epigram on GM). See also W. E. Swinton,
"Gideon Mantell and the Maidstone Iguanodon," *Notes and Records of the Royal Society of London,* 8 (1951), 261–276.

very kind friends Horace Smith and Mr. Ricardo took upon themselves to obtain it if possible and present it to me. They therefore
mentioned their wishes to several other friends, all of whom con-
curred in their opinion. This very evening the two Mr. Bensteds arrived with the specimen and it is safely deposited in my house. Now
for three months' hard work at night with my chisel, then a lecture!
I must do something to merit such kindness.

On 4 September, Ricardo sent Mantell a list of all who had contributed
toward purchase of the specimen.

Meanwhile Gideon had been applying his hammer and steel. "I have a
grand specimen on the stocks and long to show it to you," he wrote Clift
(of the College of Surgeons) on 28 August. "I am chiselling all night and
shall have finished very soon." Two weeks later, on 13 September, Mantell's journal attested that he had just placed the Maidstone Iguanodon in
his museum – but how it would ever be got out again only Heaven knew.
To celebrate the completion of his arduous task, Gideon hosted a party on
Tuesday, the sixteenth, for those who had bought him the huge fossil. Some
three dozen persons, including other friends, heard him explain the Maidstone Iguanodon bone by bone, allude to its discovery, and discuss the interesting problem of its having been found in marine strata. It was, he told
them correctly, one of the most extraordinary specimens in Europe.[11]

The Maidstone Iguanodon soon proved to be a very popular exhibit.
As a result, large crowds visited Mantell's museum to see it, and he had to
accommodate them by remaining open on extra days. Among the host of
visitors who had besieged his house, Gideon noted proudly, was John
Martin, "the celebrated – most justly celebrated – artist, whose wonderful conceptions are the finest productions of modern art." Enjoying a
prodigious reputation that has since faded, Martin was well-known for
such large, dramatic canvases as "Sadek in Search of the Waters of Oblivion" (1812), "Joshua Commanding the Sun To Stand Still" (1816), "The
Fall of Babylon" (1819), "The Destruction of Herculaneum" (1822), "The
Seventh Plague" (1823), "The Deluge" (1826), and most recently, "The Fall
of Nineveh" (1828), which Mantell had seen and admired. As Martin had
shown deep interest in its remains, Gideon hoped he could induce this
admired visionary to portray the iguanodon in its actual prehistoric

11. J (16 July, 14 Aug 1834). M. Ricardo–GM, 4 Sept 1834 (ATL); see *Petrif*, p. 303n;
    GM–WC, 28 Aug 1834 (RCS, in WC's diary). J (13, 16 Sept 1834); GM–J. J. Masquerier,
    10 Sept 1834 (Dr. Williams's Library, London); GM–BS, 4 Oct 1834 (Yale): "I am now
    certain that the hind feet of the iguanodon were very large, flat, and enormously strong."
    Medals, II, 899.

environment, being convinced that "no other pencil but his should attempt such a subject."[12]

Because of his fossil, Mantell became even more of a social lion than formerly and spent a large number of his evenings dining out. But it was still only glory, with no money attached, and his professional prospects remained "in nubibus!" On Friday, 3 October, however, Gideon received Silliman's letter to him of 24 August, which revealed that some none-too-subtle hints in previous correspondence had, together with the Maidstone Iguanodon, been turned to good account. "It was my pleasure," Silliman informed him, "to propose your degree of LLD to our college authorities and I have the satisfaction to inform you that through the various stages in two committees – first a standing and secondly a special one – and lastly through the college senate it passed readily" on 21 August. The diploma would soon be on its way to England. Gideon was thoroughly delighted with this blatantly solicited but unexpected testimonial. He had long hoped to receive or buy an honorary M.D., but such professional distinction was never to be his, and though he chose ever after 1834 to be known as "Dr. Mantell," it is important to stress that Gideon was only an honorary doctor of laws, never of medicine, and remained solely a surgeon rather than a physician.[13]

Dr. Mantell replied to Silliman's letter the next day, on 4 October, thanking his friend profusely for the high honor he had received and wishing himself more worthy of it. "The only return I can make to your college for enrolling my humble name in its archives," Gideon pledged, "is by patiently persevering in the path of science, in the hope that sooner or later I may add something to the mass of facts already known, and prove that I had the desire if not the ability to advance the cause of knowledge." The honor had come at a very good time, he admitted, "for my prospects here are at present so hopeless that I had almost abandoned my family motto, 'Nil desperandum.'"

He had given no more lectures, as doing so for money would have involved "a loss of caste," and the two he had donated to benefit institutions had been profitless to him. "It is most strange," he meditated once again,

---

12. For John Martin, see J (30 May 1828, 27 Sept, 29 Oct 1834); J. Martin–GM, 7 Nov 1834 (ATL); Thomas Balston, *John Martin* (London, 1947); Ruthven Todd, *Tracks in the Snow* (New York, 1947); and William Feaver, *The Art of John Martin* (Oxford, 1975).

13. J (28 Sept, 3 Oct 1834). BS–GM, 24 Aug 1834 (Yale). The most obvious earlier hints: GM–BS, 24 Nov 1831, 3 Oct 1833 (Yale). In 1833 GM had also attempted to obtain an honorary M.D. degree from the University of Glasgow (GM–T. Brown, 18 Aug 1833; Hunterian Museum, Glasgow). GM–BS, 4 Oct 1834. The Mantell family motto, *Nil desperandum* ("Despair of nothing"), appearing in GM's correspondence at this time is from Horace (Odes, I, vii, 27).

Figure 7.3. Two cartoons: *Top:* "Dr. M[antell] in Extasies at the Approach of His Pet Saurian" (Henry De la Beche, 1834; ink on paper, 185 × 229 mm; Alexander Turnbull Library, Wellington, New Zealand). Previously unpublished, this crude but engaging bagatelle satirizes Mantell's newly awarded L. L. D. (note the laurel crown and stratified Parnassus) while congratulating him for acquiring the Maidstone Iguanodon, which appears in restored form as a playful reptilian puppy. *Bottom:* "A Sawrian" (Thomas Hood, ink on paper and published in his works, 1830s; Alexander Turnbull Library, Wellington, New Zealand). With puns, Mantell–mantle, saurian–sawrian–sorry 'un. The setting is Mantell's museum, and his fears may include some for its survival. Hood later contributed a comic geological expedition in prose to Mantell's *Medals* (1844).

"that while I enjoy a reputation both in my profession and in science far, very far beyond my deserts, yet I am reaping no advantages from it. I fear my next letter will tell you that I must abandon this place and sell off my museum." That would be horrible, of course, but if Mantell did not succeed in Brighton, he would have to look out for a practice in London, where he could not afford to take his collection, as it would require too large a house. Regarding his present museum, Gideon still allowed visitors to come every Tuesday between 1 and 3 P.M. and approximately fifteen hundred had done so, but no more than three of them subsequently became his patients.

Mad Hawkins

Mantell's anxiety regarding the possible sale of his collection had been aggravated earlier that year by his evaluation of Thomas Hawkins' specimens for the British Museum. Hawkins, whom Gideon had gotten to know in 1832 (as we have seen), was the twenty-four-year-old author of *Memoirs of Ichthyosauri and Plesiosauri* (1834), an extravagantly written treatise illustrated throughout with beautifully drawn plates. When Mantell first announced Hawkins' then-forthcoming book to Silliman, in February 1833, he described its author as "a very young man, not versed in science and knowing but little of anatomy," though with time and money at his command; he had spared neither to obtain outstanding specimens of *Ichthyosaurus* and *Plesiosaurus*.

Initially attracted to Hawkins' romantic personality, Mantell later became more than a little convinced that the young collector was genuinely mad. "His drawings," Gideon predicted in January 1834, "will be beautiful, but unless Buckland (to whom it is to be dedicated) looks over it, the letterpress will completely mar the work. The fact is, Mr. H. is a very young man who had more money than wit, and happened to take a fancy to buy fossils." Hawkins' were, to be sure, the finest specimens in the world, and when he first showed them in London – eliciting Mantell's enthusiasm, for example – the savants very injudiciously praised him to the skies. But Hawkins subsequently "became so inflated, and behaved so extravagantly, that fears were entertained for his sanity." Gideon did not then know whether Hawkins would ever get his mind together long enough to bring forth his book.

Finally, on 18 June, it did appear. Mantell sent a copy to Silliman the same day, together with this detailed assessment:

> Of Mr. Hawkins' book I need say but little. You will agree with me
> that it is very splendid; the plates are accurate, and the specimens quite

as perfect as they are represented. Mr. H. is a young man of 22 [*sic;* he was 24]: he has lost his father and mother, and his guardians left him to take care of himself. He was designed for the medical profession, but his father's death left him at liberty, and as he is very deaf he abandoned it, imagining that his few thousands would be sufficient. Poor fellow! He is very romantic, very weak, very good natured, and I fear very headstrong. He has spent nearly *four thousand* pounds upon his collection and now wishes to dispose of it to the British Museum.

"Our government," Gideon added, "ought to buy it – first, because it is a British collection and the finest in the world; and next, to save a man who has devoted (although injudiciously) his all to science."

Only eight days later, on 26 June, Buckland wrote from Oxford to Lord Farnborough, one of the trustees of the British Museum, explaining that Hawkins wished to see his collection placed in the Museum and would sell it to them for a price set either by Buckland alone or conjointly with any other person the Oxford professor might name. Fully supporting national acquisition, Buckland had agreed to designate a second referee; as Sedgwick was in Wales, he suggested Mantell, a choice Hawkins endorsed. Having received preliminary approval from Lord Farnborough, Buckland then addressed the trustees as a group. In a formal letter of 7 July he strongly urged the Museum to acquire Hawkins' collection, proposed to evaluate it for them with Mantell, and pointed out that Hawkins' book was actually a beautifully illustrated catalog of his fossils. It would be "a national discredit," Buckland concluded, "if these productions of England should be purchased for public museums in other countries." Further letters to the trustees from Conybeare, Clift, and Mantell were just as insistent.

The trustees met on Saturday, 12 July, to consider Hawkins' formal petition of the ninth, offering his specimens. The supplementary letters from Buckland, Mantell, Conybeare, and Clift were also read, each of them attesting to the importance of Hawkins' collection. Buckland himself appeared to answer the trustees' questions. He thought the illustrated specimens worth about £1,000, and the others £250, exclusive of the cases which contained them; he and Mantell should examine the collection more particularly and report back upon its value. After hearing a further endorsement of this proposal from Charles König, the Museum curator, its trustees authorized Buckland and Mantell to proceed.

That result had been a foregone conclusion. At 11 A.M. the same morning Gideon boarded a coach for town and was already en route when the authorizing vote was taken. He reached London soon after five, met Buckland at the Salopian Coffee House by prior appointment, dined there, and then proceeded with him to Adelaide Street for their evaluations of Hawkins' fossils. Afterward, he returned to the Salopian and sat up with

Buckland until 2 A.M. comparing figures. Gideon arose four hours later to breakfast with Hawkins and leak him the news; he left London at eleven, returning to Brighton at five.

On 1 August a trustees' committee met to read Buckland's affirmation that he and Mantell had separately estimated every item in Hawkins' collection, agreeing on a figure of £1,277; his specimen cases, separately evaluated, were worth £60/5s. The trustees duly petitioned Parliament for a grant of £1,310/5s, which was eventually forthcoming. Hawkins' very showy collection then moved to the Museum early in 1835, and much of it is still on display.[14]

Fossil Fishes

Once the Maidstone Iguanodon had been chiseled out, Mantell's primary scientific concern throughout 1834 and 1835 became fossil fishes. Abundant in the Chalk, they had interested him since 1817, when he first began to collect vertebrate fossils in earnest. In *The Fossils of the South Downs* (1822), he had devoted ten of his forty-two plates to fishes; by the time of *South-East England* (1833), in which many of them reappeared as line drawings, Mantell was prominent among the few British writers who dealt with fossil fishes at all. He had, moreover, published a large folding plate of his most remarkable fossil fish in April 1825 and another of his four best fishes in 1826.

The study of paleoichthyology, as it would later be called, was not then well organized and, during the 1820s, still awaited its systematizer. One arose in Louis Agassiz (1807–1873, the son of a Protestant minister), a precocious Swiss naturalist who had been interested in living fishes since childhood. After a university career that had already included Zurich and Heidelberg, Agassiz enrolled at Munich in November 1827 and soon established himself as the school's most promising researcher. At twenty-one, the next year, he was commissioned to describe a large number of previously unknown Brazilian fishes collected a decade earlier – and

---

14. Mantell helped Thomas Hawkins (1810–1889; *DNB*) to find his publisher, Relfe and Fletcher. TH–GM, 18, 29 Jan, 6 Mar 1833 (including a prospectus), 20 June 1834 (all ATL); GM–BS, 10, 28 Feb 1833, 18 Jan, 18 June 1834, 25 Sept 1835 (Yale); *AJSA*, 24 (1833), 212; GM–SGM, 18 June, 4 Dec 1834 (APS). Hawkins also wrote a second volume on fossil reptiles, *The Book of the Great Sea Dragons* (London, 1840) and a number of pretentious literary works. Regarding the sale of his collection, see *Memoirs of Ichthyosauri and Plesiosauri,* pp. 53–58 (WB–TH correspondence), and Thomas Hawkins, "Statement Relative to the British Museum" (London, 1848), the latter including many documents. J (12, 13 July 1834). His specimens are now at BMNH.

succeeded brilliantly. Agassiz then sent the resulting book (1829) to Cuvier, its dedicatee. Together with his colleague Achille Valenciennes, Cuvier had already begun to publish a comprehensive survey of living fishes that would eventually require twenty-two volumes. He was also collecting specimens and drawings of fossil fishes for an intended sequel. Unaware of Cuvier's intentions, Agassiz himself envisioned a richly illustrated magnum opus on fossil fishes and began to spend what little money he had toward realizing it, sending his hired Munich artist Joseph Dinkel to collections in various museums while going as far as Vienna himself.

Eventually – inevitably – Agassiz went to Paris, ostensibly to observe the cholera then raging, for he was still officially a student of medicine, but actually to study collections of fossil fishes. Meeting on 17 December 1831, he and Cuvier mutually astonished one another. A closer relationship soon followed, as Agassiz (now twenty-four) joined Valenciennes in working under the great anatomist. With incredible generosity Cuvier then relinquished his own intention of systematizing the fossil fishes and in February turned over to Agassiz an extensive file of information and drawings, much of it based on specimens in England. The younger men grieved deeply when, on 7 May 1832, Cuvier was suddenly stricken with paralysis and died six days later (at age sixty-three) from the still-prevalent cholera.

Agassiz left Paris that fall, having accepted a professorship at Neuchâtel, a small Swiss town of perhaps five thousand persons; from there, belatedly in 1834, he published the first *livraison* (installment) of *Recherches sur les poissons fossiles (Researches on Fossil Fishes;* 5 vols., Neuchâtel, 1833–1843). Issued piecemeal as a disorderly accumulation of text and plates, the book would eventually depict about seventeen hundred specimens, though Agassiz's classification of them, based on the structure of scales, was not our modern one. Ever loyal to Cuvier, moreover, he would always explain the fossil record catastrophically. Among his British counterparts, Agassiz naturally identified most closely with William Buckland, whose works embodied an outlook similar to his own.

The same insatiable curiosity concerning fossil fishes that had taken Agassiz to Paris in 1831 lured him to Britain three years later, Buckland having proposed he come. "I should very much like to put into your hands what few materials I possess in the Oxford Museum relating to fossil fishes," the English geologist wrote on Christmas Day 1833, "and am also desirous that you should see the fossil fishes in the various provincial museums of England, as well as in London." Among those provincial resources, Buckland particularly emphasized "the very rich collection of fossil fishes in the museum of Mr. Mantell, at Brighton." To see these various assemblages, as well as to obtain some further, much-needed subscribers for his expensive book, Buckland suggested, Agassiz should plan to visit England in September 1834, when he could attend that year's meeting of the British Association for the Advancement of Science in Edinburgh.

Agassiz accepted his friend's advice. Arriving in August, he accompanied Buckland and Lyell to Edinburgh, seeing collections along the way, and obtained Lyell's promise to go with him to Brighton. "He has found about a hundred new species of English ichthyolites," Lyell then informed Gideon on 1 October, "making seven hundred fossils in all." Together with many other prominent geologists, Lyell thought Agassiz a strange combination of superb intelligence, great energy, brilliant capacity, and occasionally foolish ideas. "His readiness and knowledge are surprising," Lyell forewarned, "and you will find him very skilful in reptiles, though he does not profess to know anything about them, except that he maintains he can prove the pterodactyls to have been swimming, and not flying, animals!" Lyell himself, however, had already felt the impress of Agassiz's commanding personality and now embarked on the study of "fossil ichthyology" under his influence.

Shortly after receiving this news, Mantell relayed it to Silliman, characterizing Agassiz as "a very clever man" who had made the most of his several opportunities. Like Lyell, Gideon looked forward to learning from the Swiss, who was far more experienced than himself in dealing with recent fishes and could therefore identify fossil ones more accurately. "I expect he will throw much light upon my collection," Gideon concluded, "for my additions in fishes have been very great." Swimming pterodactyls, however, he wisely declined to accept.

Following some preliminary correspondence, Agassiz, Buckland, and Lyell arrived in Brighton on Wednesday, 29 October 1834, to inspect Mantell's collection. They were soon joined by Moses Ricardo, Robert Bakewell, and Michael Faraday. Agassiz spent the next two days rearranging his host's fossil fishes and making notes toward his great work. Though Gideon had his own tasks (a few patients, at last), he, Buckland, and Agassiz dined together both days, adding Lyell on Thursday and Bakewell on Friday. Mary Ann, unfortunately, was very ill, a fact Gideon resented, as it kept him from enjoying his distinguished visitors more fully. On Saturday, 1 November, Buckland left for Oxford and Lyell for London, but Agassiz remained one further day.

"The week before last I had a visit from M. Agassiz," Mantell then wrote Silliman on the eighteenth. "He is a professor of natural philosophy in the University of Neuchâtel and has for years been engaged in the study of recent fishes with a view to the arrangement of fossil ones. Two livraisons of his splendid work are out and he complied with my wish most readily to send them to you in my box." Gideon also enclosed prospectuses in the hope that some American institutions or men of science would subscribe. Bakewell, he added, had written an account of Agassiz's visit to the Mantellian Museum and published it in the *Athenaeum*. Gideon then similarly advised Woodward on 2 December. Though Agassiz "had seen every public museum in Europe," he confided, "yet he was astonished at the beauty and perfection of my Chalk fossils, and particularly the fishes." Agassiz himself would echo this opinion in his book.

When the fourth *livraison* of Agassiz's text reached England, in September 1835, it contained an extensive tabulation of Mantell's Chalk fishes. Unfortunately, Agassiz had changed not only generic designations but specific ones as well – names in which Gideon had commemorated his friends. "Surely the original discoverer is entitled to the slight privilege of naming his own discoveries," he complained angrily to Silliman. Agassiz himself returned to Brighton on 26 October, to complete arrangements with his artist Joseph Dinkel, whose drawings of Gideon's Chalk fishes later appeared prominently in atlas segments four and five of *Poissons fossiles*. During the three months it would take Dinkel to do them, he and Gideon became very good friends. An able but not a speedy worker, Dinkel also befriended Mantell's elder daughter, Ellen Maria, giving her drawing lessons and encouragement that would determine the course of her life. On Wednesday the twenty-eighth, Agassiz, Gideon, and Ellen Maria went by coach to London, with Gideon expounding local geology all the way. Despite the impact of his visits, however, Agassiz did not significantly retard Mantell's progressive abandonment of Cuvierian catastrophic geology in favor of Lyellian uniformitarianism.[15]

---

15. Agassiz. *Recherches sur les poissons fossiles* (1833–1843) is, as noted in my text, an unusually complicated book. Probably the finest, most authoritative copy available is that prepared by Agassiz himself for the Earl of Enniskillen, a noted collector. I wish to thank Bernard Quaritch, Ltd., for allowing me to spend several hours at their premises in Golden Square examining the Enniskillen copy. The primary statement concerning Mantell appears in "Feuilleton Additionnel" (volume one, but missing in most copies), pp. 54–56, with further references and plates throughout the ten-volume set. See, for example, I, Chap. 1, pp. 5, 23, 28–28, 50 (FA, following p. 188, emends p. 23); II, plates 30, 65bis, 66, 69; III, plates 8, 10a, 10b, 22, 25a&b, 27, 36, 40, 40a, 52, 53, 56, 57a; IV, Chap. 14 (pp. 114–122); V, Part 1, pp. 64–65, 99, 101–103, plate 25; V, Part 2, pp. 105, 108–110, plate 60a–c. Gideon was of course a subscriber. Dinkel's original drawings of his fishes are now at GS.

WB–LA, 25 Dec 1833 (in E. C. Agassiz, I, 232–233). CL–GM, 1 Oct 1834 (ATL); GM–BS, 4 Oct 1834 (Yale; note WB, *G&M*, 1836, I, 224). J (25, 29 Oct–2 Nov 1834). GM–BS, 18 Nov 1834 (Yale); RB, "Visit of Prof. Agassiz to Mr. Mantell's Museum at Brighton," *Athenaeum*, 15 Nov 1834, pp. 841–842; *AJSA*, 28 (July 1835), 194–197. GM–SW, 2 Dec 1834 (Norwich; see also RB above and LA, *Poissons*, I, FA, 54–56).

GM–SGM, 20 Aug 1835 (APS); GM–BS, 25 Sept 1835 (Yale). J (15, 26–28 Oct 1835);

Lyell's Ascendancy

When Lyell was awarded the Royal Society's medal of recognition in December 1834, specifically for his *Principles of Geology,* no one congratulated him more warmly than Gideon, who pointed out that this new honor would strengthen Lyell in his struggle against the entrenched power of theological conservatism, a force they both opposed. Mantell moved even closer to Lyell in December and January when, in a medical capacity, he ministered to Lyell's mother and sisters at Hastings. Lyell's gratitude toward him during this anxious time revealed itself fully in a letter of 7 January. "Mantell, whom I visited lately at Brighton," he wrote to Dr. John Fleming at Aberdeen,

> has made a bold professional stroke in removing there, which you will be glad to hear is likely to succeed, in spite of the misgivings of many of his friends, who had not the confidence which I always had in his genius. He is, in fact, a man of great medical skill, and tact so great, as to triumph over the drawback of his having so fine a museum and so much fame in certain branches of geology.

Such commendation, optimistic though it was, became all the more satisfactory to Gideon in that Lyell was currently enjoying great prestige.

Apart from the Royal Society's medal, another unmistakable sign of Lyell's ascendancy was his becoming president of the Geological Society of London in 1835. Seven years earlier that society had instituted its Wollaston Medal, funded by William Hyde Wollaston a few days before his death, to honor outstanding geological contributors. First bestowed in 1831, it had gone to William Smith, the stratigrapher. No further award took place until 1835, when on 21 January the Council of the Society, which included Lyell, "*Resolved:* That the Wollaston Medal together with the balance of the annual produce of the Wollaston Fund be given to Gideon Mantell, Esq. of Brighton (late of Lewes in Sussex) for his various publications and continued labors on the comparative anatomy of fossils, especially for the discovery of two new genera of fossil reptiles, the Iguanodon and the Hylaeosaurus." The gold medal was worth ten guineas; and the balance that year, twenty-two pounds. As incoming president, Lyell

---

also *Petrif,* pp. 215 (misdated May), 419, 441; *Medals,* II, 587ff. As after his meeting with Cuvier in London (1830), Gideon emended copies of his earlier works to reflect Agassiz's identifications: *Illus* (ATL) and *Geol SE Engl* (GS), for example. ATL preserves some LA–GM letters, 1835–1850. See also AS–RIM, 6 Nov 1835 (GS; proposed issuing that part of *Poissons* dealing with GM's fishes as a separate work); CL–GM, 9 Nov 1835 (ATL; same idea); and Joseph Dinkel, autobiography (ms, Harvard), p. 25.

was chosen to make the presentation, which took place at the anniversary dinner on Friday, 20 February, at the Crown and Anchor Tavern.[16]

Richard Owen

Throughout much of 1835 Gideon remained sick, listless, and despondent. Despite having just been awarded the Wollaston Medal, he disliked appearing publicly in London and even when presenting papers arranged for others to read them in his stead. He was also provocatively contentious, having minor tiffs throughout the year involving such various acquaintances as Richard Owen, Henry Brougham, William Kirby, and even his Brighton friends, who, like Silliman and Lyell, were trying hard to please him.[17]

Comparable in several ways, the careers of Gideon Mantell and Richard Owen (1804–1892) remained intertwined even after Mantell had died. Like him, Owen had been apprenticed to a surgeon and soon developed an interest in comparative anatomy. He matriculated at Edinburgh University in 1824, studied under Abernethy at St. Bartholomew's Hospital in London the next year, and attained his certification in 1826. Afterward, through Abernethy, Owen became William Clift's assistant at the Hunterian Museum, of the Royal College of Surgeons (where Gideon probably met him), and was soon engaged to Clift's daughter, Caroline. By 1828, rising brilliantly, Owen had begun to lecture at St. Bart's himself. He was publishing significant papers on comparative anatomy by 1830. That same year Cuvier visited the Hunterian Museum while in England and was shown through by Owen, who spoke fluent French. In return, Cuvier in-

16. CL–GM, 10 Dec 1834, 3, 5 Jan, 20 May 1835 (ATL); CL–Fleming, 7 Jan 1835 (*LLJ*, I, 446); Wilson, pp. 410–412. Wollaston Medal: *Hist GS*, pp. 89–92, 317; *Proc GS*, 2 (1835), 124, 142–143; CL–GM, 18, 21 Feb 1835 (ATL); GM–BS, 14 Apr 1835 (Yale); *AJSA*, 28 (July 1835), 391–393; *Petrif*, pp. 229, 488–489. Other materials at ATL include an extract from the GS Council minutes of 21 Jan 1835 (when the award was determined) and two newspaper accounts of Lyell's speech on 20 Feb.

17. Henry Peter Brougham (1778–1868) had been Lord Chancellor from 1830 to 1834; his *Discourse of Natural Theology* (London, 1835; see pp. 45–49) provoked numerous published objections, and GM–Brougham, 15 July 1835 (draft, ATL): "The remarks on the Universal Deluge and comparatively recent origin of the Earth cannot fail materially to retard the progress of those correct views which authors more modern than the eminent philosophers quoted by your Lordship entertain on these subjects." Mantell then cited Lyell's *Principles* specifically. At some later time, GM recorded on his draft that "No notice whatever was taken of this letter by Lord Brougham, but his Lordship was half mad at this period, and this probably was the cause." Kirby's Bridgewater Treatise, attacking GM for his age of reptiles idea, also appeared in 1835.

vited Owen to Paris. Rather unexpectedly, the young Englishman arrived in July 1831, while Cuvier and Valenciennes were deeply involved with their massive catalog of fishes. Though Owen attended the usual Saturday evening soirees at Cuvier's home regularly, he was substantially ignored. Nevertheless, Owen carefully went through parts of the Museum of Comparative Anatomy by himself and must have absorbed a great deal.

Owen's brilliant "Memoir on the Pearly Nautilus," which significantly modified Cuvier's classification of the cephalopods, appeared in August 1832, too late for Cuvier to see. But the Frenchman's demise undoubtedly helped Owen, who was now free to publish on comparative anatomy without a world-famous supreme authority to controvert him. A second fortuitous death followed in September when Clift's only son expired in London, leaving the conservator with no successor. At the time, ironically, the elder Clift had been visiting Gideon at Lewes; he then went on to the Isle of Wight and did not learn for some days that his son had been spilled onto his head by an overturned cab. Taken to St. Bartholomew's, young Clift was received by Owen, who could do nothing for his fractured skull. Following the death of William Home Clift, Owen became William Clift's only assistant, then professor of comparative anatomy at St. Bart's in 1834, and a fellow of the Royal Society in January 1835. Having by now published a series of important papers, Owen became securely established within the scientific and medical community of London and could therefore marry Miss Clift in July. Like Agassiz, he was associated closely with William Buckland, who shared Owen's predilections without in any way endangering his genius.[18]

Geological Disputes

A distinguished group of geologists and comparative anatomists, including Owen, Buckland, Agassiz, Silliman, and others, strongly opposed aspects of the new geological beliefs that Lyell had promoted in his *Principles of*

---

18. Owen. Brilliant, prolific, and heartily disliked by almost everyone in the scientific community of his own time, Richard Owen (1804–1892) has frequently been disparaged. Even *The Life of Richard Owen* (2 vols., London, 1894) by his grandson, the Rev. Richard Owen, includes a pejorative essay by T. H. Huxley regarding Owen's contributions to science. *DSB* (Wesley C. Williams); Adrian Desmond, *Archetypes and Ancestors* (Chicago, 1982); and N. Rupke, *RO, Victorian Naturalist* (New Haven and London, 1994) present more recent, but not necessarily sounder, perspectives. I review the GM–RO relationship in Chap. 12 below. RO, "Memoir on the Pearly Nautilus" (London, 1832). Death of William Home Clift: Owen, *Life*, I, 67–68; C. Dick–GM, 20 Sept 1832 (ATL); GM–CL, 29 Sept 1832 (ATL).

Geology. First among several related issues was, very broadly, the relationship of science and religion, particularly the extent to which the book of Genesis should be regarded as historically valid. Though definitions of geological time were fundamental to this dispute, by the 1830s it was widely acknowledged that the biblical six days of Creation could not be literal. (Parkinson had correlated them with geological evidence in 1811 but only as vague periods.) If the geological past *were* of vast duration, however, then the slow changes postulated by Lyell might be possible. While not all catastrophists were biblicists, those biblicists who utilized geological opinion to any extent were almost necessarily anti-Lyellian.

A second major dispute concerned the biblical Flood (or Deluge) of Noah, which had been commonly associated with the origin of fossils in the late seventeenth century and throughout the eighteenth. As it developed, however, the science of stratigraphy soon proved that sea had invaded land not once but dozens of times; at most, therefore, one could claim only that the Deluge was the last of a series of incursions. Since this position was obviously tenuous, biblical scholars (and geologists who discussed the Bible) increasingly characterized the Deluge as a divine act of moral retribution that had come and gone miraculously without significant physical effects. Throughout the 1830s, a number of important geologists in England and America formally recanted their earlier belief in the once-popular school of Deluge geology.

A third major dispute concerned the validity of natural theology, or the empirical verification of divine attributes through study of God's creation. Despite the several difficulties it opposed to literal acceptance of the book of Genesis, geology had long been regarded by clergymen as particularly compatible with their vocation. Natural theology, moreover, was a required study at British universities. William Paley's finely reasoned *Natural Theology* (1802), the necessary text, had impressed Lyell with its argumentation and strongly influenced almost every other naturalist of the time who attended either Oxford or Cambridge, including Darwin. As scientific knowledge increased, however, Paley's solutions became less and less applicable, so that the case for natural theology had to be reargued every few years – with the evidence, as many feared, going increasingly against any sign of direction or purpose in nature.

The fourth and fifth major disputes, therefore, regarded the creation and direction of life. As scholars are now well aware, Darwin's theory of 1859 was by no means first in dealing with the origin of species. Evolutionary speculations, for example, had been fairly commonplace in eighteenth-century Europe; they lacked adequate factual substantiation, however, and failed initially to provide a more convincing explanation of biological variety than did such prevalent theological concepts as plenitude and design. But stratigraphers established beyond doubt that life on earth had been historically sequential, and to many theologians in the earlier nineteenth century this was welcome verification of organic *progress,* which seemed to them (and to most geologists) evidence of purpose, plan, or – if you will – Providence in nature. Thus, Cuvier had interpreted the fossil record as a progressive series of distinct creations, though he was neither so literally biblical as some of his British followers had made him out to be nor so devastatingly catastrophic as Agassiz. Owen's opinion was very nearly Cuvier's own, while Lyell remained adamantly antiprogressive and favored special creations of species, largely to protect the unique identity of man.[19]

Mantell had already been concerned with these issues for some time, as was evident from the donated chapter on natural theology that began *The Fossils of the South Downs,* his pious conclusion to *The Geology of the South-East of England,* and the explicit disclaimers of religious implications in his geological lectures of February 1834. He was subsequently attacked in public nonetheless. Gideon responded that same year by anonymously exposing the "unaccountable blunders" of fact in George Fairholme's book on scriptural geology (1833). Writing privately to Silliman on 18 June, he also criticized some "perfectly absurd" remarks in the fourth edition of his friend Bakewell's *Introduction to Geology.* While discussing the Stonesfield Slate, Bakewell had been hard-pressed to explain how mammalian bones (the tiny fragments reported by Buckland in 1824) could have been found among reptilian ones, and he was reduced to postulating an improbable island on which the "higher class of animals" had lived, their later bones then mixing in with earlier. Following Lyell, however, Mantell regarded the simultaneous appearance of mammals and the great saurians as definitive evidence that the creation of life had not been progressive. "The theory of the creation of the simplest forms of organization [being] the earliest," he informed Silliman with unusual certitude, "is not tenable."[20]

19. Charles C. Gillispie emphasized most of the same disputes in his *Genesis and Geology* (Cambridge, Mass., 1951) and has since been augmented by numerous scholars.
20. GM on Fairholme: "News in Science," *BGaz,* 15 May 1834, p. 4. On Bakewell (1833, pp. 275–276): GM–BS, 18 June 1834 (Yale). Nonprogressive views are apparent in GM, "On the Bones of Birds," *Proc GS,* 2 (1835), 203; *Trans GS,* ns 5 (1840), 175–177. Also: GM, "Remarks on the Coffin Bone," *Proc GS,* 2 (1835), 203–204; and GM, "Anatomical Description of the Fox," *Trans GS,* ns 3 (1835), 291–292. See also GM–RO, 28 May 1835 (Spokes, p. 75); and CL–GM, 13 Apr, 14 May, 28 June 1835 (all ATL).

8

Wonders of Geology

By now, of course, Gideon Mantell had become thoroughly appalled at the public's ignorance of natural history and how shamelessly that ignorance was being exploited by the clergy and other self-serving groups. In the fall of 1835, therefore, he resolved to educate the masses and, with his medical practice in shambles, saw no reason to delay further his appearance as a professional expounder of modern secular knowledge.

At 3 P.M. on Friday, 13 November 1835, a cautious but unsuperstitious man of science appeared in the Assembly Room of the Old Ship Hotel to lecture on geology and organic remains for the benefit of a blind florist named Phillips. Almost 350 persons attended, securing a profit of twenty-five pounds. Dismiss from your minds, Gideon advised his audience, any idea that the universal Deluge caused every geological change on earth. Whatever may have been its effects, such a catastrophe was irrelevant to the phenomena he was about to consider: How shells had gotten into solid rock; how wood had been changed into stone; how delicate leaves and plants had been preserved as fossils; how coral was enclosed in marble; how the bones of animals and remains of fishes had become entombed in stony sepulchres; and how the Weald of Kent and Sussex had once been the delta of an ancient river. Here, and again in Chichester on the twenty-seventh, he explained basic geological processes in the uniformitarian manner of Lyell.[1]

Having received accolades on both occasions, Mantell lectured once more, "On Corals and the Animals Which Form Them," at the Central National School in Brighton at 2 P.M. on Boxing Day, 26 December, the proceeds going this time to a proposed association for destitute fishermen. Now he defended scientific endeavors against charges of frivolity and godlessness; affirmed that corals were animals, not plants; explained the formation of coral reefs, with interesting drawings; described fossil corals; and even said a few words about crinoids. Afterward, he announced that

1. *BGaz,* 19 Nov 1835, p. 4; J (13 Nov 1835); broadsheet (ATL). *BGaz,* 3 Dec 1835, p. 3; J (27 Nov 1835); broadsheet (ATL).

the increasing number of applications (more than six thousand since it opened) had compelled him to close his museum to visitors for a time, but that arrangements were being made to associate it with a new institution – promoted, as he did not say, by friends hoping to keep him and his collections in Brighton. Thus, Gideon capitalized on his recent successes as a lecturer to firm up his museum against the enforced sale he so often miserably predicted.[2]

The New Institution

A series of organizational meetings followed, beginning on 31 December, when several gentlemen convened at Mantell's home to discuss proposals by Ricardo, Smith, and himself toward opening the museum for public exhibition. After a subcommittee of this as yet informal group met again on New Year's Day, 1836, Gideon wrote the Earl of Egremont ("now, alas, 84 years of age!") to report their progress and problems. Ricardo, Smith, and Sir Richard Hunter then visited Egremont at Petworth on the fourth, securing his name as patron of the institution and a pledge of one thousand pounds. Despite these auspicious beginnings, Gideon remained more impatient than grateful. "For the last fortnight," he grumbled in a journal entry of 19 January,

> scarcely a day has passed without my time being engrossed by meetings concerning the projected scientific institution. I am already tired of the eternal changes of opinion which the gentlemen engaged in it are constantly evincing. I see but too clearly that I shall be made a mere stepping stone for the accomplishment of the principal object with most of them – a gossiping club.

Their Majesties, moreover, had refused to sanction the nearby organization by gracing it with a royal name. So much, then, for the advancement of science and the encouragement given its cultivators!

Though January turned into March, no obvious solution to Mantell's problems forthwith appeared. "I have had an immense deal of trouble with all these affairs," he complained to Silliman, "and even now nothing certain is arranged." If the proposed institution did materialize, Gideon would still not be released from difficulties unless his practice significantly increased, of which he saw no chance. Since moving to Brighton he had been forced to spend two thousand pounds in savings and was now quite

2. *BGaz*, 31 Dec 1835, p. 4; J (26 Dec 1835). Broadsheet and 23-page ms ("Enough for three lectures at least! Far too lengthy!!" [ATL]). That corals were animals, not plants, had been discovered in 1752 (published 1755) by John Ellis, a Brighton resident.

at a loss regarding what to do. "If I could have a good practice in London, I should prefer it," he continued, "or I would sell off my collection or attach it to any public museum if by so doing I could promote the interests of my family." Mantell seriously considered housing his wife and children in some little cottage near London and then offering himself as surgeon or anatomist to a voyage of discovery – but his health would not allow it. He thought also of going to America. "In fine," Gideon concluded, "I am still floating on the sea of circumstances and know not where the billows may convey me."[3]

At 2:30 P.M. Saturday, 12 March, Mantell lectured to an extremely large audience at the Old Ship "On the Extraordinary Fossil Remains of a Crocodile" recently discovered at Swanage. This was a fine specimen of *Goniopholis* that Robert Trotter, the generous "poacher" of Borde Hill, had purchased from quarrymen in Dorsetshire and then presented to the speaker. Recalling his first public lecture on the science, in the same room two years before, Gideon regretted the continuing prejudice against geology, which was still suspected of discrediting God's revealed word. Nothing, however, was more rash than to identify theories in physical science with particular interpretations of the sacred text. In actuality, geological researches, like all inquiries into the created world, confirmed the essence of Scripture. After quoting a favorite passage from Bishop Sumner's *Records of Creation* (which upheld the compatibility of geology and *Genesis*), Gideon dealt in turn with Trotter, methods of the comparative anatomist, and Sussex during the Age of Reptiles. In closing, he announced the foundation of the Sussex Scientific and Literary Institution and Mantellian Museum, with Davies Gilbert as president and the Earl of Egremont as patron. His museum had been engaged for three years. Soon open to the public, it would be augmented with a library and reading room as well as pertinent lectures by himself and others. Annual subscriptions of two pounds included free admission to the museum; those of five pounds added the library, reading room, and other functions. Ladies, moreover, were eligible to join. Otherwise, general admission (to the museum only) was one shilling per entry. A modest broadsheet to the same effect circulated later that month.[4]

3. J (31 Dec 1835 ["What misery have I not endured in this year!"], 1, 4, 19 Jan 1836); GM–SGM, 16 Jan 1836 ("Alas! I see nothing but ruin before me" [APS]); GM–BS, Jan & 1 Mar 1836 (incl. Egremont quote [Yale]).

4. *BGaz*, 17 Mar 1836, p. 3; broadsheet (ATL). In this lecture Gideon also publicly identified the Rev. Henry Hoper as having authored the first chapter of *Foss SD*. The Swanage Crocodile and other gifts from Trotter are described in GM's museum catalog (note seven below). SSLI&MM broadsheet, with annotations by GM (ATL). John Bird Sumner, *A Treatise on the Records of the Creation* (2 vols., London, 1816) included a well-known

Figure 8.1. The Swanage Crocodile (*Wonders,* third edition, pp. 387–389; Plate I). Size: Three feet ten inches by three feet. Mantell identified: (1) the left side of the lower jaw, with two teeth remaining in their natural position; (2) detached vertebrae of the back and tail, showing the transverse and dorsal processes; (3) dermal, or skin bones; in life, arranged in parallel rows along the spinal column; (4) ribs; (5) chevron bones; (6, 7, 8) pelvic bones.

Revealingly, even Mantell's innocuous observations on science and religion drew vitriol, as the Rev. Robert Fennell, of Brighton, replied to them with "Geology. Remarks on Bishop Sumner's Appendix to His Work Entitled *Records of Creation.*" Though nominally addressed to Dr. Sumner, a liberal clergyman who had persuasively defended geologists' right to free inquiry, Fennell's pamphlet appeared just after Gideon's lecture and specifically attacked him, accusing Mantell and other geologists of falsifying Scripture and bringing the Bible into contempt. "The cackling of a goose once saved the Capitol," Fennell pointed out unwarily, "and if by my shouting I can but arouse attention to the sappers and miners beneath our walls, I shall be no 'vain babbler'" (ix–x). Some mildly laudatory remarks about Gideon's soon-to-be-opened museum in no way compensated for what one Brighton newspaper called "a tissue of personal and direct insult." All who knew Dr. Mantell or had heard his lectures or read his books, the reporter added, were aware of the falsity of Fennell's charges. Public opinion agreed, and Fennell (whose name was equated with religious fanaticism for a time) found himself obliged to support the new institution with a five-pound subscription.[5]

The Institution's public operations began on 4 April when its reading room and library opened to subscribers at No. 20 The Steine. In addition to books, the collection (acquired by purchase and donation) offered learned journals, periodicals, and newspapers. It had not been possible to relocate Mantell's fossils and antiquities, so his expensive house was being redesigned to accommodate the Institution's needs. "I have *lent* my

---

appendix arguing "That the Mosaic History Is Not Inconsistent with Geological Discoveries" (I, 267–285). He relied heavily on the Kerr-Jameson edition of Cuvier's preliminary discourse (*Essay on the Theory of the Earth*; Edinburgh, 1813).

5. The Reverend Robert Fennell, "Geology" (London, 1836). "Of geology," he admitted,

> I know but little. I once endeavored to understand something of its outlines, but I soon found myself in a maze; nor could I find a single writer to guide me from the labyrinth. It appeared to me a science of opinions, and scarcely one of those opinions could I find supported by two authorities. . . . Hence, I returned with double zest to the Bible, resolving not to *endeavor* to be wise beyond what was therein written. (v–vi)

Like many clerics, he had been greatly upset to learn from *Quarterly Review* (56 [Apr 1836], 321–364) that William Buckland's forthcoming work on natural theology would deny the Deluge while affirming the Age of Reptiles and "development"; Lyell, Buckland, and Mantell were his chief targets. GM, Notice in reply to Rev. R. Fennell's attack (ms, ATL). William Buckland, *Geology and Mineralogy* (2 vols., London, 1836); for GM, see I, 86, 168n (actually discovered by MAM), 191n, 198&n, 216, 234n, 240–242, 290, 334; II, 35, 67. See also Buckland, *Life,* Chapter 8, and D. W. Gundry, "The Bridgewater Treatises and Their Authors," *History,* 31 (1946), 140–152.

museum to the Institution for two years and a half, ending at Christmas 1838," he explained to Silliman on 14 July, "and I have let my house for the purpose, reserving for myself only a little parlor and a bedroom." For this he would receive £450 per annum – £250 for his museum, £150 for his house, and £50 for giving five lectures – but this meant losing £100 every year on the house and having to relocate his family to a rented cottage at Southover, near Lewes, where his younger son Reginald was in school. Science was unappreciated in Brighton and Gideon's medical practice there had not been equal to his expenses. Alone and discouraged, he therefore resolved to struggle through one more season while attempting to buy a practice in London.[6]

Gideon's first lecture on behalf of the Sussex Institution took place in the commission room of Brighton's town hall at 3 P.M. Monday, 16 May. The museum, he announced then, would open on Wednesday, the eighteenth; conversazioni, for members, would be held at his former home every other Tuesday evening at 8 P.M., beginning on the twenty-fourth. His topic on the present occasion, known to him only the week before, was "Local Antiquities." Though very interested in such relics as a boy, he had since gone on to other pursuits, not to mention his professional obligations, and had been called away unexpectedly this last week by the illness of his younger daughter, Hannah Matilda. Nevertheless, Gideon thrilled his audience as he traced the development of civilization and then detailed the respective funereal practices of Britons, Romans, Danes, and Saxons, illustrating the subject with drawings of various barrows excavated by himself in the vicinity of Lewes and a large collection of artifacts. Finally, he explained the new arrangement of his museum and regretted that its sponsoring institution had not attracted the public support to which it was entitled. A closing tribute publicly acknowledged the bounty of Lord Egremont, who was in Gideon's large audience, as were Mary Ann and his children.

Gideon's museum opened privately to members on 17 May, with about forty of them attending, and to the public the next day. It consisted now of three rooms, each filled with wall and standing cases, arranged on two floors of the tall, narrow house. The drawing room floor, little changed, included large collections of iguanodon and megalosaur remains, together with individual cases for the Maidstone Iguanodon and *Hylaeosaurus*. Another important exhibit was a group of six iguanodon tail bones presented

---

6. *BGaz*, 19 May 1836, pp. 3, 4; *BH*, 21 May 1836, p. 3. *BH*, 28 May 1836, p. 3; J (25 Apr 1836). GM–BS, 14 July 1836 (Yale). On 27 Aug, GM's wife and daughters (the sons remaining at school) left Southover and moved to 114 Western Road, Brighton, to be near him.

by Robert Trotter. Further cases held relics of megalonyx, mastodon, Irish elk, and various local fossils. Gideon's bust of the Earl of Egremont, by John Edward Carew (a sculptor subsidized at Petworth), adorned the mantelpiece, with his stuffed iguana by the fireplace on the left. Two large displays of fossil fishes, arranged in Agassiz's manner, and the Swanage Crocodile dominated the upper front room. An adjacent back room contained antiquities primarily, as well as local shells that had long been in Gideon's possession.[7]

The Mantellian Museum remained open to visitors every day except Sunday from ten till five and therefore required a full-time curator – George Fleming Richardson – salaried by the Institution. Richardson's father, a silk and lace mercer on Castle Square, had tried unsuccessfully to interest his brilliant son in shopkeeping. Finding the stupidity of customers unbearable, however, young Richardson devoted every opportunity to the study of languages, including Greek, Latin, Hebrew, French, Italian, and especially German, the literature of which he soon came to love. His own literary productions, written in a variety of modes and styles, included *Poetic Hours* (1825), a prose tale (1827), a translation of the life of K. T. Körner (1827), and *Sketches in Prose and Verse* (1835). Desperate for approval, he would try almost anything in public – singing, dancing, and acting being among his failures. What probably impressed Gideon and other members of the Institution, however, were Richardson's prodigious memory and his incredible aptitude for acquiring knowledge. Once hired, therefore, he took up geology from scratch and by the time Gideon's museum reopened had mastered enough of it to become an excellent cicerone.[8]

---

7. J (14–17 May 1836). GM, "A Descriptive Catalogue of the Objects of Geology, Natural History, and Antiquity (Chiefly Discovered in Sussex) in the Museum Attached to the Sussex Scientific and Literary Institution at Brighton" (London: Relfe and Fletcher, 1836). J (15 Sept, 20, 26 Oct 1836). GM, "Thoughts on a Pebble" (London: Relfe and Fletcher, 1836), written for his son Reginald, expanded in its various editions from eighteen pages to more than one hundred.

8. For Richardson, see *DNB;* Charles Fleet, *Glimpses of Our Sussex Ancestors* (Lewes, 1882), pp. 136–145; and (despite some flaws) Hugh S. Torrens and John A. Cooper, "George Fleming Richardson (1796–1848)," *The Geological Curator,* 4 (1985), 249–268. His most important scientific contribution was Hermann von Meyer, "On the Structure of the Fossil Saurians" (1832), trans. GFR, communicated by GM, *The Magazine of Natural History,* ed. Edward Charlesworth, ns 1 (June 1837), 281–293; (July 1837), 341–353, with frequent mentions of GM and interesting editorial comments throughout. GFR, *Sketches in Prose and Verse, Second Series* (London, 1838), dedicated to GM, includes "Visits to the Mantellian Museum," pp. 1–27, 189–222, the best contemporary description of it that we have. His *Geology for Beginners* (London, 1842 and later editions) was a major irritant to GM, who regarded it as a plagiarism of his own and others' works (*Medals,* Preface). Overcome by financial embarrassments, GFR died by his own hand, a demise GM genuinely regretted.

By fall, the new Institution seemed to be working well. Its conversazioni, in particular, had begun to attract significant contributions and a good deal of local interest. When Richardson, for example, spoke on the language and literature of Germany (9 and 23 August), the *Brighton Herald* published him in full. Similarly, a lengthy essay by Horace Smith on steam locomotion and another by Richardson on Lewes Priory were again reported verbatim. At each session Gideon would extemporize for a time on donated specimens and other topics.

The Museum, too, was drawing as many as 150 visitors per week while rapidly expanding its collections. In October, Mantell wrote a new catalog for its holdings as presently arranged. All the famous specimens were described fully, and Gideon reaffirmed his previous reconstructions of both *Iguanodon* and *Hylaeosaurus;* many new specimens had also been donated, by Robert Trotter and others. An unusually elaborate section on fossil fishes quoted Agassiz in praise of the collection and utilized his terminology, together with Joseph Dinkel's drawings.[9] Feeling a good deal more settled now, Gideon resumed lecturing once again.

On 19 December the Sussex *Royal* Institution and Mantellian Museum held its first annual meeting. Affiliates and friends, some six or seven hundred strong, met at 2 P.M. in the Town Hall, where Moses Ricardo read a brief annual report. Later published, it reviewed in turn the reading room and library, the museum (three thousand visitors), the lectures (seven by Gideon and four others), and the fortnightly conversazioni. A list of donors and subscribers for 1836 included 168 names, but more than half the money had come from the Earl of Egremont alone; even with his munificence, little more than three hundred pounds remained. "The present funds of the Institution," Ricardo stated hopefully, "are in a more flourishing state than was anticipated." His brother David would probably have been more brutal.[10]

---

9. Fall 1836: J entries; GM–JH, 20 Aug ("My professional expectations have been completely disappointed" [WSRO]); *BGaz*, 11 Aug, p. 2; *BH*, 13 Aug, p. 3; 27 Aug, p. 4. *BGaz*, 8 Sept, p. 2; 22 Sept, p. 3; *BH*, 24 Sept, p. 4; *BGaz*, 6 Oct, p. 2; *BH*, 8 Oct, p. 2; 15 Oct, p. 4, 2 Oct, p. 4, 29 Oct, p. 4 (HS's essay in three parts). Museum: GM's catalog (note seven above); GFR's description (note eight above). *BGaz*, 27 Oct, p. 4; *BH*, 5 Nov, pp. 3, 4; *BGaz*, 10 Nov, p. 4; 17 Nov, p. 4. "I am still in a state of the utmost uncertainty," Gideon then wrote John Hawkins; "I have been and am still anxiously seeking for a practice at the west end of London" (GM–JH, 6 Nov 1838 [WSRO]).

10. Annual meeting. *BGaz*, 22 Dec 1836, pp. 2, 3; *BH*, 24 Dec 1836, p. 4. J (19 Dec 1836). "First Annual Report of the Sussex Royal Institution and Mantellian Museum" (4 pp., Brighton, n.d.; includes GM, WBDM, and the Rev. Robert Fennell among its "List of Donors and Subscribers for 1836"); "Sussex Royal Institution and Mantellian Museum" (18 pp., Brighton, n.d.), quoting p. 2 (Ricardo); both ATL.

Omens

By reason of the weather, Christmas 1836 was unusually miserable for everyone in Brighton. "A snowstorm, begun last night, has continued through the day," Mantell recorded then, "and everything is most dreary and wretched." He had returned to his meager den that evening wet through from walking in snow up to his knees – only to find the fire out and smoke coming down his chimney instead of going up. Drifts increased until the roads became impassible. No coach left Brighton for three days, and those coming from London were stopped at towns fifteen or twenty miles away. At Lewes, moreover, an avalanche had fallen from the cliff in South Street on Christmas Eve, overwhelming several cottages in Boulder Row and burying fifteen inhabitants, eight of whom were killed. As Gideon learned from the London papers (on the twenty-ninth, when mail finally got through), most of those slain had been among his former patients.[11]

Gideon had already suffered an even more personal loss in 1836, his brother Joshua having been taken from him most cruelly. Precisely five and three-fourths years younger, Joshua was born on 3 November 1795, though (because of his hunchback) not baptized until 7 September 1811. A promising chemist, in December 1817 he published a letter on the reduction of silex. Two years later, Joshua lectured before the Lewes Philosophical Society on the physical and chemical properties of silver. In another lecture of January 1821 he discussed the nature and properties of gases. Joshua often gave fossils (and natural history specimens of other kinds) to Gideon, who cited him three times in *South Downs* for having done so. Together with Mary Ann, Gideon also looked at Jupiter and other planets through his younger brother's telescope. Joshua was therefore a naturalist of some consequence.

Having no regular occupation, Joshua conducted a stationer's shop in Brighton for a time. In September 1822, however, he gave up that unrewarding work, binding himself to Gideon for five years as a surgical apprentice. Beginning in January 1823 Joshua studied obstetrics in London; he was then visiting patients with Gideon at Lewes in October 1825. Two years later, in October 1827, Joshua braved London again to attend the medical lectures of John Epps. Subsequently admitted to the Apothecaries' Company in 1828, he next practiced surgery at Newick, north of Lewes. Throughout these years Joshua remained passionately interested

11. J (25, 29, 31 Dec 1836); S. Warren Lee–GM, 18 Jan 1837, with sketch (ATL); S. C. M., "An Avalanche in the South Downs," *The Sussex County Magazine*, 1 (1927), 70–74 (from *SA*, 2 Jan 1837).

in flora, devoting himself to horticulture as fully as Gideon had to geology; he published a short book on flowers in 1832.

Four years later, in March 1836, gentle, deformed Joshua fell from his horse, permanently injuring his brain. He would imagine, for example, that whatever was said to him had been said before, and any disclaimers to the contrary would enrage him. Though cared for solicitously in both London and Lewes, Joshua had become incurably deranged. On Monday, 26 September, therefore, his elder brother Thomas auctioned off Joshua's effects at the Star Inn, Lewes. They included literary, medical, and scientific works, some of them rare; electrifying machines, blowpipes, glass tubes, telescopes, thermometers, air baths, and various other chemical apparatus, together with a fine collection of minerals and fossils. It was let out, and commonly believed, that Joshua had died.[12]

Predictions

"Still surrounded by snow – the cold intense," Gideon lamented at year's end. "Suffering from toothache, and disappointed, I close the last day of 1836 [words deleted]." As if fearful of divine retribution for what he had just written, however, he then expressed contrite gratitude to the Eternal for blessings permitted him and "with renewed confidence and hope in His goodness and mercy" hailed the coming year. Consciously the Roman stoic now, but also defensively pious, Gideon had become convinced by his brother Joshua's fate, the randomness of disease, and the blind slayings of his intemperately hostile environment that the future could no longer be controlled through his own efforts. "A gypsy foretold me that 1837 would make or mar my fortunes!" he remembered. "Be it so."[13]

The gypsy's year of 1837 began favorably for Mantell, who lectured for a third time on the nervous system to a very attentive audience at the Old Ship, then gossiped on the structure of the eye at a related conversazione. As they were intended to do, these *medical* lectures brought him patients;

---

12. Joshua: J entries; *Foss SD;* documents at SAS. *SWA,* 22 Dec 1817, p. 2; *SA,* 26 Feb 1827, p. 3 (both supplied by Dr. Colin Brent); John Epps, *Diary,* ed. Mrs. Epps (London, 1875), pp. 195–197; 279 (supplied by Charlotte MacKenzie). Joshua Mantell, "Floriculture, Comprising the General Management and Propagation of Stove, Green-House, and Hardy Herbaceous Plants, Hardy Trees, and Shrubs" (Lewes: J. Baxter, 1832; 36 pp.). CL–GM, 16 Apr, 6 July 1836 (ATL). Auction: *SA,* 19 Sept 1836, p. 1 (adv.); J (26 Sept, 31 Dec 1836). See also MAL, *The Worthies of Sussex,* p. 159n; and *DNB.* The apprenticeship of a younger brother to his older was not unusual.

13. J (31 Dec 1836).

when an epidemic of influenza struck southern England the following week, he had many cases to attend. Gideon also lectured on the disease in public and at his usual Tuesday evening soiree. His forty-seventh birthday, on 3 February, was the happiest in some years, in part because his elder son Walter (now an apprentice surgeon) came voluntarily from Chichester to be with him. Gideon then lectured regularly on geology, sometimes repeating topics, but his offerings became less popular as their frequency increased. The conversazioni, too, appealed mainly to a small but loyal following of only five or six, with Gideon the primary attraction. Religious opposition to science, moreover, had by no means disappeared.[14]

Some of the Institution's difficulties were attributable to continuing poor weather, for Brighton's unusually harsh winter had been followed by a disappointing spring in which frosts were common, with a late snowfall in mid-April. "This town has therefore suffered dreadfully," Mantell explained to Silliman on the twenty-ninth, "the fashionable visitors hurrying away as fast as possible." The countryside had been sorely afflicted also, with not a blade of grass in sight and the price of hay alarmingly high. Inevitably, Gideon's practice had suffered as well, for the rush of business he experienced during the influenza epidemic had not compensated for the emptiness of the town that followed. "Altogether the prospects of the Institution are bright," he continued, "but I much fear they will be marred by the principal managers, who are not men of science but of literature, and I fear it will all end in a mere book club." He would not, however, dwell on this. "A few months," he predicted, "will decide my fate, or rather that of the Museum," adding that he was "prepared for the worst."[15]

For a time, the portents were encouraging. On 12 May, to begin with, Mantell went to London, where Lyell (who had mentioned him quite prominently in the just-published fifth edition of the *Principles*) introduced him to Charles Darwin. That evening Gideon lectured gratis on "The Iguanodon and Other Fossil Remains Discovered in the Strata of Tilgate Forest" to an audience of perhaps seven hundred at the Royal Institution. On Saturday, the thirteenth, he attended a glittering soiree given by the

14. Nervous system lectures: J (30 Nov, 5 Dec 1836, 7 Jan 1837). *BGaz*, 8 Dec, p. 3; *BH*, 10 Dec, p. 3; *BGaz*, 12 Jan 1837, p. 2; *The Lancet*, 1 (1837), 758. Geological lectures: J (7 Jan 1837); *BGaz*, 12 Jan 1837, p. 2; J entries and broadsheets, ATL; *BH*, 14 Jan, p. 4; *BGaz*, 19 Jan, p. 2; *BH*, 28 Jan, p. 3 (poem on GM); 25 Feb, p. 3; J (18 Feb). The 18 Feb lecture as a whole (in which GM quotes Charles James Blomfield, Bishop of London, *Sermons* [1829], p. 116) suggests how adverse public reactions to geology and paleontology were becoming, as the fuller import of Buckland's concessions in *B&M* (1836) appeared. Conversazioni: J (18 Feb, re the 14th; 29 Mar, re the 28th); *BGaz*, 20 Apr, p. 2. Conversazioni resumed on Tuesday, 6 June.
15. GM–SGM, 20 Mar, 27 Apr 1837 (APS); GM–BS, 29 Apr 1837 (Yale).

Duke of Sussex (president of the Royal Society) at Kensington Palace, presenting the Duke with an elegantly bound copy of the Mantellian Museum catalog. Gideon then gave two more lectures in Lewes and Brighton, and sat on 3 June to J. J. Masquerier, who would work for months on the best-known portrait of him (issued 1 November as an engraving). At Gideon's conversazione of 20 June, seventeen-year-old Walter debuted as a geological lecturer, displaying fossil specimens found by himself at Chichester and pleasing his father very much. Unfortunately, William IV had died the same day, leaving the Sussex Royal Institution without its official patron.[16]

By now Mantell had resolved to sell his practice and buy another, a move he had contemplated almost since arriving in Brighton. "By purchasing a practice," he had explained to Silliman on 29 April, "I mean the payment of a sum to a person about to retire, or desirous for a partner. This is the usual mode of transferring a medical practice, and is the only certain mode of success in London." The sum required was generally equivalent to two years' profits. There would then be a transitional period, during which the retiring surgeon would treat the incoming one as a partner, ingratiating him with patients gradually. "My practice at Lewes was disposed of in this manner," Gideon pointed out, "and my successors are doing very well." He had been looking for such an opportunity in the west of London, but had heard of none as yet. On 8 September, however, with an arrangement in view, he agreed to sell his Brighton practice to a surgeon named James Pickford, and for only £105 (which suggests how minimal his "profits" had been).

Writing to Silliman the next day, Mantell described his situation. He would leave Brighton within a few months and take up the practice of a retiring surgeon at Clapham Common, which was about three and a half miles from Westminster Bridge, on the Surrey side. "The gentleman to whom I succeed has been in extensive practice there about fourteen years," Gideon advised, "and will remain till next summer to introduce me to his patients and then retire in my favor." But of course he would have to pay a great deal of money for this and was therefore obliged to sell his museum. Gideon had offered it to the members of his Institution for three thousand pounds immediately and the rest – whatever it might be valued at, even if no more than five thousand pounds in all – to be paid off at 3 percent. It had been proposed to raise the three thousand pounds through proprietary shares of twenty-five pounds each, bearing 3 1/2 percent interest. Anxious to preserve the museum for Sussex, Lord Egremont

16. J (11–13, 22, 30 May; 3, 5, 20 June 1837): *BGaz*, 8 June, p. 2; *BH*, 10 June, p. 4. There was also an SRI excursion from Brighton to Lewes: J (22 June 1837); *BH*, 24 June, p. 3 (eight-page reprint, ATL); *BH*, 1 July, p. 3 (excursion poem by GFR).

prepared to invest heavily in the scheme, but not until others had joined him in financing it.[17]

Six Lectures

Though its eventual demise now seemed assured, the Sussex Royal Institution experienced a last hurrah, Mantell having agreed to give his long-desired course of six lectures on geology. Beginning 16 September, they were offered on consecutive Saturday afternoons at 3 P.M. in the Town Hall, with individual admissions of two shillings sixpence or twelve shillings for the course (less for schools). A published eight-page syllabus described topics, illustrations, and specimens. Each lecture was then transcribed by Richardson on behalf of Gideon's next book, *The Wonders of Geology*, and summarized independently in the press.

His overall approach was historical and recessive, but Mantell began with some introductory remarks about geology and astronomy. After reviewing nebular theories by Herschel and Laplace, he described the present earth and its evidence of successive geological epochs. Even the most recent of these, our own, included species now extinct – the dodo and Irish elk, for example. Human fossils had also been found. Gideon next discussed geological changes presently in operation, virtually summarizing Lyell, and concluded with a defense of comparative anatomy, which he thought the only guide to follow in paleontology (as the study of fossils was now called).

Mantell's second lecture, on 23 September, dealt with the epoch of the large mammalia, which had included not only the familiar long-haired pachyderms but also *Megatherium, Megalonyx,* and *Dinotherium*. Thus, Gideon reviewed the frozen carcasses of mammoth and rhinoceros found in Siberia, mammoth and mastodon bones from England and the United States, the discovery of *Megatherium* by Sir Woodbine Parish in Argentina, and assemblages from India and Burma found by Captain Cautley and John Crawfurd, respectively. William Buckland had explored and described a number of ossiferous caves in Germany and England, in which relics of hyena and bear were common; some of them also contained pottery, a brash discrepancy suggesting an older age for humankind than contemporary estimates allowed.

17. GM bought the practice of William Hyde Pearson (d. 1849), who was elected F.R.S. on 9 March 1826 and would be knighted at St. James's Palace on 18 July 1838. GM–BS, 29 Apr 1837 (Yale). J (20 July, 11 Aug, 11 Sept 1837); GM–James Pickford memorandum, 8 Sept 1837 (ATL); GM–BS, 9 Sept 1837 (Yale); Kell and Son's itemized bill to GM for legal services, 8 Jan, paid 13 Apr 1838 (ATL).

The third, on 30 September, described Tertiary formations situated between the Chalk and more recent layers. Lyell had concentrated on these strata, his division of them into Eocene, Miocene, older and newer Pliocene being widely accepted. Gideon displayed a number of shells, both freshwater and marine, then described some early mammals and other fauna. He also introduced his audience to the London, Hampshire, and Paris basins, emphasizing his own researches and those of Cuvier and Brongniart in France.

Lecture four, 7 October, summarized the Age of Reptiles, comprising Chalk to Carboniferous strata. After reviewing the entire sequence, Gideon particularly emphasized the Wealden, a fluviatile formation first determined by himself. Iguanodons, hylaeosaurs, megalosaurs, plesiosaurs, crocodiles, and turtles were duly conjured up in turn, to be followed by the marine Oolite; the deltaic Stonesfield beds; and the lowermost Lias, with its ichthyosaurs, plesiosaurs, and pterodactyls. For many in his audience, this lecture was the highlight of the series.

Mantell's fifth, 14 October, continued this investigation of the Secondary strata, concentrating on the New Red Sandstone and the Coal Measures or Carboniferous group. Of vegetable origin, coal preserved numerous fossils, many of them plants. The Mountain Limestone, an extensive marine deposit, included crustacea, trilobites, crinoids, and especially corals, which Gideon described at length.

The final lecture, on 21 October, began with further remarks on the Old Red Sandstone and went on to the oldest stratified rocks then known, the Silurian (about which Murchison would publish a famous book fifteen months later). This left only lifeless Primary rocks, including granite and basalt. In discussing the latter, Gideon evoked such standard localities as Fingal's Cave and the Giant's Causeway. He then described the production of ores and gems, ending with earthquakes, volcanoes, and a brief review of his five earlier lectures in the series.

Like all its predecessors, this last performance was very well received. But an item in the *Brighton Herald* that same day affirmed local reluctance to secure Mantell's museum for the initial three thousand pounds so desperately needed. By this time, the Earl of Egremont had subscribed for twenty shares (five hundred pounds); the Earl of Munster, one hundred pounds; Moses Ricardo, the same; his brother Samuel and a Miss Wright, fifty pounds each. Many others had invested half as much. But thirty to forty shares, at twenty-five pounds each, were still to be disposed of. The effort, then, lagged shy of the mark by something like a thousand pounds and gave every indication of failing. Whatever outward appearances Gideon may successfully have maintained, he became seriously depressed and would remain so for almost three years, during which he failed to maintain his journal, writing little more than a few lines per year. This run of bad luck culminated on 11 November 1837, when the Earl of Egremont died, obliterating any last remaining possibility that Gideon's museum would be preserved in Brighton. The gypsy's fateful prediction, therefore, had come true.[18]

Necessary Expedient

His Brighton project irretrievably smashed, Mantell turned now to the completion of his long-delayed agreement with Sir William Pearson, who had decided to retire. On 3 January 1838 Pearson, a surgeon, apothecary, and accoucheur in partnership with John Parrott, agreed to sell his Clapham Common practice to Gideon for £1,500. During the next three months, Pearson would introduce Mantell to his patients. On 4 April, £1,000 was due; all later profits would be Gideon's and Pearson would no longer practice in the vicinity, but till 4 July would recommend Mantell through means other than personal introduction. The final £500, with interest, was due a year later. (Receipts signed by Pearson show that Gideon paid him £1,000 on 3 April 1838 and a further £525 on 1 April 1839.) Finally, a means of arbitration was established in case violations of the agreement were alleged by either party. Having shortly to reduce himself by a further thousand pounds, Gideon now moved quickly to liquidate his museum.[19]

In January 1838 he wrote Charles König at the British Museum regarding negotiations with the trustees. "Tell me who they are," he urged, "and through whom I shall best succeed." Elsewhere, Lord Northampton was already working on his behalf. Gideon planned to have Henry Stutchbury come down in a week or two and help him number the principal specimens but thought that König, on behalf of the Museum, should see them also. "I am besieged with applications from local institutions," he added (perhaps untruthfully), "but I am now resolved to have it in the British, if I can obtain anything like a return for what I have expended." König replied on the fifteenth, regretting that Mantell had just missed the

18. Six lectures. Sussex Royal Institution, "Syllabus of a Course of Six Familiar Lectures on Geology by Gideon Mantell, Esq., L.L.D., F.R.S." (Brighton, 1837 [ATL]). After 14 Sept, GM virtually abandoned his journal for two and a half years. *BGaz*, 21 Sept, p. 3; *BH*, 23 Sept, p. 4; 30 Sept, p. 4; 7 Oct, p. 4; 14 Oct, p. 4; 21 Oct, p. 4; 28 Oct, p. 3; 4 Nov, p. 4; 11 Nov, p. 6; 18 Nov, p. 4; 25 Nov, p. 4. Other newspapers in Brighton and London noticed the lectures more generally. Conversazioni, also held, were reported briefly. *BGaz*, 16 Nov 1837, p. 2 (obituary of Lord Egremont); *BH*, 23 Dec 1837, p. 3 (gloomy SRI second anniversary).

19. GM–Wm. Pearson agreement, 3 Jan 1838, with receipts (ATL) See also J (Dec 1840).

trustees' meeting two days before and advising him as requested. In particular, he should have a number of important geologists write the Museum to recommend that it acquire his collection – and the sooner the better, as other offers might well arise to compete with his.

Gideon immediately relayed this information to his good friend Lord Northampton, who then solicited several geologists of his acquaintance. Thus, Murchison wrote König on the eighteenth. "Such is the real value of this collection," he thought, "that it would honor the nation if placed at Somerset House or the British Museum." Murchison emphasized the "grave imputation upon our national character" that would result if this opportunity were lost, for "Mantell's Museum has deservedly acquired a great name among foreigners and, containing as it does so many of the originals of Cuvier and Agassiz, must always be an invaluable source of *reference*." Place *livraisons* nine and ten of Agassiz's great work on fossil fishes before the trustees, Murchison suggested, and they would see how much of their superior interest derived from the quality of the Mantellian Museum.

Not surprisingly, the terrible uncertainty of Gideon's fortunes afflicted him. "I have long been very ill," he wrote Silliman on 4 February, and was still much broken in health and spirits, surrounded as he was by difficulties and trials, of which the most aggravating was the base ingratitude and poor conduct of his professed friends. "My museum," he continued, "is thrown upon my hands; the Brighton people have refused to have it upon any terms. My only chance of getting back some portion of what I have expended rests with the British Museum," which was yet to be heard from. He asked Silliman to ascertain whether some American city, perhaps Philadelphia or Boston, might not be interested in acquiring his specimens. Meanwhile, Gideon was working hard on his new book, *The Wonders of Geology*, expected to be out by April.

When the trustees of the British Museum met on 17 February, Mantell offered them his collection of a lifetime for five thousand pounds. He also agreed to negotiate – specifying, however, that his museum would remain attached to the Sussex Royal Institution until December 1838, when their lease on it would expire. Enclosing the most recent catalog of his collection, Gideon suggested that the trustees consult Northampton, Buckland, Murchison, Lyell, or any other of his most reliable friends for guidance. "I beg to apologize for this hurried communication," he concluded disingenuously, "but an application made to me yesterday from the agent of a foreign government to purchase the collection compels me to lose no time in soliciting, my Lords and Gentlemen, your favorable consideration."

The trustees also lost no time. That same day, they directed König to examine the catalog of Mantell's collection and report to themselves the scientific and pecuniary value of his specimens. König soon decided, however, that only a thorough, firsthand inspection would do. The upshot was that Henry Stutchbury and some others spent several days toward the end of March trying to catalog the full collection in some reliable way. Gideon cannily pleaded professional engagements to escape most of that heartbreaking task and could offer even less assistance after 1 April, when he moved to temporary quarters in Clapham.[20]

The Wonders of Geology

On 2 April *The Wonders of Geology* appeared, its first edition consisting of one thousand two-volume copies selling for fifteen shillings the set. These went within the month, and a "Second Thousand" was likewise virtually exhausted by September. The revised third edition of May 1839 then included a new preface, more woodcuts, and additional material; from it, American and German editions were drawn, as well as a special large-paper British one. There would be eight British editions in all, making *Wonders* the best selling popular geology in Victorian England; through his book Gideon achieved broad cultural impact and specific influence on such important men of letters as Bulwer-Lytton, Tennyson, and Hardy.[21]

20. The sale of Gideon's collection to the BM is extensively documented by numerous letters preserved at ATL (CK, CL, GFR, RIM, H. Stutchbury, Lord Munster, Sir Philip Egerton, the Rev. Josiah Forshall [for the Trustees]) and BMNH (CK, Trustees). For an extensive listing, see R. J. Cleevely and S. D. Chapman, "The Accumulation and Disposal of Gideon Mantell's Fossil Collections and Their Role in the History of British Palaeontology," *Archives of Natural History*, 19 (1992), 307–364, pp. 346–347. GM–CK (and CK–GM), 15 Jan 1838 (BMNH); RIM–CK, 18 Jan 1838 (BMNH). GM–Northampton, 19 Jan 1838 (Castle Ashby); GM–BS, 4 Feb 1838 (Yale). Petition, and direction to CK, 17 Feb 1838 (BMNH). H. Stutchbury–CK, 26, 28 Mar 1838 (BMNH). Ms catalog of GM's collection (BMNH). WBDM–BS Jr., 1 Apr 1838 (Yale); J (re 1 Apr 1838).

21. *Wonders*. Announcement: *BH*, 31 Mar 1838, p. 1 (adv.). GM, *The Wonders of Geology* (with a second title page adding the subtitle, "A Familiar Exposition of Geological Phenomena" and "From notes taken by G. F. Richardson"); 2 volumes, London: Relfe and Fletcher, 1838. Re the title page vignette, see GM, "On the Remains of Man" (1850), pp. 242–243. Dedication to George Augustus Frederick Fitzclarence, first Earl of Munster (1794–1842; *DNB*), eldest son of William IV and son-in-law of the Earl of Egremont. As chairman of the Sussex Royal Institution and Mantellian Museum, he wrote frequently to GM (ATL), who thought him an excellent friend and was shocked by the Earl's suicide (J, 20 Mar 1842). Gideon had been contemplating a book called *Wonders of Geology* since 1834 (GM–BS, 4 Oct, 18 Nov 1834 [Yale] and J, 8 Dec 1836). Editions one through four acknowledge GFR's reportage (Preface).

  Edward Bulwer-Lytton cited *Wonders* in Book VI, Chapter 7, of his popular novel *Zanoni* (1842); for Tennyson's reaction to GM's frontispiece, see my "Tennyson and Geology" (Lincoln, 1985); Patricia Ingham, "Hardy and *The Wonders of Geology*," *Review of English Studies*, 31 (1980), 59–64. John Martin, the artist, kept *Wonders* in his library.

Figure 8.2. "The Country of the Iguanodon" (John Martin, 1838; engraving on steel from a painting now lost; the frontispiece of Mantell's *The Wonders of Geology*). See p. 369, first edition. Later editions added (with textual variants and page numbers): "The painting represents a country clothed with a tropical vegetation, peopled by colossal reptiles, and traversed by a river, which is seen to empty itself into the sea in the distance. Oolitic [i.e., Jurassic] rocks form the heights and cliffs with which the landscape is diversified. The vegetation consists of the trees and plants whose fossil remains have been discovered in Tilgate Forest, namely arborescent ferns, zamias, and coniferous trees, the lesser species being distributed over the foreground.

"The greater reptiles are the Iguanodon, Hylaeosaurus, Megalosaurus, and Crocodile. An Iguanodon attacked by a Megalosaurus and Crocodile constitute the principal group; in the middle distance an Iguanodon and Hylaeosaurus are preparing for an encounter; a solitary Pterodactyl, or flying reptile, with its wings partly expanded, forms a conspicuous object in the foreground while tortoises are seen crawling on the banks of the river. Ammonites and other shells of the Portland Oolite, which is the foundation rock of the country, are strewn on the shore" (*Wonders*, third and seventh editions, condensed).

The most immediately striking feature of *Wonders* was its steel-engraved frontispiece, "The Country of the Iguanodon," John Martin's remarkable visual manifestation of Mantell's researches in Tilgate Forest. Though somewhat contrived (as his work tended to be), Martin's originally nine-foot depiction of the Age of Reptiles was intended to be accurate. Despite a great deal of artistic license, only some freshwater ammonites outrightly contradicted available knowledge. The genre itself – a special kind of history painting – was new, and only a few paleontological localities were sufficiently well investigated to be reconstructed in such detail. Gideon's commissioned frontispiece therefore became the first well-known attempt to re-create the actual appearance of dinosaurs (not yet so called) and their environment. It effectively awakened Victorian imaginations to a saurian past that has continued to fascinate us ever since.[22]

Mantell had contemplated some kind of work to be called *The Wonders of Geology* since 1834. Once his lecture series of 1837 had been arranged, he planned from the beginning to make it the basis of that long-anticipated book. As presented in September and October, his lectures were largely extemporaneous – given from notes rather than a full text – but G. F. Richardson transcribed all of them verbatim (or nearly) and may have done even more. An "Advertisement" to the first and second editions of *Wonders* acknowledged him as "Editor." Richardson's name appeared on the title page as well but was dropped from the third and subsequent editions. Conceivably, Richardson may have provided so much as a first draft, worked out between September 1837 and, say, January 1838. Yet, in both matter and style, the final version of *Wonders* was almost wholly Gideon's. It took him, as he told Silliman, about fourteen weeks, his dedication (to the Earl of Munster) being signed on 27 February. By all indications, then, Richardson had been little more than a temporary amanuensis.

A comparison of *Wonders* with newspaper accounts of Mantell's Brighton lecture series reveals several important differences, chief among which is that the book was organized into eight lectures rather than six; Gideon added material from his nonseries lectures on corals and the Chalk

22. "Country of the Iguanodon" (for dimensions, see GM–BS, 4 Aug 1842 [Yale]). Ancestral to Martin's frontispiece was the museum habitat group, invented by William Bullock around 1812. An early try at recreating an Age of Reptiles environment pictorially was the lithograph by Henry De la Beche entitled "Duria Antiquior." Presenting a view of Dorset in Liassic (now Jurassic) times, it appeared in 1830 (CL–GM, 13 May 1830 [ATL]). Gideon's unpublished "Reptiles Restored" of 1833, commissioned by him from George Scharf, was the earliest attempt ever to restore the appearance and environment of dinosaurs. He used it and similarly conjectural restorations in his public lectures, beginning on 8 Feb 1834; given to Yale at his death, these historically important attempts have since disappeared. The first book to feature such a frontispiece was Thomas Hawkins' *Memoirs of Ichthyosauri and Plesiosauri* (1834). Tiny illustrations of individual prehistoric creatures, including the earliest *published* reconstruction of an iguanodon, accompanied a colored geological chart in Buckland's *Geology and Mineralogy* (1836). Gideon's *Wonders* then popularized the saurian frontispiece as a genre. Other examples soon followed, including G. F. Richardson, *Sketches . . . Second Series* (1838) and *Geology for Beginners* (1842, 1843). Thomas Hawkins, meanwhile, had used another in his *Book of the Great Sea-Dragons* (1840). Two additional early tries at depicting *Iguanodon* (supplied by J. B. Delair) appeared in A. Riviere, *Elements de Géologie* (Paris, 1839), and Joshua Trimmer, *Practical Geology and Mineralogy* (1841).

primarily. The opening remarks of his first lecture and the concluding remarks of his last were better organized and more elaborate than their oral
counterparts. Since *Wonders* retained the lecture motif, however, alluding
to specimens and diagrams displayed, parts of the manuscript may simply
have been Richardson's draft emended. Mantell himself arranged for the
illustrations, which included not only Martin's frontispiece but another
(to Volume 2) in color on corals by his elder daughter, Ellen Maria, and
numerous woodcuts throughout; these were further increased in the third
edition. By then, Gideon's original lectures and Richardson's assistance
had become increasingly vestigial, as *Wonders* gained independence from
its origins.

   Reviewing the first edition, Silliman praised both substance and style.
"In point of science," he wrote, "it is precise, accurate, condensed and cu-
mulative in proof . . . ; no important facts are omitted and none are un-
duly expanded" (390). He thought the book's illustrations very useful and
extolled its prose as lucid, flowing, simple, and elevated. In private, however, Silliman suggested a long list of stylistic and grammatical corrections,
many of which revealed Mantell's compositional haste and the survival of
Richardson's less fastidious calligraphy. When Gideon then utilized these
discerning emendations fully, his Connecticut friend delightedly praised a
"greatly improved" style in the third London/first American edition.

   But Silliman had also been sufficiently astute to point out that, however successful *Wonders* might be, it was still only a popularization, fur-
thering public understanding rather than knowledge itself. "Although we
have to regret that the rich fields of research so assiduously and successfully cultivated by the acute and indefatigable author in the southeast of
England can be explored by him no more," Silliman wrote in 1840,

> we conceive that a post of usefulness is now assigned him of not less
> importance, although it may be in a more humble sphere. His highly
> gifted mind, endowed with all the auxiliary sciences, disciplined at
> once to accuracy and taste, and prompt to observe and secure every
> discovery and improvement, will vigilantly watch the progress of ge-
> ology and post it up, as we trust, by such constant revision that the
> successive editions of the *Wonders,* improved and even enlarged as
> occasion may require, will continue to reflect, as from a perfect and
> beautiful mirror, the full face and form of this lovely and splendid
> science. (10)

Gideon himself saw both face and form clearly, but only in retreat, like
those of an unsatisfied spouse about to abandon him forever. "My new
work, *The Wonders of Geology,* is making a sensation," he assured Silliman in April 1838; "Lyell writes me that it will do more to popularize the

science than any work that has yet appeared." Alas, he continued, "it is my farewell to geology!"[23]

23. *Wonders,* editions one to three. The newspaper accounts of GM's six original lectures are cited in note eighteen above. GM–BS, 4 Feb 1838 (misdated 1837), 8 Apr 1838 (Yale). BS, rev. of first edition, *AJSA,* 34 (July 1838), 387–392 (quoting *BH,* 31 Mar 1838); 35 (Jan 1839), 384; rev. of first American edition, *AJSA,* 39 (Oct 1840), 1–18. BS–GM, 10 July 1838 (Yale); GM–BS, 8 Apr 1838 (Yale). See also William Sydney Gibson, *The Certainties of Geology* (London, 1860), pp. 217–218.

9

Crescent Lodge

Though lonely and unsettled in Clapham, Gideon Mantell complacently assumed that his recent problems had been overcome and was again enjoying the bustle of a genuine practice. By 28 April, however, all optimism had been crushed. Solely because of unexpectedly low tax revenues, Parliament had refused to fund the British Museum's purchase of his collection. Influential friends offered to remonstrate with the Chancellor of the Exchequer on Mantell's behalf, but there seemed to be no chance of success in that. "How contemptible," he thought, "is that legislature which, with the enormous riches of this country, cannot afford the petty sum of five thousand pounds to obtain what is peculiarly British and unique." Time was of essence, as Gideon's lease on No. 20 The Steine would soon run out, his collection thereafter being without a home.

Lyell and others, meanwhile, were exerting themselves unremittingly to secure governmental reconsideration of the purchase. On the twenty-ninth Mantell enlisted still further support at a soiree given by the Duke of Sussex for fellows of the Royal Society. Once again, his friends did not fail him; within a week the government changed its mind, agreeing to buy his museum if he would wait till 1839 for the money. Stipulating only that persons be named to value and transport his specimens, Gideon accepted.

For several weeks, nothing further was done – largely because the lavish coronation of young Queen Victoria on 28 June engrossed all England. The trustees then ordered König alone to Brighton, where his evaluation was completed on 28 July. Mantell wrote Silliman immediately to announce the price, a sum far below his collection's real value, no doubt. But as the trustees undertook to pack and transport everything at their own expense, the Museum's four thousand pounds seemed almost equivalent to what Gideon had anticipated. Capably orchestrated by König and Richardson, the overland removal to London then took place on 10 December, precisely as scheduled.[1]

---

1. The sale and removal of GM's collection. GM–BS, 8 & 28 Apr 1838 (ATL); GM–CK, 18 Apr, 5 May; CK–GM, 6 May 1838 (BMNH); GM–SGM, 1 June 1838 (APS); CL–GM,

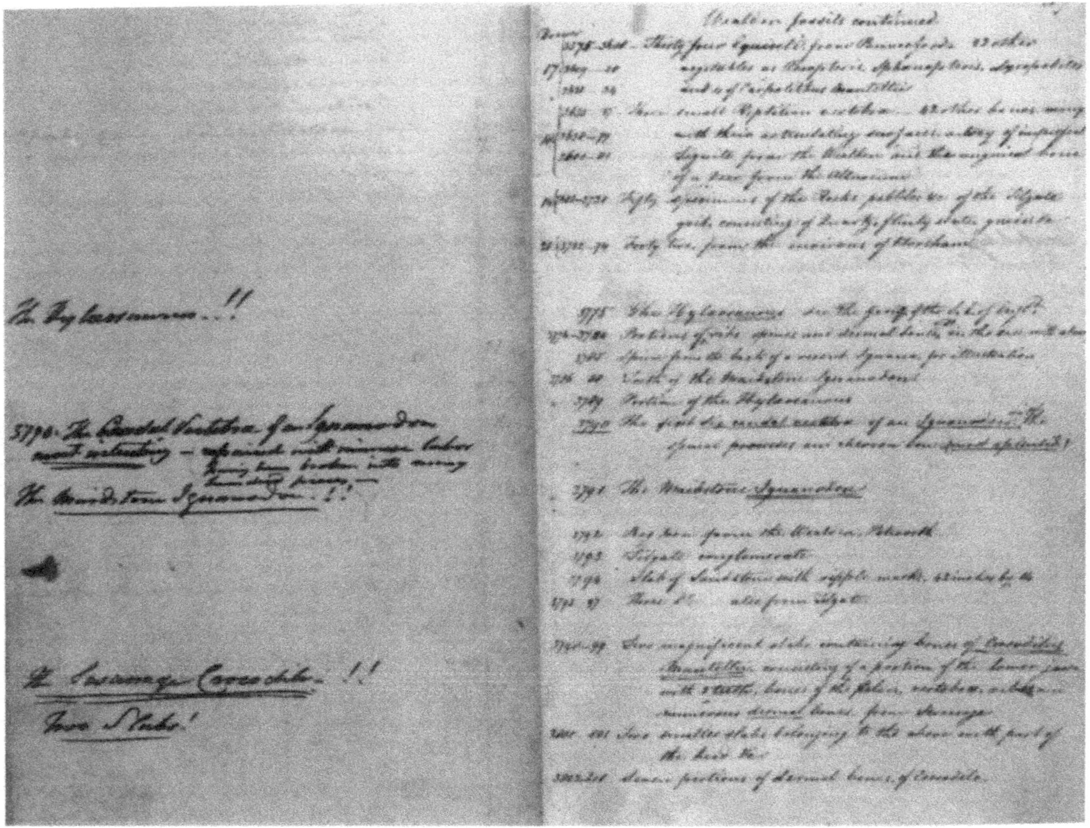

Figure 9.1. Sale catalog of Mantell's former collection, with poignant annotations by him regretting the loss of some of his finest specimens (BMNH).

Meanwhile, Mantell himself had moved, and after 29 September took up residence at Crescent Lodge, one of a pair of three-story houses on the south side of Clapham Common, at the entrance to Crescent Grove.[2] He was situated on the main road from Brighton to London, about three miles from Westminster Bridge, and only twenty minutes' drive from the Geological Society, the meetings of which he could now attend more regularly. Having already held two soirees at home, Gideon spent the rest of the year developing his medical practice.

On 1 May 1839 the third edition of *Wonders* appeared, with a new preface alluding to Mantell's misfortunes in Brighton and an extensively rewritten text including the most recent geological discoveries. The Old Red Sandstone of his first edition was now established as the Devonian system, while work by Murchison and Sedgwick had transformed discussion of the Silurian and Cambrian strata. A number of minor flaws having been meticulously corrected as well, the third edition was so much improved that Gideon felt unusually certain of its success. "There is a very

---

5 June 1838 (ATL). Further details then obscured and endangered the outcome of this complicated sale till 13 June, when Lyell, Buckland, Northampton, Sedgwick, Fitton, Owen, and especially Murchison (among other members of the Geological Society) once more urged the government to acquire GM's collection. GS–Trustees, 13 June 1838 (ATL; published in Thomas Hawkins, *Statement*, 1848, pp. 24–25). RIM–CK, 17 July 1838 (BMNH); GM–Northampton, 18 July 1838 (Castle Ashby); Trustees (J. Forshall, secy)–CK, 18, 20 July 1838; GM–CK, 20 July; Forshall memo, 28 July 1838 (all BMNH). GM–BS, 28 July 1838 (Yale); J (30 July); GM–CK, 31 July 1838 (BMNH); GM–RIM, 31 July 1838 (GS); J. Forshall–CK, 4 Aug 1838 (BMNH); GM–CK, 10 Aug 1838 (BMNH); GM–BS, 18 Sept 1838 (Yale).

SRI wrap-up. Earl of Munster–GM, 26 Oct 1838, with release (ATL); Mantellian Museum Visitors Book: 18 May 1836–1 Nov 1838, listing 7,270 names (subscribers not included), presented to GM 14 Nov 1838 (ATL).

The move. Horace Smith, "A Vision. On the Removal of Dr. Mantell's Collection from Brighton to the British Museum, 1838" (poem, ATL). GFR–GM, 5 Dec 1838 (ATL); movers' bill of 10 Dec 1838 (Edinburgh University Library); R. J. Cleevely and S. D. Chapman (Chap. 8, note 20 above), note 77. On 8 Jan 1839 CK acknowledged receipt of ninety variously sized cases from Brighton (BMNH). GM received his four thousand pounds from the trustees on 1 Aug 1839, and immediately resolved to "begin de novo!" (J).

A final indication of what his collection included at the time of its dispersal was GM's eighty-page "Catalogue of the Mantellian Collection in the British Museum" (1839, ATL), compiled for his own use from the official one made earlier by König and Stutchbury (BMNH).

2. Clapham Common. GM–BS, 18 Sept, 25 Oct, 31 Dec 1838 (Yale); J (19 Sept 1838, 3 July 1843, 29 Sept 1844). GM leased Crescent Lodge from a Mr. Northcote for five years, later extended to precisely six. Which of the twin houses at the entrance of Crescent Grove was his has been disputed, but a labeled drawing by George Scharf (BM) seems to me conclusive. GM's home is now called Denmark Lodge; its unfortunate twin (dental offices) presently masquerades as Crescent Lodge. Eric E. F. Smith, *Clapham* (London, 1976), and Arthur H. Noble, *The Families of Allnutt and Allnatt* (Aylesbury, 1962) are helpful.

great demand for it," he wrote proudly, "and my bookseller thinks it will sell very rapidly." American and German editions followed.[3]

Breakups

Though now fully and profitably engaged in his profession, Mantell himself fared poorly in both health and spirits. While his collection had arrived more or less safely at the British Museum (the Maidstone Iguanodon, unfortunately, having been damaged), it was to be broken up and distributed throughout theirs – plants with plants, fishes with fishes, and so on. Greatly distraught, he implored Northampton, Buckland, Fitton, and others to join him in a petition to the contrary, but they refused to intervene; Silliman tried, unsuccessfully. As a last hope, Gideon then wrote to urge that his Wealden fossils (at least) remain together. Though Northampton (now president of the Royal Society) supported him in that, König did not and would prevail. "Your collection at Yale College," Gideon then advised his American friend, "will now be the only Mantellian Museum."[4]

Mantell wrote even more poignantly to Silliman on 4 March 1839, a date that would haunt him for the rest of his life. In return for the sale of his collection, he had been overwhelmed with professional engagements, including a number of important cases. Working often with Sir Astley Cooper and other prominent surgeons, he had been highly successful – but was now "almost dead from fatigue, anxiety, and that sorrow which language cannot describe!" Throughout the past few weeks, despite noticeable kindness and sympathy from many of his patients and friends, Gideon had experienced severe fainting spells, excruciating tic douloureux (neuralgic facial pains), and other symptoms of nervous breakdown. Yet were it not for —— (something deleted from this letter by a later hand), he would be happy in Clapham. The deleted passage concerned his wife, Mary Ann, who after twenty-three years of often stormy marriage and several extended separations that day left her husband forever.[5]

---

3. *Wonders*, 3rd. GM–BS, 18 Sept, 31 Dec 1838, 18 & 22 May 1839, with details in BS replies (Yale); GM–SGM, 21 May 1839 (APS); Joseph Burkart (of Bonn)–GM, 19 Sept 1839, 14 Mar 1840 (ATL). For the British edition of twenty-five hundred copies, GM eventually received £250.

4. GM–Northampton, n.d. (Dec 1838, Castle Ashby); GM–BS, 31 Dec 1838 (quoted; Yale); BS–GFR, 7 Feb 1839 (Yale). GFR had been named curator at the BM under CK, so GM's collection remained in his care.

5. MAM walks out. GM recorded the date (not previously ascertained) in his own copy of William Berry, *County Genealogies . . . Sussex* (London, 1830), p. 20 (ATL). J (4 Mar 1839; a one-and-a-half line entry thoroughly deleted, with a despairing note on the loss of his museum added). GM–BS, 4 Mar 1839 (Yale; MAM passage deleted). In a memo-

The Mantell's longtime housekeeper, Hannah Brooks, left with her. She and Mary Ann then began living together in a small cottage near Exeter. When her spring term at Dulwich ended, Ellen Maria returned home to Gideon, Reginald, and sick Hannah Matilda. Walter probably visited as well and was certainly exchanging some heated opinions with his father regarding future plans, now that his apprenticeship at Chichester was ending. He refused to enter into partnership with his father and had no interest in surgery at all. On 15 September, despite Gideon's unavailing efforts to dissuade him, Walter sailed for New Zealand as a pioneering immigrant and would never see his father again.[6]

Like Walter, Ellen Maria also left Gideon in September 1839, to live with Mrs. Richardson in Dulwich and attend her school. Ellen had been twenty-one since May and could therefore not be restrained from doing as she pleased, though her father probably continued to support her, as he did Mary Ann. Since Gideon had almost abandoned his journal, we know his state of mind only from one partially deleted entry: "September – My son Walter and my daughter Ellen [obliterated, but probably something like 'have both deserted me']." On 30 October, Bakewell then wrote nervously to Silliman: "I am anxious about Dr. Mantell. I have not seen or heard of him since his son sailed for New Zealand in September." Lyell did what he could for Gideon, recognizing the cruel disappointment that Walter's departure necessarily involved while radiating sanguine assurances that prospects for New Zealand were most auspicious.[7]

randum preserved at Yale, BS noted that he had "faithfully removed or obliterated every trace of that subject [MAM's desertion] in Dr. Mantell's letters" (Private Journals, IV, 194; about 1857). Deletions in GM's journal are generally WBDM's, made posthumously in obedience to his father's instructions. By leaving her husband, MAM gave up legal control of both her property and her children; there was never a divorce, which would have required an act of Parliament.

6. WBDM and New Zealand. G. Duppa–GM, 20 July 1839 (ATL). For the Wakefield colonization scheme, see [Edward Gibbon Wakefield and Harry George Ward], *The British Colonization of New Zealand* (London: John W. Parker, 1837); Edward Jerningham Wakefield, *Adventures in New Zealand* (2 vols., London, 1845); Louis E. Ward, *Early Wellington* (Auckland, 1929); and Marjorie Appleton, *They Came to New Zealand* (London, 1958). Biographical sketches of WBDM and his fellow colonist George Duppa appear in Ward; G. H. Scholefield, ed., *A Dictionary of New Zealand Biography* (2 vols., Wellington, 1940); A. H. McLintock, ed., *An Encyclopedia of New Zealand* (3 vols., Wellington, 1966); and W. H. Oliver, ed. *The Dictionary of New Zealand Biography*, I: 1769–1869 Wellington (1990).

7. J (Sept 1839); RB–BS, 30 Oct 1839 (Yale); CL–GM, 2 Nov 1839 (ATL). GM's brother Joshua also contributed significantly to his dejection. After hiding him for thirty months, brother Thomas finally took Joshua to the enlightened asylum in Ticehurst, Sussex, where he was admitted as a patient on 12 April 1839. Throughout the 1840s his care required annual payments of £54/12/0; it then rose to £60 (financial and observational records

With Walter and Ellen Maria gone, Gideon had only his two younger children, of whom he cherished poor, suffering Hannah Matilda the more. Throughout her unlucky existence, she was his "sweet angel" and most tragic loss. In 1836, at age fourteen, she began experiencing problems with her health and had more than once to drop out of school because of them. From the spring of 1837 onward, Hannah Matilda remained at home, unable to walk. "My sweet girl is still wholly confined to her bed," Mantell wrote Silliman in May 1839, "but she is better than she was and does not suffer. She is obliged to lie constantly on her back, but still she is able to draw, paint, knit, write, and work, and her sweet temper and disposition make everything delightful around her." During that summer and early autumn she continued to improve, the tuberculosis in her hip being so much diminished that Gideon had renewed his hopes of her ultimate recovery.

When Walter and Ellen Maria left, Hannah Matilda relapsed. She continued to be heroically cheerful, however, and even so late as February 1840 was able to take airings in an invalid carriage that Gideon had constructed for her. In March he dined at the Athenaeum, a prestigious literary club in London to which he had been elected the previous month. Coming home early, Gideon was abruptly summoned to his fainting daughter's side; sudden, profuse bleeding surged from the abscess on her hip. Though soon improved, she was watched over constantly thereafter by Gideon's sister Mary (Mrs. West) and a niece.

After a good night Hannah Matilda awoke on Thursday, 12 March, to breakfast on lobster from Brighton. While sipping her tea, she began to faint again and whispered her father's name. When Mrs. West called out in alarm, Gideon rushed in from an adjoining room just in time to hear his favorite daughter's final sigh. "And thus," he sorrowed, "[words blotted out] one whose sweetness of disposition and affectionate heart endeared her to me beyond even the natural ties that united us, is taken from me! Before the Chastener humbly let me bow, o'er hearts divided and o'er hopes destroyed!" Five days later, Hannah Matilda was interred at Norwood Cemetery, near her Dulwich school, in a grave that Gideon Mantell would visit faithfully for the rest of his days and now lies beside.[8]

from Ticehurst, as researched for me by Charlotte MacKenzie, Wellcome Institute for the History of Medicine, London). See also Chap. 8, note 12 above; J (*passim*); Mrs. Epps, ed., *The Diary of John Epps* (London, 1875), pp. 196–197 (a visit to Joshua at Ticehurst); and Charlotte MacKenzie, *Psychiatry for the Rich: A History of Ticehurst Private Asylum, 1791–1917* (London, 1993).

8. Hannah Matilda. GM, poems to her (July 1823, Nov 1825, Dec 1826; ATL and, in *Illus*, SAS). GM–BS, 18 Sept, 25 Oct 1838, 4 Mar, 18 & 22 May 1839 (Yale); J (12, 17 Mar 1840); GM–BS, 17 Mar 1840 (Yale); GFR, "To the Memory of Miss Hannah Mantell" (poem, ATL); J (Apr 1840); GM–BS, 23 May, 2 & 26 Sept 1840 (Yale). In 1840 tuberculosis caused an estimated one-sixth of all deaths in England.

Though close friends sent obligatory condolences, and G. F. Richardson a poem, Mantell was desperately unable to reconcile Hannah Matilda's death with his previous belief in Providence or to face the future with any kind of remaining optimism. "Very ill, and in a state of depression almost unbearable," he wrote to himself in April. "My little boy Reginald [now thirteen] is the only one that remains to me." He himself was fifty; despite an unchanged hairline, Gideon's once-youthful features, marked now by excruciating care and sorrow, revealed all too plainly that for him the vigor of manhood had departed forever.[9]

Recovery

By September, a fundamental tenacity of spirit began to reassert itself. Returning to geological activity, Mantell was actively collecting fossils once again, participating regularly in the Geological Society, and reviving some public and literary ambitions. Quickly establishing himself as a local spokesman for the scientific outlook, he even agreed to lecture for charity in Clapham, a conservative suburb still famous for its devotion to Evangelical religion.

Mantell's first effort, on Friday, 30 October, stressed the "Interest and Importance of Geological Researches," which enable us to examine and value productions of the Creator. An imaginary geological stroll over Clapham Common allowed him to discuss nearby phenomena. Several members of the Established Church surrounded Gideon throughout his hour-and-a-half address; at its conclusion, one of them (the Reverend Mr. Murray) then stepped forward and begged him to allow them the privilege of announcing a sequel. There were in fact two more, with audiences of more than five hundred attending each. The Reverend Murray then wrote Mantell to thank him for having inaugurated the Clapham Boys' Parochial School lecture program so successfully; he had raised more than £110 on its behalf.[10]

9. J (Apr 1840). "But for my kind friends Miss Foster and Mr. and Mrs. Allnutt," he added, "I must have sunk under the accumulated evils that have pressed upon me." The well-to-do wine and brandy merchant John Allnutt (1773–1863) and his second wife Eleanora Brandram Allnutt (1789–1866) welcomed GM to their elegant mansion (see Smith, note 2 above). Catherine Foster (a subscriber to *South Downs*) conducted a seminary for young people. Her brother, the Reverend John Foster, was RNM's schoolmaster; she herself served as his godmother. Louisa Broadhurst, another good friend, was remembered in GM's will.

10. GM–BS, 2 & 26 Sept, 24 Oct, 27 Nov 1840 (Yale); broadsheets and newspaper clippings, incl. *Sussex Press,* 9, 23 Nov, 7 Dec; *Morning Post,* 30 Nov 1840 (ATL). Rev. W. Murray–GM, 17 Dec 1840 (ATL).

Mantell and Owen

In January 1841 Mantell reopened his journal, ostensibly to record his usual share of vexations but actually to admit that his outlook had improved. He was, Silliman learned, "fairly embarked in the London world of science," having been elected to the councils of the Geological and Linnaean societies and secretary to the Geological Committee of the Royal Society while remaining active in the Medico-Chirurgical Society and the Athenaeum as well. Gideon had been elected to the latter (with help from Greenough and Murchison) on 25 February 1840 – as had Richard Owen, the anatomist. The two men were being frequently compared, and this as-yet friendly rivalry was stimulating for a time, a useful prod to further original research.

Among other projects, Mantell contributed two long papers on paleontology to the *Transactions* of the Royal Society. The first recalled how in his *Iguanodon* paper of 1825 he had hoped eventually to find a jaw, so that his inferences from teeth alone could be confirmed or modified. Now, in this "Memoir on a Portion of the Lower Jaw of the Iguanodon," he announced that a specimen found by himself near Cuckfield in 1837 was probably the long-sought evidence. After discussing it, Gideon dealt also with further news of *Iguanodon* and other Tilgate saurians. Since his previous memoir he had collected more than 250 iguanodon teeth, of which 120 were now in the British Museum. Since then also, the best of these teeth had been examined microscopically by Professor Owen, who pointed out that they differed fundamentally in structure from those of the iguana. ("Iguanodon" – having iguana-like teeth – was therefore a misnomer, but Mantell did not acknowledge this.) In any case, he left such technicalities to be elucidated more fully by Owen, who was now at work on his *Odontography,* a three-part investigation of vertebrate teeth that Gideon thought a work of genius.[11]

Iguanodon's hand and finger bones also failed to correspond with those of the iguana; instead, they more nearly resembled those of large herbivorous mammalia. From the structure and condition of its teeth, Mantell

---

11. J (Jan-May 1841, *passim*). GM–BS, 29 Mar & 4 Apr 1841 (Yale). For Mantell's opinion of the new Ice Age hypothesis, see this letter, GM–BS, 14 June 1842 (Yale); and *Wonders,* sixth edition, I, 73. Neither GM's "First Lessons in Geology" nor his proposed treatise on the nervous system ever appeared. For his interest in microscopics, see J (27 Aug 1833, 26 June 1837); GM–BS letters, Nov 1840–Aug 1842; and later references below. RO, *Odontography; or, A Treatise on the Comparative Anatomy of the Teeth* (2 vols., London, 1840–1845); *Iguanodon* is at I, 246–253, and Plates 70, 71. For Mantell's opinion of RO's book, see GM–BS, 29 Mar, 4 Aug 1841, 4 Aug 1842, 6 July 1845 (Yale); *Medals,* II, 589–590; and *Petrif,* pp. 489–490.

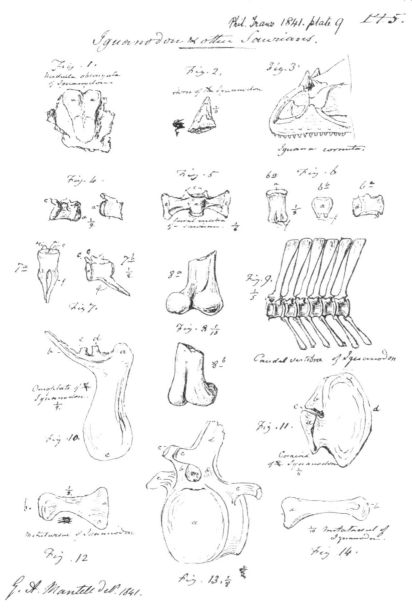

Figure 9.2a. Iguanodon & other Saurians (original drawings by Mantell affixed to a single sheet; later published in *Philosophical Transactions of the Royal Society*, 131 (1841), Plate IX; RS).

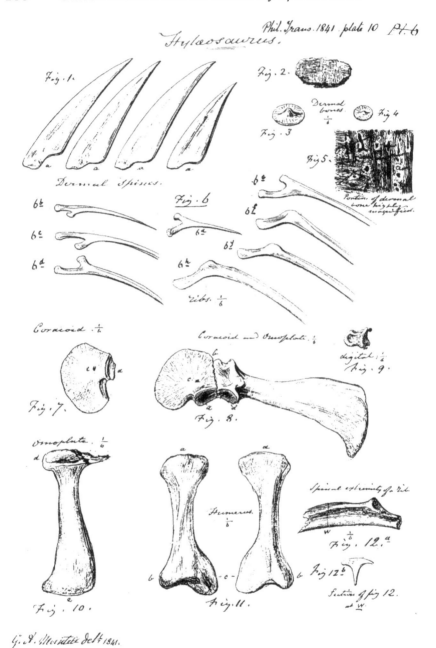

Figure 9.2b. Hylaeosaurus (original drawings by Mantell; later, *Philosophical Transactions of the Royal Society*, 131 (1841), Plate X; RS).

reaffirmed, *Iguanodon* was clearly herbivorous also. He now imagined it as standing upright on enormous hind legs and using slender, prehensile forelimbs to pull down foliage from trees and arborescent ferns. Thus, Gideon had by now reconstructed not only *Iguanodon's* environment but something of its predominant ecological role as well. In twenty years he had found the remains of no fewer than seventy individuals, from specimens "just burst from the egg" to fully grown giants. Yet evidence of at least three times as many had probably been destroyed by the quarrymen. *Iguanodon,* therefore, must have been the ruling animal of its time.

In the same paper Mantell reviewed his original discovery of *Hylaeo-saurus* in 1832, which had since been augmented by two further major specimens, both unfortunately damaged by the workers before he had seen them. No certain examples of the skull, jaws, or teeth of *Hylaeosaurus* had yet appeared. In the same stratum, however, a number of distinct teeth had been discovered – and were incorrectly identified in Gideon's *Geology of the South-East of England* as phytosaurian. The one tooth of this kind still in his possession had distinctly reptilian structure (as Owen had defined it), but nothing could be decisively concluded until a tooth-bearing jaw was discovered in association with other bones. Unfortunately, it would have to be discovered by someone else. "Removed from the field of my former labors, having disposed of my collection, the fruit of twenty-five years assiduous research, and being engaged in the duties of an arduous profession," Gideon concluded unexpectedly, "this memoir is, in all probability, the last contribution which it will ever be in my power to offer in this interesting department of paleontology" (144–145). Six (V–X) plates depicted various parts of *Iguanodon* and *Hylaeosaurus,* but on several of his identifications, Gideon would subsequently change his mind.[12]

Fossil turtles enjoyed something of a vogue in 1841 as both Owen and Mantell wrote papers on them. The latter's specimen had been found for him in March by W. H. Bensted, the discoverer of the Maidstone Iguanodon, and promptly displayed at one of Northampton's soirees. Written in April, Gideon's paper was then dated, received, and presented in May. Though turtles are no longer indigenous to Great Britain, Mantell pointed out, they were once common. Numerous examples of their remains, for example, occurred in the strata of Tilgate Forest, together with bones of

---

12. GM, "Memoir on a Portion of the Lower Jaw of the Iguanodon, and on the remains of the Hylaeosaurus and other Saurians discovered in the Strata of Tilgate Forest in Sussex," *PT,* 131 (1841), 131–151; read 18 Feb 1841. J (Feb, 14 Sept, 9 Dec 1841, 3 Feb 1842). Noticed in *AJSA,* 43 (Oct 1842), 189–190. GM later reidentified the jaw; see *Petrif,* pp. 242, 280–281, 286, 333–335. GM's copy of this paper (at ATL) records several changed identifications. It actually includes parts of four dinosaurs: *Iguanodon, Hylaeosaurus, Cetiosaurus,* and *Regnosaurus.*

the giant saurians. Higher up, however, in the Chalk strata, turtles were extremely rare. Gideon himself had formerly owned more than ten thousand Cretaceous fossils from the South Downs, but only one was definitely chelonian. In a paper lately read before the Geological Society, Professor Owen had described the first turtle fossil ever found in the British Lower Chalk (at Burham, Kent); Gideon's specimen, from the same place, was much finer and had been identified for him by Thomas Bell, a turtle expert, as being probably freshwater. Like the Maidstone Iguanodon, therefore (also found in marine deposits), the dead Burham turtle had been carried far out to sea by some large river.[13]

In recent years Owen had published several very competent papers correcting egregious mistakes by his predecessors in fossil comparative anatomy. For example, in 1838 he revealed that the bent tails so often seen in ichthyosaur skeletons were not coincidentally damaged but rather consistent indications of a large caudal fin. The next year Owen untangled a mistake by his father-in-law, William Clift, who had confused armadillo and sloth remains from South America; he also contradicted Dr. Richard Harlan regarding some fossil teeth. Not surprisingly, it was Owen who then identified the mammalian fossils brought from South America by Darwin and the *Beagle*. His forthcoming *Odontography* (on fossil teeth, and likewise undertaken in 1839), seemed one of the most promising advances in vertebrate comparative anatomy since the death of Cuvier. Owen, therefore, was widely regarded as the dead Frenchman's equally brilliant successor, a title once seriously attributed to Gideon. Fully aware of his rapidly advancing eminence, Owen grasped opportunities which permitted him to accelerate it, particularly by undermining the credibility of his rivals.[14]

Owen on Fossil Reptiles

On Saturday, 31 August 1839, Owen had given a two-and-a-half-hour presentation before the British Association meeting in Birmingham. Read-

13. GM, "On the Fossil Remains of Turtles, discovered in the Chalk Formation of the Southeast of England," *PT*, 131 (1841), 153–158; read 20 May 1841. J (17 Apr, 3, 11, 20 May). Noticed in *AJSA*, 41 (1841), 205; 43 (Oct 1842), 189–190. GM–RO, 4, 5 Oct 1841 (BMNH). RO, "Description of the Remains of a . . . Tortoise," *Proc GS*, 3 (1840), 298–300; RO, "Description of the Remains of Six Species of Marine Turtles," *Proc GS*, 3 (1841), 570–578. *Medals*, II, 766.
14. RO, "On the Dislocation of the Tail . . . of Many Ichthyosauri," *Proc GS*, 2 (1838), 660–62; *Trans GS*, ns 5 (1840), 511–514 (but his explanation was inexact). RO, "Description of a Tooth and Part of the Skeleton of the Glyptodon," *Proc GS*, 3 (1839), 108–113; *Trans GS*, ns 6 (1841), 81–106. RO, "Observations on the Teeth of the Zeuglodon," *Proc GS*, 3 (1839), 24–28; *Trans GS*, ns 6 (1841), 69–80. RO, *The Zoology of the Voyage of H.M.S. "Beagle" . . . Part I: Fossil Mammalia* (London, 1841). Owen, *Life*, Chaps. 4–6.

ing the first part of his landmark *Report on British Fossil Reptiles,* he dealt only with what Conybeare had called *Enaliosauria,* or marine lizards. Summarizing and clarifying all the information then known concerning them, much of it original to himself, Owen distinguished sixteen species of *Plesiosaurus* and ten of *Ichthyosaurus.* Since relics from these creatures of the Oolite were foreign to Sussex, Owen mentioned Gideon Mantell only once, regarding some ichthyosaurian remains the latter had found at Dover.[15]

Two years later, however, on 2 August 1841, Owen spoke at equal length to the British Association meeting at Plymouth, reading his Part II (of 144 printed pages eventually), a document important to Gideon's subsequent reputation. As Owen reminded his audience, the British fossil reptiles described in the first part of his report had been characterized by modifications of spine and limbs particularly adapting them to a marine life. Though formerly referred to a broad class of "saurian" reptiles by Cuvier, they were a group by themselves, serving to link the reptiles with fishes and cetaceous mammals. In his present report, Owen would discuss the most remarkable amphibious and terrestrial sauria yet found in England, as discovered and described by Buckland and Mantell. These, Owen had now classified as Crocodilian, Dinosaurian, Lacertian, Pterodactylian, Chelonian, Ophidian, and Batrachian reptiles. Through this superior classification, Owen proposed to supersede not only Cuvier's but also that of Hermann von Meyer, a distinguished German paleontologist who had written importantly about saurians in several recent papers.

A close reading of Owen's *Report,* II, illuminates the extent to which Gideon's discoveries in the field had enlarged the study of fossil reptiles in England – how significant his collection and researches were – as well as the symbiotic relationship of Mantell and Owen. To begin with, Owen was significantly indebted to Mantell in his section on *Chelonia,* or fossil tortoises. Thus, he cited *Illustrations of the Geology of Sussex* (1827) and three of its drawings to establish Gideon's discovery of a freshwater tortoise in the Wealden strata of Tilgate Forest. A second and similar discovery was *Platemys;* found by Mantell, and discussed by Cuvier, it was then named *P. Mantelli* by J. E. Gray. There were still difficulties with the genus *Trionyx,* however. One specimen so described in *Illustrations,* for example, was actually crocodilian, as Gideon himself had subsequently recognized.

Among the marine turtles, genus *Chelone,* Mantell had discovered and figured in his "valuable" *Illustrations* parts of a large species with a carapace nearly three feet long. The specimens were fragmentary, however,

---

15. RO, "Report on British Fossil Reptiles," [Part I], *Report of the Ninth Meeting [1839]* of the British Association for the Advancement of Science (London, 1840), pp. 13–126; on p. 126 GM is credited with the find actually made by MAM.

and Gideon had been sure of their nature only after consulting with William Clift. (He remained wrong, Owen noted, on one specimen listed in his catalog as the "Sternal plate of a marine turtle," which was actually from a land tortoise.) Mantell had certainly discovered a new species of marine turtle, but more complete specimens were necessary to designate it properly. Earlier in 1841, Owen had announced the first discovery in Britain of a chelonite from so far down as the Chalk formation. Gideon then described a new find from Bensted's quarry; after consulting others, he identified it as a freshwater species whose remains had subsequently been swept out to sea. Owen now pronounced it marine, a true *Chelone*, and perhaps the first example of a previously unknown subgenus.

Owen's discussion of the fossil crocodile *Suchosaurus*, renamed by himself, began with a passage from the third edition of Mantell's *Wonders* and relied throughout on data and specimens from him. *Illustrations of the Geology of Sussex* (1827) and *Geology of the South-East of England* (1833) were also cited, as well as two important specimens – numbered by König and Stutchbury – from Gideon's former collection. A second fossil crocodile, *Goniopholis*, was Mantell's Swanage specimen, for which *Wonders* remained authoritative. Distinguishing between ordinary crocodiles and gavials (the latter having long, slender muzzles), Owen went on to describe *Poekilopleuron*, for which his evidence was entirely Gideon's. The supposed lumbar vertebra of an iguanodon in *Illustrations*, however, was neither lumbar nor iguanodontian, but the only indication thus far of an otherwise unknown crocodilian genus. Mantell had also been in error regarding the gavial *Streptospondylus*, having failed in his *South-East England* to identify a specimen correctly, while confusing anterior and posterior vertebrae as well. "The determination of the true nature of the convexo-concave vertebrae of the Wealden," Owen pointed out, "and of the affinities of the reptile to which they belonged . . . removes one of the chief difficulties attending the determination of the true vertebral characters of the *Iguanodon*" (93–94).

Mantell, Owen continued, had long ago endeavored to separate the various forms and types of vertebrae found in Wealden strata. Owen would now further this enterprise by distinguishing those of *Cetiosaurus*, a new creature defined by himself earlier in 1841. What had been labeled a "Gigantic vertebra of *Iguanodon*" in Gideon's sale catalog of 1838 was actually cetiosaurian. Mantell had published this mistake earlier, in *Illustrations* (1827), and his ongoing confusion had led to an erroneous discussion of vertebral characteristics in *South-East England*. Von Meyer, Buckland, Lyell, and Sedgwick had then been guilty of derivative errors. Owen thought *Cetiosaurus* very like a whale (as his name for it suggested) and strictly aquatic, but was pleased to add its bulk to "that period of the earth's history which has been aptly termed the 'Age of Reptiles.'"

Owen on Dinosaurs

The most famous and memorable section of Owen's *Report* sought to establish a new classification of fossil reptiles called Dinosaurians. This proposed group, which included at least three well-established genera, was characterized by a large sacrum (termination of the lower spine) made of five united vertebrae, some specialized aspects of the dorsal vertebrae, twofold articulation of the ribs to the vertebrae, and other similarly technical distinctions. The limb bones of this group, moreover, tended to be unusually large; resembling those of elephants, they proved that dinosaurs were land dwellers. "The combination of such characters," Owen believed, "all manifested by creatures far surpassing in size the largest of existing reptiles, will, it is presumed, be deemed sufficient ground for establishing a distinct tribe or suborder of saurian reptiles, for which I would propose the name of *Dinosauria*" (103), a Greek derivation meaning "fearfully great lizards." Owen also coined the English word "Dinosaur" and used it freely within his paper.

His "distinct tribe" consisted only of *Megalosaurus, Hylaeosaurus,* and *Iguanodon,* "the worthy fruits of the laborious researches of Prof. Buckland and Dr. Mantell" (103). Regarding the first, *Megalosaurus* had never been found in any extensively connected form, the evidence for its existence consisting only of isolated teeth, one lower jaw, a number of vertebrae, various limb and toe bones, and some others. In general, these remains proved *Megalosaurus* to have been a lizard-like animal with significant crocodilian affinities. Its trunk was broader and deeper than that of modern reptiles, however, and its limbs were larger and longer, like those of a large mammal. Its teeth proved *Megalosaurus* to have been carnivorous. The most extensively articulated *Megalosaurus* fossil consisted of five ankylosed [fused] vertebrae, which both Buckland and Cuvier had discussed without realizing that they were the *sacrum,* analogous to that in birds, a skeletal feature central to Owen's conception of the dinosaur. It was this structure, "beautifully exemplified in the sacrum of the young ostrich, which Creative Wisdom adopted to give due strength to the corresponding region of the spine of a gigantic saurian species, whose mission in this planet had ended probably before that of the ostrich had begun" (106). And how large *was Megalosaurus?* Though some earlier estimates had suggested lengths of as much as sixty or seventy feet, Owen's sophisticated analysis could affirm no more than half of that. *Megalosaurus,* he thought, had a five-foot head, a twelve-foot trunk, and a thirteen-foot tail, for a total length of approximately thirty feet. While the most important specimens of *Megalosaurus* had been found at Stonesfield, Owen acknowledged that Mantell had discovered a fine vertebra, a large femur, and many teeth in the Ironsand of Tilgate Forest.

Figure 9.3. *Megalosaurus* and *Pterodactyl,* as restored by Benjamin Waterhouse Hawkins (ca. 1854) in accordance with ideas derived from Richard Owen (BMNH).

Figure 9.4. *Iguanodon* and *Hylaeosaurus,* as restored by Benjamin Waterhouse Hawkins (ca. 1854) in accordance with ideas derived from Owen and Mantell (BMNH).

Owen's second dinosaur, *Hylaeosaurus,* was entirely Gideon's discovery. Its assignment to the dinosaurs depended primarily on a supposed sacrum; dental features placed *Hylaeosaurus* closer to the lizards and farther from the crocodiles, but other evidence suggested the reverse. Unlike *Megalosaurus, Hylaeosaurus* was known almost exclusively from a single, well-articulated specimen including ten anterior vertebrae, a small part of the skull, the two coracoids, various ribs and vertebrae, and remarkable plates or spines that had possibly ranged (as Mantell had said) down the center of its back. The most striking feature of hylaeosaurian vertebrae, for Owen, was their unusual development of the neural arch. Its ribs, as with other of Owen's dinosaurs, were bifurcated at the spinal end. In his memoir on Tilgate saurians published the same year (1841), Gideon had argued that this feature and the expanded arch of *Hylaeosaurus'* ribs were probably related to the enormous development of its dermal spines. But this, Owen retorted, was precisely where *Hylaeosaurus* differed most from presently existing spined lizards and most closely resembled *Megalosaurus* and *Iguanodon.*

Though Owen had assigned *Hylaeosaurus* to the dinosaurs on the basis of its sacrum, he did not have a reliable specimen. A partial sacrum in the Mantellian collection, evidently belonging to "a small or young" dinosaur, could not be assigned either to *Megalosaurus* or *Iguanodon;* Owen therefore bestowed it upon *Hylaeosaurus* and proceeded to a detailed analysis. He also analyzed a further specimen of Gideon's found at Cuckfield in 1837 – a section of tail almost six feet long, consisting of about twenty-six vertebrae – and was allowed to analyze a small portion of bony scute microscopically. Further discussion of Mantell's ingenious identification of the striking dermal spines followed, and while Owen did not outrightly contradict him, he suggested rather plainly that they were more probably abdominal ribs that had been displaced from their natural position during the body's decomposition. Yet another of Gideon's specimens, a nearly complete scapula, proved that the original *Hylaeosaurus,* on which Mantell had based his announcement of 5 December 1832, was not fully grown. Though Gideon had listed some possible *Hylaeosaurus* teeth in his 1841 paper, Owen quoted only from Mantell's earlier description of the same teeth in *South-East England,* where they had been otherwise identified, without revealing his subsequent self-correction. Finally, Owen reassessed one more specimen from the Mantellian collection, a jawbone Gideon had identified as iguanodontian. While Owen had accepted that designation in his *Odontography,* he now withdrew it in favor of *Hylaeosaurus.* This section of Owen's *Report* should be compared in detail with Gideon's paper of 1841.

Owen's third and final dinosaur was *Iguanodon Mantelli,* an enormous reptile admittedly discovered by Gideon (but with exaggerated "aid" from

Cuvier and Clift), named by Conybeare (the naming was actually a collaboration), and mistakenly placed by Owen – who was ignorant of stratigraphy – within the Oolitic formation (120). Among several osteological peculiarities (which proved it unquestionably terrestrial), *Iguanodon* was particularly distinguished by its teeth. They resembled those of the iguana in shape but differed from the teeth of every other known reptile and proved *Iguanodon* herbivorous. Owen then quoted Mantell's narration of their discovery, including Cuvier's remarks, from his famous paper of 1825, and summarized his own analysis of the teeth in *Odontography,* which had emphasized their dissimilarity, in certain respects, from those of the iguana. Though Gideon had identified two fragmentary specimens in his collection as parts of an iguanodon's jaw, Owen did not specifically accept either, having already assigned one more probably to *Hylaeosaurus.* Mantell had also discovered in Tilgate strata (and figured in *South-East England*) a very remarkable specimen, identifying it – with assistance from Thomas Hodgkin – as the tympanic bone of an iguanodon. Owen accepted that, and added nothing to Gideon's remarks, which he quoted.

His analysis of iguanodon vertebrae was necessarily more independent because Mantell (Owen declared) had confused valid examples with those of *Cetiosaurus.* A specimen identified in Gideon's collection, and in his memoir of 1841, as the vertebra of a young iguanodon, moreover, was actually crocodilian, probably of unknown type. These unfortunate mistakes, introduced into Mantell's discussions of iguanodon vertebrae in his *Illustrations, South-East England,* and memoir of 1841, were then repeated by Hermann von Meyer in good faith; there being no reliable previous account to which readers could be referred, Owen necessarily proceeded to a lengthy discussion of his own, based largely on the Maidstone Iguanodon and other Mantellian specimens. Regarding further bones, however, usually identifiable beyond question, Owen routinely agreed with Gideon, citing his specimens and publications regularly. Other significant examples had been collected by G. B. Holmes of Horsham and W. D. Saull. In an important summary, Owen stressed that "the hind-legs at least, and probably also the fore-legs, [of *Iguanodon*] were longer and stronger in proportion to the trunk than in any existing saurian" (142). Exaggerated resemblances of the iguanodon to the iguana, however, had misled Mantell (not named, but cited), who had grievously overestimated the size of *Iguanodon;* for Owen, it was not one hundred feet long but a more believable twenty-eight.[16]

16. RO, "Report on British Fossil Reptiles," Part II, *Report of the Eleventh Meeting [1841]* of the British Association for the Advancement of Science (London, 1842), pp. 60–204: turtles, pp. 160–180; crocodiles, pp. 65–102; dinosaurs, pp. 102–144 – with GM prominent throughout. RO's creation of the *Dinosauria* grouping was announced only in the published version of his paper (i.e., in 1842, not 1841). See H.S. Torrens, "When Did the

Even at this time, no one really *knew* what dinosaurs looked like. There was fairly substantial agreement concerning *Megalosaurus,* though Buckland's original amphibian model had been discarded. Whatever its lapses in detail, Gideon's conception of *Hylaeosaurus* (as an armored dinosaur) was both more daring and more accurate than Owen's, which attempted to reduce dorsal spines to misplaced ribs. Except in matters of size, where Mantell's imagination had risen to heroic proportions, the same remained true of *Iguanodon.* Though Owen was an exacting and latter-day comparative anatomist, privileged to correct the errors of the pioneers, he saw no reason to challenge Gideon's original vision of *Iguanodon,* ignored the vexing problem of its "horn," and plausibly but quite erroneously attributed to *Iguanodon* (without evidence) some pondrous forelimbs that would distort reconstructions of it for a generation. Mantell, on the other hand, had preceded Owen and surpassed him in describing a more nearly bipedal *Iguanodon* similar to that which later evidence confirmed. But he had done so in a paper that was "in all probability, the last contribution" to vertebrate paleontology Gideon would ever make. Isolated and sick, he felt in no position to confront his brilliant rival with inadequate evidence. For most persons in Victorian England, therefore, Owen's subsequent misidentification of certain fossil foot-tracks found in western Massachusetts as having been made by gigantic prehistoric *birds* settled the issue of possible reptilian bipedalism. As a result, conceptions of the dinosaur remained wholly quadrupedal and elephantine until bipedality was rediscovered in the latter nineteenth century, largely through a new find of *Iguanodon.*[17]

Dinosaur Get Its Name?" *New Scientist,* 134 (1992), 40–44. The Greek *deino* is more usually translated as "terrible" or "monstrous" rather than "fearfully great," as Owen had it. In any case, the name "*dinosaur*" would appear to exclude docile or smaller reptiles, irrespective of their anatomical affinities. See also *Lit Gaz* (14 Aug 1841), pp. 513–519 (a long summary of RO's Part II, with no mention of dinosaurs); GM's reply (28 Aug 1841, pp. 556–557); RO–GM, 30 Sept 1841 (ATL); GM–BS, 11 Sept 1841, 30 Apr, 7 June, 4 Aug, 18 Oct 1842 (Yale). *Petrif* (1851), GM's fullest response to RO's paper, is discussed below. In "A Description of . . . the Cetiosaurus," *Proc GS,* 3 (1841), 457–462, RO confidently interpreted that animal as an aquatic carnivorous crocodile; we know it to be a terrestrial herbivorous dinosaur.

17. I discuss the bird tracks controversy below. Some apparently bipedal iguanodon tracks were found in southeast England by Samuel Beckles during the 1850s. Joseph Leidy, in America, described his newly discovered *Hadrosaurus* as probably bipedal in 1858, and later conjectures by Thomas Henry Huxley supported him, but decisive evidence was not at hand before 1878, when the Bernissart iguanodons were discovered. See Edgard Casier, *Les Iguanodons de Bernissart* (2nd ed., Brussels, 1978); and David Bruce Norman, *On the Ornithischian Dinosaur Iguanodon Bernissartensis* (Brussels, 1980). Less technical are William Edmonds, *The Iguanodon Mystery* (Harmondsworth, 1979); and Pat Shipman, "How a 125 Million-Year-Old Dinosaur [*Iguanodon*] Evolved in 160 Years," *Discover* (Oct 1986), 94–102.

While impressive in its way, Owen's famous new suborder of saurian reptiles was a mistake from the beginning, primarily because his "Dinosauria" amalgamated two distinct classes of saurians having substantially different features. The osteological definitions on which Owen based his concept have now been entirely discredited. Though Owen inconsistently allowed for the possibility of "small" dinosaurs, size was also a criterion for him – and we know it to be irrelevant. Owen realized that his saurians differed from lizards in the structure of their limbs but failed to suggest (as Mantell had) the bipedality since found characteristic of many dinosaurs. He substantially misunderstood the real appearance of both *Hylaeosaurus* and *Iguanodon*. More damagingly, he failed to realize that *Cetiosaurus, Streptospondylus, Thecodontosaurus,* and *Poekilopleuron* (all of which he discussed) were as fully dinosaurs as any of the three genera he chose. Two others – *Cladeiodon* and *Paleosaurus* – were known only from teeth, but Owen was expert on teeth. He overlooked *Macrodontophion* (1837) and *Plateosaurus* (1834; a tooth only) entirely. Thus, of the eleven possible dinosaurs available to him, Owen included three, misclassified six, and omitted two.

In discussing *Megalosaurus, Hylaeosaurus,* and especially *Iguanodon,* Owen combined ostensible good will and professional courtesy with conspicuous ungenerosity and snide distortions, particularly where Gideon was concerned. Like other brilliant men, he had swelled too much to see the laborious foundations on which he stood. Though he refined and confused much about dinosaurs in his *Report,* Owen discovered nothing. *Megalosaurus,* for example, was the combined discovery of Cuvier, Buckland, and Mantell, while *Iguanodon* and *Hylaeosaurus* were Mantell's alone. That an Age of Reptiles had once existed, moreover, was first glimpsed by Cuvier and then significantly elaborated, through ecological analysis foreign to Owen at this time, by Gideon, who had also publicized his significant concepts of *terrestrial, herbivorous,* and at least partially *bipedal* saurians to professional and lay audiences alike. In all justice, therefore, Gideon Mantell was both the primary discoverer and designer of dinosaurs.[18]

18. In a survey of the animal kingdom published in 1829, Cuvier had noted that three recently discovered prehistoric saurians (*Megalosaurus, Iguanodon,* and *Geosaurus*) were insufficiently well known to classify. But Hermann von Meyer's *Palaeologica* of 1832 paired *Megalosaurus* and *Iguanodon* as "Saurians with Limbs Similar to Those of the Heavy Land Mammalia." When this important suggestion had been translated by G. F. Richardson, forwarded by Mantell, and published by Edward Charlesworth (Chap. 8, note 8 above), the latter – or Gideon speaking through him – added *Hylaeosaurus,* a later discovery. By 1837, then, the three major saurians had already been grouped, clearly anticipating Owen.

In his own more elaborate but erroneous grouping (1842, pp. 189–191), Owen classified *Poekilopleuron* (misspelled), *Streptospondylus,* and *Cetiosaurus* as crocodiles; *Megalosaurus, Hylaeosaurus,* and *Iguanodon* as dinosaurs; and *Thecodontosaurus, Paleosaurus,* and *Cladeiodon* (misspelled) as lizards. The latter's nature is still problematical but all the others are definitely dinosaurs. He also reclassified Gideon's "very young Iguanodon" (possibly *Hypsilophodon*) as a crocodile. Compare Edwin H. Colbert, *Men and Dinosaurs* (New York, 1968), pp. 31–32; and David B. Weishampel, Peter Dodson, and Halszka Osmólska, eds., *The Dinosauria* (Berkeley, 1992), *passim.*

Today (following Thomas Henry Huxley), many experts regard limb structure as perhaps the most significant of all dinosaurian characteristics. Though Owen (pp. 103, 108, 200) accepted Von Meyer's point, and also speculated about heart structure (204n), his formal definition of the dinosaur on pages 102–103 was solely osteological and thoroughly erroneous. According to a leading authority, all of Owen's proposed distinctions, including his criterion of size, "must now be regarded as unsatisfactory" (Alan Charig, "Problems in Dinosaur Phylogeny," *Geobios, 6* [1982], 113–126; p. 119). So far as individual dinosaurs are concerned, Owen corrected Gideon on size but otherwise delayed and obscured the latter's superior conjectures regarding anatomical details. For an alternative point of view, see Adrian J. Desmond, "Designing the Dinosaur: Richard Owen's Response to Robert Edmond Grant," *Isis, 70* (1979), 224–234.

With Owen out of the picture, the two remaining figures of importance are Cuvier and Mantell. In 1800, 1808, and 1812 Cuvier described saurian bones found by others as crocodilian. In 1812 he briefly proclaimed what we now call the Mesozoic an age of reptiles. On visiting England in 1818 he decisively identified megalosaurian remains from Stonesfield as reptilian rather than mammalian. By 1821 he had mentally reconstructed from such bones a generalized gigantic lizard, of which subsequent descriptions by Parkinson and Buckland were only a reflection. Cuvier unquestionably invented the first dinosaur. Except as a reporter of British and others' accomplishments, however, he had no further role in dinosaurian discoveries.

Under Cuvier's more general influence, Mantell had independently begun to collect dinosaurian remains from Cuckfield by 1817. Writing to WDC in 1821 (a footnote to the latter's *PT* paper), he was first to describe *Megalosaurus* – or any other dinosaur – in print. *Foss SD* (1822) then included an unprecedented assemblage of megalosaurian remains and a fuller description (p. 54) clearly derived from them. It also offered some preliminary indications of *Iguanodon,* on which Mantell elaborated in 1824 (in Horsfield) and 1825. As I have already remarked, *Illus* (1827) then became the first book in which dinosaurian remains were foremost. Unlike Cuvier, Gideon persisted in his dinosaurian researches, finding additional dinosaurs and adding parts to (while refining his conceptions of) those already discovered. Though Cuvier had been the inventor, Gideon Mantell was certainly the first major hunter, discoverer, and designer of dinosaurs.

10

Medals of Creation

Despite inclement weather, Gideon Mantell gladly drove to Craven Cottage, Fulham, on 10 October 1841 to dine with the novelist Sir Edward Bulwer-Lytton, a distinguished correspondent he had never met. Charmingly situated on the north bank of the Thames between Chelsea and Hammersmith, Craven Cottage featured an imaginative Egyptian Hall with a dining room on one side, a drawing room on the other, and a library study at the rear, all of them richly furnished in admirable taste. The elegant Sir Edward received Mantell cordially, sitting next to him at a stag dinner for five. They conversed that evening on a great variety of topics, including animal magnetism, geology, Shakespeare's plagiarisms, the microscope, the telescope, literature, drama, and still other subjects he could not remember. Gideon then left at ten, much gratified with his visit to an author whose writings had so often delighted him during long hours of sorrow and sickness.[1]

The next day Mantell was en route to a patient in his carriage when his coachman allowed the reins to become entangled. Impulsively leaping out of the still-moving vehicle to extricate his horse, Gideon fell violently, the wheels just grazing his head. Not feeling seriously injured, he continued with his duties, attending a case of brain concussion that could easily have been his own, and afterward walked home, on an intensely cold night. Pains already in his lower back intensified, and his left foot numbed, with paralysis of his legs and pelvis following. Mantell remained at home for the rest of the month, continuing to experience numbness and immobility. In succeeding weeks the paralysis came and went, sometimes extending to his right arm. Though feeling thoroughly incapacitated, he forced himself to deliver a promised lecture (on corals and animalcules) at the Parochial School, Clapham Common, on 7 December. "A numerous audience," Gideon noted; "passed off very well, but almost killed me." His

---

1. Edward Bulwer-Lytton (1803–1873) published volumes of poetry and a long series of popular novels, including *The Last Days of Pompeii* (1834) and *Zanoni* (1842). He inherited Knebworth in 1843. J (10 Oct 1841, 23 Feb, 4 June 1842).

paralysis had lessened but was still distressing. On the twenty-first he addressed a second audience of about 260 persons at the Parochial School. When this equally successful effort left him "almost dead" from pain and fatigue, he resolved never to lecture again.[2]

Mantell's illness preoccupied and confined him for the next several weeks, effectively restricting his activities to correspondence, brief visits, a few meetings, and the lectures of others. By 12 March 1842, however, he felt well enough to enjoy a soiree given by Murchison, now president of the Geological Society, in Belgrave Square and managed to lecture himself (on corals) four days later at the London Philosophical Institution, Finsbury Circus. "My health is much better," he subsequently advised Silliman, "but I am still quite an invalid, the affection of the spine and its consequent symptoms remaining very much the same; I cannot stoop or use any exertion without producing loss of sensation and power in my limbs." Though physical debility curtailed his interactions with the scientific world, reactions to his recent lectures had been extremely gratifying. On 29 March the local barrister Henry Hopley White presented Gideon with a valuable Ross microscope worth ninety guineas as a testimony of respect from himself and more than a hundred other admirers in Clapham.[3]

Mantell continued to restrict his normal activities throughout April and most of May, kept often to his bed with severe neuralgia and some paralysis; he attempted to treat both with increasingly stronger doses of laudanum and prussic acid. In May, however, Gideon was roused to attend a series of lectures on the nervous system – a topic of his own – by Richard Owen at the Royal College of Surgeons. He also managed some unstrenuous socializing with close friends like the Allnutts of Clapham and a short rail excursion to Brighton and Lewes. By late May Mantell was visiting patients again, and after hearing himself praised in Owen's final lecture on the twenty-eighth, exhibited his microscope that evening at a soiree given by Lord Northampton, as president of the Royal Society, for members of his Council. Prince Albert arrived soon after nine; taken

2. The accident. J (11, 27–31 Oct, 1, 5, 6, 9, 15 Nov 1841). GM–BS, 7 Nov 1841 (Yale). J (7, 9, 15, 21, 31 Dec 1841, 25 Jan 1842). Lecture prospectuses, 7 and 21 Dec 1841 (ATL). See also Joseph Burkart (of Bonn)–GM, 4 May 1844 (ATL).

3. Ross microscope. The idea of giving it to Mantell (as a belated birthday gift) had originated with Rosina Zornlin, author of *Recreations in Geology* (1841). She, her sister Georgiana, the John Allnutt family, Miss Foster, and other friends then spent the evening with Gideon as White showed all of them how to use the splendidly sophisticated instrument. J (3, 27 Feb, 6, 12, 16, 29, 30 Mar 1842). GM–BS, 30 Apr 1842 (Yale). "Testimonial to Dr. Mantell," incl. inscription; GM–Rev. Dealtry, 4 Feb 1842; and a list of the 110 subscribers (ATL). R. Zornlin–GM, 10 Feb 1842 (ATL); *The Lancet*, 2 (1842), 32; *AJSA*, 43 (Oct 1842), 206–207.

immediately to Gideon, he spent more than half an hour viewing animal-cules through the Ross microscope and, in his usual, courteously ingrati-ating manner, made a number of pertinent observations.[4]

New Topics

Two days before Christmas, Mantell lectured at a private soiree of John Allnutt's on three geological topics currently of interest to him. The first was the so-called Temple of Serapis at Puzzuoli, near Naples, whose columns had been pierced by boring marine mollusks within historical times. This could only mean that the entire site (adjoining the sea) had been first de-pressed and then uplifted a number of feet in a geologically short time. Gideon, who had earlier been delighted to receive a fine specimen taken from one of the perforated columns by Sir Woodbine Parish, and later a model of the ruin as a whole, was now presented with a sketch of it made on the spot by Mrs. Allnutt. Lyell, it will be remembered, had made the Temple of Serapis his frontispiece, emblematic of his reliance on geologi-cal causes now in operation to explain phenomena of all ages.

Mantell's second topic, fossil infusoria, combined his interests in pale-ontology and microscopics. Stimulated by the researches of Christian Ehrenberg, Joseph Jackson Lister, and Andrew Pritchard, Gideon, his son Reginald, his daughter Ellen Maria, and his assistant Hamlin Lee had all been pursuing microscopical researches for some time. What further knowl-edge he now displayed was largely attributable to his own finds, his new microscope, and a swiftly developing friendship with the Reverend Joseph B. Reade, president of the Microscopical Society, who had visited the Man-tells that afternoon. Gideon's remarks at the Allnutts' soiree preceded an exhibition of microscopic specimens.[5]

His third and most complex topic was the supposed fossil footprints of birds. Though long known locally, these curious tracks from the Con-

---

4. J (Apr–May 1842, *passim*); "Confined to my bed – very ill with neuralgia of the heart. Took seventy-five drops of laudanum and twelve drops of prussic acid before any relief was obtained" [4 May]. J (28 May, 2–4 June, 8–15 Aug 1842). In June he paid a first visit to Joshua at Ticehurst and in August attempted an ill-advised excursion to Clifton and Wales (notebook, ATL). J (Aug–Dec, *passim*); GM–BS, 18 Oct 1842 (Yale).
5. J (23 Dec 1842). Serapis: J (23 Mar, 1 July 1842); CL, *Principles,* frontispiece (all edi-tions). Eleanora Allnutt's drawing (as copied by EMM) is at ATL. Fossil infusoria: J (27 Aug 1833, 26 June 1837, 28 May, 2 June 1842); GM–BS, 27 Nov 1840, 29 Mar, 4 Aug, 11, 29 Sept 1841, 4 Aug 1842 (Yale). GM–J. B. Reade, 20 Nov 1841 (Alan G. Nichol-son). GM, *Medals,* Chap. 7; *Animalcules* (1846); and several papers. ATL preserves his copies of microscopical books and papers by others; also slides, pill boxes, and vials filled with fossiliferous earths.

necticut River Valley of Massachusetts escaped scientific attention until
1835, when Dr. James Deane of Greenfield saw them and notified the state
geologist, Edward Hitchcock, at nearby Amherst College. The latter responded immediately, collected numerous examples, and then concluded
in a hasty publication of January 1836 that the puzzling trifid impressions
had undoubtedly been made by the feet of giant prehistoric birds. Hitchcock's theory became well known in England the same year when publicized by William Buckland at the Bristol meeting of the British Association for the Advancement of Science – where he grotesquely imitated the
ungainliness of a gigantic hen getting her feet stuck in the mud – and more
seriously in his Bridgewater Treatise (1836).

Gideon alluded briefly to Hitchcock's bird prints in his fifth geological
lecture of 1837 and in *The Wonders of Geology* (1838). "These impressions are supposed by Professor Hitchcock to have been made by the foot-
steps of birds," he wrote cautiously in the latter. "If this opinion be correct, it is clear that the feathered tribes of that ancient epoch were the
iguanodons of their race, for the dimensions of one kind of foot-mark are
fifteen inches long . . . twice as large as those of the ostrich!" (424). But
since Hitchcock had no skeletal evidence whatsoever, Mantell thought
that any identification of the prints should remain extremely tentative until bones were discovered in association with them. In his oral report on
British fossil reptiles (1841), Owen adopted a very similar position. Hitchcock, however, strongly reasserted his avian theory in a *Final Report on
the Geology of Massachusetts* (1841) and was firmly supported by Lyell,
then on an American visit.

By November 1841 Mantell had received a copy of Hitchcock's *Final
Report* from its author; illness prevented him from responding properly,
with a letter and a box of fossils, until March 1842. Hitchcock then showed
his reply to Silliman and Lyell at a meeting of the American Geological Association. Because Mantell had expressed significant reservations about
Hitchcock's supposed birds, Benjamin Silliman, Jr., replied to him in May:
"I fancy you would be convinced of the truth of Prof. Hitchcock's bird
tracks if you could see slabs containing six or eight continuous steps," this
being the evidence that had persuaded Lyell. Gideon, however, remained
skeptical and stood his ground. "Notwithstanding what your son assures
me," he responded coolly in June to the elder Silliman, "and Mr. Lyell's
opinion, I still think Prof. Hitchcock's bird tracks will be found to be reptilian." (They are now attributed to very early bipedal dinosaurs.)[6]

6. "Bird" footprints. DRD, "Hitchcock's Dinosaur Tracks," *American Quarterly,* 21 (1969),
   639–644, covers the early history (and its literary influence). Edward Hitchcock (1793–
   1864; *DAB*) published several footprints papers in *AJSA,* beginning in Jan 1836 (29:
   307–340). Gideon received his *Final Report* (2 vols.; Amherst and Northampton, 1841) on

Figure 10.1. *Top:* Hitchcock's "Bird" Tracks (from Hitchcock's *Final Report* [1841], Plates 39 and 49). *Bottom:* the trackmaker, as restored by Professor O. C. Marsh, after 1893, when the first bones were found (from George P. Merrill, *The First One Hundred Years of American Geology* [1924], p. 56).

The Moa

In the meantime, another line of argument was developing. On 28 February 1837 in Australia – half the world away – a New Zealand trader named John Williams Harris had sent Dr. John Rule, his nephew by marriage, a curious bone (later identified as the shaft of a femur with both ends broken off) which, according to Maori interpretation, was that of a gigantic bird. Two years later, Rule took the bone with him to England and the Hunterian Museum, where Owen reluctantly agreed to identify it. He first thought the unfossilized bone that of an ox, but comparisons with a series of mammalian skeletons soon proved him wrong. Abandoning such unfruitful analogs, Owen next compared Rule's specimen with the femur of an ostrich and was surprised to find striking similarities. "In short," he later wrote, "stimulated to more minute and extended examinations, I arrived at the conviction that the specimen had come from a bird." On 12 November, Owen courageously read before the Zoological Society a paper in which he announced, on the basis of Rule's bone only, that "there has existed, if there does not now exist, in New Zealand a struthious [ostrich-like] bird nearly, if not quite, equal in size to the ostrich, belonging to a heavier and more sluggish species" (171). Gideon later called Owen's pronouncement the most brilliant example he knew of "successful philosophical induction" in comparative anatomy. But the Hunterian Museum refused to buy Rule's specimen, and though the Zoological Society agreed to publish Owen's paper, it noted alongside that responsibility for conclusions rested exclusively with the author.

Hoping to learn of further bones, Owen commissioned one hundred extra copies of his paper and, with the assistance of Colonel William Wakefield – a colonizer who had arrived in Port Nicholson only a few weeks before – got them distributed throughout the British settlements in New Zealand. (The *New Zealand Journal* of 1840 also abstracted Owen's

16 Nov 1841; within it, see II, 520–521, 766 for GM; they met in London on 15 June 1850. WB, *G&M*, II, 39–42. Owen, 1842, p. 203. CL, *Travels in North America* (2 vols.; London, 1845), chaps. 12, 13, 23; "Lectures on Geology . . . in the City of New-York," reported by H. J. Raymond (2nd ed., New York, 1843), Lecture 6.

J (16 Nov 1841, 20 Mar, 3, 4 Nov 1842). BS Jr.–GM, 14 May 1842; GM–BS, 7 June, 30 Dec 1842 (Yale). Dr. James Deane (1801–1858; *DAB*) published footprints papers in *AJSA* from 1843 to 1848 (and others elsewhere). [BS, ed.], "Ornithicnites of the Connecticut River Sandstones and the Dinornis of New Zealand," *AJSA*, 45 (Oct 1843), 177–188, incl. J. Deane–GM, 20 Sept 1842; GM–J. Deane, 13 Feb 1843; RO–BS, 16 Mar 1843; and RIM–GS, 17 Feb 1843. GM–RIM, 20 Feb 1843 (APS). Through Gideon's efforts, the British Museum acquired three large slabs of Deane's "bird" footprints: J (30 Oct, 16 Nov 1843, 21 Jan, 27 Feb, 18 May 1844); J. Forshall–GM, 18 Mar 1844 (ATL); *Medals*, II, 808–816.

Figure 10.2. *Dinornis Maximus,*
as reconstructed by Owen (1854).

paper.) This strategem succeeded, for on 28 February 1842 the Reverend William Williams, of Poverty Bay, New Zealand, wrote Buckland at Oxford, telling him that Maori searchers, who believed the giant birds still existed, had found a large number of moa bones. Williams shipped forty-seven of the latter to Buckland, and they were soon forwarded to London. The first box arrived at Owen's on 19 January 1843; only five days later, he displayed some of these bones before the Zoological Society as the remains of a gigantic struthious bird, *Megalornis Novae-Zealandiae,* and created a sensation, having thus vindicated his daring prediction of November 1839 brilliantly. On 14 February (before his paper was published) Owen changed "Megalornis" to "Dinornis." When Williams' second box arrived, Owen further emended his paper to designate four additional species of *Dinornis,* which was the largest of all known birds, though probably extinct. He also alluded briefly to the supposed bird prints in America, but only to point out that *Dinornis* was larger and surely unidentical.[7]

7. The moa. A good brief history is A.G. Bagnall and G. C. Pietersen, *William Colenso* (Wellington, N.Z., 1948), pp. 464–467. See also T. Lindsay Buick, *The Mystery of the Moa* (New Plymouth, N.Z., 1911); Buick, *The Discovery of Dinornis* (New Plymouth, 1936); and J. R. H. Andrews, *The Southern Ark: Zoological Discovery in New Zealand, 1769–1900* (Honolulu, 1986), chaps. 7–9. Herries Beattie, "The Moa: When Did It Be-

By 4 January 1843, meanwhile, Mantell had received a box of actual footprint specimens from Dr. Deane, together with a long letter that reviewed the history of their discovery, emphasized their avian characteristics, and affirmed the "absolute certainty" of their bird origin, as established by comparative anatomy. Gideon immediately presented slabs of the footprints, and Deane's valuable letter, to the Geological Society. At the same meeting, Lyell communicated a similar letter from William C. Redfield of the United States, who had discovered additional footprints in the New Red Sandstone of New Jersey. When Owen announced his *Megalornis* three weeks later, the avian nature of these huge fossil footprints from America easily became more plausible. Knowing that his own opinion would now be of great interest to Silliman's readers, Gideon sent Deane a belated reply (via New Haven) in mid-February:

My dear Sir,

I have deferred replying to your highly interesting communication, and acknowledging your kindness, until an opportunity occurred of submitting the specimens of ornithoidicnites [fossil bird footprints] to the examination of the Geological Society of London. At the last meeting I placed the specimens before the Society and read the letter with which you had favored me, and afterwards gave a *vive voce* description of the fossils, illustrating my remarks by drawings showing the position and relative distances of the footprints when in situ. A brief notice of footprints by Mr. Redfield was read on the same evening, and Mr. Lyell, who communicated it, gave a graphic account of the appearance of the impressions of feet seen by him in various localities of the United States in company with Professor Hitchcock.

come Extinct?" (Otago, N.Z., n.d. [ca. 1950]) pursues a related question. A. Hamilton's very useful bibliography of nineteenth-century scientific papers on the moa appeared in *Trans NZ Institute,* 16 (1893), 229–257; several of the most important were GM's. RO collected his own numerous papers (technical anatomy, but with valuable historical information) as *Memoirs on the Extinct Wingless Birds of New Zealand* (2 vols.: London, 1879); preface, p. iii quoted; 73–76 (Rev. Williams–WB, 28 Feb 1842). RO, "On the Fragment of the Femur of an Unknown Struthious Bird from New Zealand," *Proc ZS,* 7 (1839), 169–171; *Trans ZS,* 3 (1849), 29–32. GM, "On the Fossil Remains of Birds," *QJGS,* 4 (1848), 225–238. Owen, *Life,* I, 144–151, 207–212; RO, "On Dinornis Novae-Zealandiae," *Proc ZS,* 11 (1843), 8–10, 144–146; RO–ZS, 14 Feb 1843, changing *Megalornis* to *Dinornis, ibid.,* p. 19.

In future years RO would be particularly dependent on specimens discovered by WBDM in NZ, sent to England by him, and then placed at Owen's disposal by Gideon. Walter's sister Ellen attended Owen's lecture of 2 Feb 1844 on the moa, was unexpectedly delighted, and sent a glowing report of it to her brother, who had already resolved to find a living example of the huge bird and make his fortune by exhibiting it. There being no living examples, he sent important collections of bones instead. RO acknowledged WBDM several times in *Memoirs.*

Mr. Owen (of the College of Surgeons) was not present, but the president, Mr. Murchison, read a short note from that gentleman expressing his doubts as to footprints alone being sufficient evidence to prove whether the animals which made them were birds or reptiles. Mr. Murchison was also sceptical as to these markings having undoubted claims to be considered as true footprints of birds, but Mr. Lyell stated his conviction that they were genuine ornithoidicnites. The enormous magnitude of the largest imprints seemed to present the greatest objection to some of the fellows, but this difficulty is removed by the recent discovery in the modern alluvial strata of New Zealand of some bones of a struthioid bird with trifid feet equal in size to the most colossal of the fossil footprints hitherto observed in your country. This New Zealand bird is stated by Mr. Owen (who has described the bones already arrived in England, in the *Zoological Transactions*) to belong to a new genus allied to the ostrich and emu.

Maori tradition, Gideon added, held that such birds were alive not more than 150 or 200 years ago. They must therefore have been annihilated by human agency.

Among others, Murchison had been convinced by the new evidence from New Zealand and at the Anniversary Dinner of the Geological Society on 17 February retracted his previous skepticism regarding the avian footprints. It was Mantell, Murchison stressed, who had brought Dr. Deane's original discovery of the footprints before the Geological Society. He also cited Mantell's discovery of bird bones in the Wealden strata. Owen, one of the few major scientific figures in London who still opposed the American triumph, then capitulated to Silliman and his compatriots in a gracious letter of 16 March, after having seen some further specimens of the footprints exhibited by Gideon at a Royal Society soiree, and being convinced primarily by his own birds from New Zealand.

Though Lyell, Buckland, Murchison, and now Owen were all in accord with Hitchcock, Mantell remained dubious. Despite further wheedling by Silliman, Deane, and Hitchcock, he was extremely reluctant to include hypothetical American birds among his various wonders of geology. In 1844, it is true, Gideon tentatively accepted what was by then almost unanimous scientific opinion. Four years later, however, he drew back from any such affirmation, and his final statement, in 1851, reemphasized the insufficiency of present evidence, despite overwhelming professional support by others for Hitchcock's erroneous surmise.[8]

8. GM, "Notice on a Suite of Specimens of Ornithoidicnites, or Footprints of Birds on the New Red Sandstone of Connecticut," *Proc GS*, 4 (1843), 22–23; CL, "Extract of a Letter from W. C. Redfield," *Proc GS*, 4 (1843), 23; J (4 Jan 1843). GM–J. Deane (via BS), 12 Feb 1843 (Yale); *AJSA*, 44 (Apr 1843), 417–418, and 45 (Oct 1843), 184–185,

The Medals of Creation

Aside from the continuing footprints controversy and two minor papers, Mantell's geological activities throughout 1843 centered on the composition of another book. Though long contemplated, it was actually undertaken only that January, when Joseph Dinkel began drawing the first of six major plates (four of them colored) and some others for what Gideon was then calling his "Introduction to the Study of Organic Remains." More than a hundred drawings would illustrate his thousand-page text; all but the most demanding were donated by his daughter Ellen Maria, who (having returned to Gideon) labored over them for five months.

By 30 October, Mantell was fully employed with his "Medals of Creation," as it was now called, persevering at his task despite "writhing in severe pain." On 22 December his new publisher, Henry Bohn, came down to Clapham and agreed to publish "the Medals of Creation or First Lessons in Geology and in the Study of Organic Remains," but in two volumes rather than one. The first edition would be of two thousand copies, costing one guinea the set, from which Gideon would receive nothing (except reimbursement for the illustrations).

By 27 December, Mantell had sent Bohn the first section of manuscript, together with some woodblocks, and drawings by Ellen Maria. The first proof sheet arrived on 8 January 1844, and by 7 February the printing had reached page 207. Volume 1, comprising 456 pages, was finished on 4 March. At month's end Richard Clay, his printer, exhausted all the manuscript on hand, so Gideon (hard at work describing fossil fishes) had to "scribble for a fortnight" until the printing could resume. Finally, on 18 May, he completed his preface and the dedication to Lyell. On 27 June, Bohn was to have at least five hundred copies prepared, but in this he failed significantly (as Gideon discovered on the twenty-ninth) and *Medals* did not become generally available until July.[9]

Following a moralistic, theological "Introduction," Mantell began his

---

185–187, 187–188. GM–BS, 7 June, 30 Dec 1842; 12 Feb, 30 Mar, 28 Aug, 26 Dec 1843; BS–GM, 26 Feb, 9 July, 4 Sept 1843; BS–RO, 27 Feb 1843 (all Yale). GM, *Medals*, II, 808–816; *Wonders*, 6th, II, 556–559; *Petrif*, pp. 64–73. See also GM, "Description of Footmarks," *QJGS*, 2 (1846), 38; *Wight*, p. 327; and J (15 Jan, 23 Apr, 27 May, 11 June 1845).

9. J (30 Jan, 26 July, 30 Oct, 29 Nov, 22 Dec 1843). Henry G. Bohn (1796–1884; *DNB*), one of the most successful publishers of his day, sought his primary audience among the middle class. GM received two hundred pounds from him on 21 Sept 1844, as reimbursement for the illustrations in *Medals*. GM–BS, 12, 28 Aug, 30 Oct, 26 Dec 1843 (Yale). J (27 Dec 1843, 8 Jan, 7 Feb, 4, 21 Mar 1844). *Medals*, p. xviii. J (26 Apr, 18 May 1844). GM–BS, 21 May 1844 (Yale). J (27 May, 4, 13, 15, 16, 24, 29 June 1844).

tripartite book with some basic definitions and a chronological survey of British strata, from the most recent downward. The Drift, including what had formerly been called Alluvium and Diluvium, was now seen to have been accumulated by a variety of causes, including land floods, glaciers, and icebergs. (Gideon did not specifically endorse Agassiz's Ice Age theory of 1840.) The Tertiary System comprised Lyell's Pliocene, Miocene, and Eocene – but would require significant modification as new discoveries were made. Among Secondary formations, the Chalk or Cretaceous System had no fewer than seven subdivisions and the Wealden Formation, four. Below the Wealden lay the Oolite Formation, the Lias, the Saliferous or New Red Sandstone, the Carboniferous, and the Devonian. Even older (despite their recent names) were the Palaeozoic formations of the Silurian and Cambrian systems. In general, Gideon observed, this series of formations appeared throughout Britain in chronological order, with Tertiary deposits in the east and southeast receding toward Palaeozoic ones in the northwest.

A chapter on the nature of fossils followed, including some interesting suggestions for preserving them and a reluctant affirmation that the North American footprints were currently regarded as avian. It was then in order to discuss coal and fossil plants. The present flora of New Zealand, in which ferns predominated over grasses, seemed to recall that of the Carboniferous (202).

Part II, beginning on page 209, dealt with Palaeontology, or Fossil Zoology, surveying in turn infusoria, or animalcules; zoophytes, echinoderms, and mollusks; barnacles, annilida, insects, arachnids, and crustaceans; then fishes, reptiles, birds, and mammals.

Two substantial chapters on fossil reptiles reflected and overpraised Owen's *Reports* of 1839 and 1841, even accepting his new (1842) classification, *Dinosauria.* "The elaborate investigation of the fossil remains of these stupendous beings, and the luminous exposition of their organization and physiological relations" in Owen's 1841 *Report,* Mantell admonished, "must be regarded as among the most important contributions to paleontology, and affords a striking example of the successful application of profound anatomical knowledge to the elucidation of the most marvellous epoch in the earth's physical history – the *Age of Reptiles*" (730–731). Much of the discussion in this eighteenth chapter concerned details in Owen's 1841 *Report,* which Gideon too often accepted as more authoritative than his own earlier publications.

Though designed in part for a popular audience, *Medals* was arguably the first modern synthesis of paleontological knowledge in English and certainly the archetype for many others. Its sources, an impressive gathering of the best available information in three languages, had been written by specialists, for whom the Palaeontographical Society (with its admirable monographs) would be founded in 1847. No longer merely a local hobby, the study of fossils had become technical, voluminous, and exacting.[10]

Change with Improvement

Despite continued hunger, economic hardship, and rapid social flux, with the advent of the railway era in England (1825ff.), evidence for technological progress became overwhelming. Naturally enough, this same equation of change with improvement soon appeared elsewhere. The evolution of stars and planets from nebulae, for example, and the progressive transformation of languages and knowledge, were largely taken for granted. Theories of biological progression, or evolution, had been common even in earlier phases of the Industrial Revolution, popularized then by such unorthodox but fascinating writers as Monboddo, Erasmus Darwin, and Lamarck. During the 1840s, however, biological evolution was opposed by two prevalent schools of scientists: the Cuvierians, like Owen and Agassiz, who regarded the history of life on earth as a series of special creations but willingly saw them as progressive; and the more rigid Lyellians, especially Lyell himself, who opposed any progressive view of nature and could account for the undoubtedly changing life of the past only by supposing an unlikely succession of distinct climates. Both schools assumed the relative fixity of species (allowing variation but not transformation) and made special provisions to insure the uniqueness of man.

In May 1827, Gideon Mantell had loaned a copy of Lamarck (1809) to Lyell, who later wrote almost the entire second volume (1832) of his *Principles* in response. Finding Lyell's arguments convincing, Gideon and others then accumulated further scientific evidence, around 1835, to support the concepts of fixity and nonprogression. But as contrary data emerged from a variety of related fields – paleontology, embryology, microscopics,

10. *Medals.* James Parkinson used "Medals of Creation" as a section title in his *Org Rem,* I (1804), Letter two; see also *Medals,* I, 254; and *The Posthumous Works of Robert Hooke* (London, 1705), p. 335. GM–BS, 4 June 1844, is filled with important details; see also GM–BS Jr., 17 June, and BS–GM, 13 Nov 1844 (all Yale). Originating with Murchison and Sedgwick, "Palaeozoic" (together with "Mesozoic" and "Cainozoic") was formally defined by John Phillips in his *Figures and Descriptions of the Palaeozoic Fossils of Cornwall, Devon, and West Somerset* (London, 1841), p. 160. Articles ("Geology," "Organic Remains," "Palaeozoic Series") written by Phillips for George Long, ed., *The Penny Cyclopedia* (27 vols.; London, 1833–1843), an important reference work more significant than it sounds (GM's set, ATL), also influenced *Medals* (see pp. 314–315n).

geography – both concepts came under strain and by the latter 1830s were increasingly debated. Mantell, for example, alluded to these issues in his conclusion to *The Wonders of Geology* (1838), maintaining a Lyellian position. Explicitly contradicting James Parkinson's *Organic Remains* (III, 1811), in which biblical days of Creation and geological epochs were equated – a passage Gideon had transcribed in 1812 – he now argued that the "*apparent* successive development of living beings" (667) was illusory. Species had come and gone in accord with climatic changes, but the creation of man remained a unique event beyond the speculations of philosophy.

As Mantell somewhat realized, however, this position was deeply flawed. In 1838 Lyellian orthodoxy faced two unanswerable objections: First, it presupposed that the same laws of nature applied to both inanimate and animate phenomena but then failed to apply them consistently; second, it ignored prevalent fossil evidence that clearly contradicted nonprogression. Special creation, then, as Gideon did not quite admit, was an externally derived assumption significantly unsupported by available data, yet maintained for its preferable religious and philosophical implications.[11]

In concluding his 1841 *Report* on British fossil reptiles, Richard Owen had also thought a response to the transmutationists necessary. Since no one doubted that reptilian genera had appeared in succession, was this evidence for the transformation of species through development? Though he mentioned several early evolutionists, including Lamarck, Owen's primary target was Dr. Robert Grant, a comparative anatomist associated with University College, who in some lectures of 1835 had supported transmutation. "A slight inspection of the organic relics deposited in the crust of the globe," Grant had argued, "shows that the forms of species, and the whole zoology of our planet, have been constantly changing, and that the organic kingdoms, like the surface they inhabit, have been gradually developed from a simpler state to their present condition." "A *slight* survey of organic remains may indeed appear to support [developmental] views of the origin of animated species," Owen retorted snidely, "but of no stream of science is it more necessary, than of paleontology, to 'drink deep or taste not.'" Those who sought evidence for evolution would never find it among his dinosaurs because "the modifications of osteological

11. For Mantell, Lyell, and J.B. Lamarck (*Philosophie zoologique*, 2 vols., Paris, 1809), see Wilson, pp. 179–182. For pre-Darwinian evolution more generally, see Bentley Glass, Owsei Temkin, and William L. Straus, Jr., eds., *Forerunners of Darwin, 1745–1859* (Baltimore, 1968); Peter J. Bowler, *Fossils and Progress: Paleontology and the Idea of Progressive Evolution in the Nineteenth Century* (New York, 1976); and Bowler, *Evolution: The History of an Idea* (Berkeley, 1989), together with biographies of Owen, Lyell, Agassiz, and of course Darwin. GM–DG, 4 Feb 1828 (ESRO); *Wonders*, II, 663–682.

structure which characterize the extinct reptiles were originally impressed upon them at their creation, and have neither been derived from improvement of a lower, nor lost by progressive development into a higher, type" (202). Comparative anatomy was still for him a servant of natural theology.

By 1844, when several prominent naturalists (including Darwin) were taking developmental theories seriously, Mantell's position – like that of others – reflected a conflict between head and heart. Evidence for some kind of evolution was increasing, but acceptance of it would have entailed the abandonment of deeply cherished beliefs. His vacillation then influenced the "Retrospect" concluding *Medals,* which emphasized how little was actually known about the history of life. Though all the earliest fossils were entirely marine, Gideon nonetheless imagined adjacent continents "teeming with appropriate inhabitants" (873). Despite the recent work of Murchison and Sedgwick, he offered no real conception of biohistory prior to the Mesozoic. From then on, however, life advanced upward through the Age of Reptiles and the Age of Mammals to that of Man. Thus, he concluded, "by slow and almost insensible gradations, we arrive at the present state of animate and inanimate nature" (875). Tentative as these remarks may be, Gideon clearly had strong reservations about special creation in the spring of 1844 and, like others, was wavering toward the developmental theory.[12]

Vestiges

Medals, then, not only achieved a responsible synthesis of available paleontological knowledge, including some original discoveries, but ended with potentially significant conjectures. Innocuously promoted by Henry Bohn, however, it was temporarily thrust aside – almost upon publication – by a similarly titled work that seemed its malignant twin. For this anonymous author, fossils were not evidence of successive special creations, as many geologists believed, but rather *Vestiges of the Natural History of Creation* (1844), proving in aggregate that "the simplest and most primitive type" of life on earth "gave birth to the type next above it, that this again produced the next higher, and so on to the very highest" (222), by which the author transparently intended man. Among his would-be proofs for this most heterodox proposal, the nameless writer theorized

---

12. Owen, 1842, pp. 196–202; quoting Grant, p. 197n (but his citation is wrong); RO's retort, pp. 196–197, my italics. Compare *Lit Gaz* (1841), p. 513. Darwin wrote out the essentials of his theory in a then unpublished essay of 1844. *Medals,* II, 872–876.

that fetuses, in their development, passed through a series of stages recapitulating the evolutionary history of their species.

The author of this daring hypothesis was a Caledonian journalist named Robert Chambers, but only a handful of persons (Lyell among them) knew for sure until 1884, when a posthumous acknowledgment appeared. By then, *Vestiges* had reached a twelfth edition. Meanwhile, its author remained a Great Unknown and the most speculated about anonymous writer in Britain since Walter Scott. Among those suspected of perpetrating *Vestiges* were William Makepeace Thackeray, George Coombe, Sir Richard Vyvyan, Charles Lyell (to his disgust, no doubt), and even Prince Albert, none of whom was more than momentarily probable. Because of its numerous mistakes, however, *Vestiges* seemed clearly the work of a scientific amateur who had misread both his natural history and his Bible.

Like Gideon's *Wonders*, *Vestiges* began with a description of the solar system and the nebular hypothesis of its origin. It discussed the composition of Earth and Moon, then reviewed the eras of terrestrial history in *ascending* order, beginning with that of the Primary rocks, before life existed. The first signs of life appeared in some early strata preceding Murchison's Silurian. The Old Red Sandstone era, with abundant fishes, gave way to a Carboniferous period in which there was dry land, with plants (but no animals) upon it. In the succeeding era of the New Red Sandstone, terrestrial reptiles and birds appeared, the latter evidenced by foot tracks in Massachusetts. The Oolitic followed with further reptiles and fishes, as well as insects, and the introduction of small mammals known only from the Stonesfield slate.

During the Cretaceous a full complement of life existed, but with species very different from present ones. Mantell, for example, had "discovered some bones of birds, apparently waders, in the Wealden" (122–123) and there were at least two other such finds, but these, Chambers admitted, "are less strong traces of the birds than we possess of the reptiles and other tribes" (123). He was vulnerable on this point because thesis and evidence were not entirely in accord. "The birds are below the mammalia in the animal scale," Chambers concluded, "and therefore they may be supposed to have existed about the time of the New Red Sandstone and Oolite, although we find but slight traces of them in those formations, and, it may be said, till a considerably later period" (124). Chambers completed his survey with a discussion of Tertiary mammals, then went on to argue for his various speculations.

In *Medals*, published only a few months sooner, Gideon had found no evidence of life's higher orders prior to his Triassic bird tracks; by the Age of Reptiles, however, mammals had also appeared and the system of animal creation was complete; finally, the Age of Mammals included monkeys, which approximated man. Thus, "by slow and almost insensible gradations" (875) the present state of life had come to be. Chambers advocated nearly identical positions more fully, but from data similar to Mantell's he derived boldly explicit conclusions that impressed the public and outraged reputable scientists.[13]

Because of his immense authority, a key figure in the *Vestiges* controversy was Richard Owen, to whom (among others) Chambers sent a copy of his first edition when it appeared in October 1844. Owen responded to the anonymous author with complimentary eloquence, criticizing only Chambers' notorious advocacy of spontaneous generation and a few other obvious mistakes. As to evolution, he observed: "The gradation of organic beings is for the most part so close and easy that we cannot be surprised at the idea of progressive transmutation of species having been a favorite one with the philosophic mind in all ages." After all, had not Lyell disavowed Lamarck's version of it in every edition of his *Principles of Geology*? And Buckland in 1836? And himself in 1841?

When William Whewell wrote him on 30 January 1845 about *Vestiges*, Owen steadfastly opposed transmutation. "Animals in general," he replied, "cannot be arranged in a series proceeding from less to more perfect in any way . . . ; much less can they be so arranged as that the more perfect in their fetal condition pass through the successive stages of the less perfect." Whewell regarded Chambers' *Vestiges* as a distinct threat to natural theology (he had written one of the Bridgewater Treatises) and hoped to enlist Owen's authority in opposition. Sedgwick and Murchison likewise sought Owen's opinion, similarly urging him to condemn *Vestiges* in public; by then, Murchison had already spoken with J. G. Lockhart, the editor of *Quarterly Review*, about a possible essay by Owen. The comparative anatomist prudently refused, however, and for months afterward scientific orthodoxy was without an effective spokesman. As of April

13. *Vestiges*. Anon. but Robert Chambers, *Vestiges of the Natural History of Creation* (London, Oct 1844; twelfth edition, 1884). Milton Millhauser, *Just Before Darwin: Robert Chambers and "Vestiges"* (Middletown, Conn., 1959); Bowler (1976), chaps. 3 and 4; and James A. Secord, "Introduction" to a reprinting of *Vestiges*, first edition, and *Explanations* (Chicago and London, 1994). As these and other scholars have stressed, *Vestiges* addressed a number of current themes and concerns. Andrew Crosse (1784–1855; *DNB*), for example, whose electrical experiments included not only the creation of crystals but (allegedly) the creation of life, was prominently mentioned (pp. 185–190). His early work had excited Mantell's circle at Brighton. *Medals*, II, 872–876, anticipates *Vestiges* to some extent. Gideon would also have been especially interested in Chambers' opinion of Adrian Leonard (p. 338), whose domino-playing dogs were "remarkable examples of what the animal intelligence may be trained to." Mantell himself had played dominoes with one of them – and lost (J, 31 Aug 1841).

1845, meanwhile, *Vestiges* had entered on its fourth edition and was clearly succeeding with the public.[14]

Gideon Mantell's evaluation of *Vestiges* was desired also, but like Owen he felt initially loathe to speak out. Having put forth some observations similar to Chambers', and very recently, Mantell appreciated the precariousness of his own authority. Sensing that any definitive statement should come from Owen, he remained unwilling in any case to risk his fragile health in unnecessary scientific and religious strife. Perhaps what shocked him most, however, was the realization of how much this new book and his latest one had in common. The history of life now seemed more plausibly the work of nature than of God – for, despite its repeated pieties, *Vestiges* was widely considered to be atheistic. If Gideon had endorsed Chambers' positions in print (as logically he might have done), the same charge would probably have been brought against his *Medals*.

Mantell's displeasure with *Vestiges* emerged fully in his letter to a Liverpool clergyman, written 24 January 1845:

> Sir [it said in part],
>
> The theory of the progressive development of organization in the animal and vegetable kingdoms, although allowable in the earlier years of modern geology, has long been either entirely abandoned or so modified as to be virtually abandoned by the most eminent observers. That this theory was not tenable I long ago imparted from discovering Tertiary species in Secondary strata, and warm-blooded vertebrates in the Wealden and Oolite. For years I have strenuously insisted upon the dangers of those hasty generalizations which assume proof from insufficient, or even negative, evidence.

Even Lyell's subdivision of the Tertiary into Eocene, Miocene, and Pliocene periods, which assumed the successive development of new genera and species of mollusca, would have to be abandoned eventually, Gideon predicted.

The author of *Vestiges*, Mantell concluded, "although a good reader, is not an original observer; neither is he acquainted with the present state of any one of the sciences upon which he presumes to base his most vague and unphilosophical speculations – which in my opinion are obnoxious to religion and philosophy alike." For Gideon, it was humiliating to think

---

14. Owen and *Vestiges*. Owen, *Life*, I, 248–256, incl. RO–R. Chambers, n.d. [1844], quoting p. 250; William Whewell–RO, 30 Jan 1845; RO–WW, 3 Feb 1845, quoting pp. 252–253; RIM–RO, 2 Apr 1845; and AS–RO, May 1845, together with suspected authors of *Vestiges*. See also Adrian Desmond, *Archetypes and Ancestors* (Chicago, 1982), pp. 210–211.

that a work of such flimsy pretensions should have obtained so much consideration, because it proves how little the intelligent public are aware of the real progress of the sciences." Unbeknownst to him, the Liverpool clergyman had also written to several other prominent geologists regarding *Vestiges.* He then published their private opinions (some incidentally supporting Chambers) in the local newspaper and later as a pamphlet. Mantell's letter, therefore, appeared prominently in each.

Embarrassed, but unwilling to let the subject alone, Gideon next wrote privately about *Vestiges* to Silliman:

> A little volume of 390 pages, *anonymous,* called *Vestiges of the Natural History of Creation,* has made a great sensation, chiefly I believe because the author cannot be detected. Two hundred copies were gratuitously delivered to the leading scientific and literary men. It is evidently the work of a very clever man who has read much and speculated more, but who is not an original observer. He embraces all the natural sciences and abounds in the most extraordinary speculations, most of them based on insufficient data or mistaken facts. His object is to prove that creation has proceeded according to a law impressed by the Creator on matter, by which organic forms arise from inorganic atoms – and that "the simplest and most primitive type, under a law to which that of like production is subordinate, gave birth to the type next above it; that this again produced the next higher, and so on to the very highest, the stage of advance being in all cases very small, namely from one species to another." In support of this theory of progressive development, geology and physiology are made to succumb to the views of the author.

But Mantell had no space for more comments. Though *Quarterly Review* had invited him to smash *Vestiges* with a response, he declined, being "too unwell and too anxious for peace to fish in troubled waters." Otherwise, he concluded, "I think the book so false in religion and philosophy that I would gladly have done my best to expose it, for all its errors are swallowed by the upper classes, to whom everything boldly asserted and in captivating style is gospel." Silliman, too, published Gideon's private remarks almost verbatim.

On 1 May, Mantell received a personal copy of *Vestiges,* fourth edition, inscribed to himself with the unknown author's compliments. A number of Chambers' initial mistakes had been corrected in this version, which was therefore less easily dismissed than his first. On the other hand, Chambers had now added a complicated series of wholly imaginative genealogies – in which, for example, pachyderms were descended from manatees and dugongs, and dogs from seals, otters, and polecats. Though the new material was ingenious, it could hardly have fooled a knowledgeable comparative anatomist. Gideon's opinion of *Vestiges*, therefore, changed little.[15]

In June, speaking as president of the British Association for the Advancement of Science, Sir John Herschel criticized *Vestiges* without naming it, but only in general terms. Transmutation, he said, had neither been established nor explained. Though improbable, it would be just as miraculous as special creation if true. Recent geological discoveries about the life of the past were amazing enough, but we must look beyond "a mere speculative law of development" in hopes of understanding them. The "official" response to *Vestiges*, appearing in *Edinburgh Review* the next month, was a lengthy denunciation by Adam Sedgwick, the able but crusty Cambridge professor who had likewise endeavored to dispose of Lyell's *Principles* when its first volume appeared. Relying on such authorities as Owen, Agassiz, and Herschel, Sedgwick easily exposed Chambers' still numerous errors and misjudgments; their chief difference, however, was ultimately a religious one.

Meanwhile, copies of *Vestiges* (and soon after, pirated editions) had begun to appear in America. Benjamin Silliman, surprisingly, had praised an American edition of *Vestiges* (New York, 1845) in the April issue of his journal, calling it "novel and interesting, . . . with many bold conceptions and startling opinions." Though unable to condone all of its furtive author's views, he "strongly recommended" the book. Perhaps because it mentioned the Massachusetts bird tracks so prominently and appealed in general to frontier optimism, *Vestiges* enjoyed an American vogue. Thus, meeting concurrently at Yale, several members of the Association of American Geologists expressed great admiration for it, believing presciently that the book represented a new scientific beginning. Mantell's remarks on *Vestiges*, when Silliman read them aloud from his friend's recent letter, caused a stir; they were then published in the *American Journal of Science*.

By this time, Gideon had received the April issue. "I was sorry to read your commendation, brief though it be, of *Vestiges*," he replied on 10 June.

---

15. GM and *Vestiges*. GM–Rev. Abraham Hume, 24 Jan 1845 (FMC); repr. (abridged) in Hume, "Remarks on the *Vestiges of the Natural History of Creation*," *The Liverpool Journal*, 8 Feb 1845, p. 2; and Hume, "Examination of the Theory Contained in the *Vestiges of Creation*" (pamphlet; Liverpool, 1845; not seen). Related items, both mentioning GM, appeared in *The Liverpool Journal*, 22 Feb, p. 5; and 1 Mar, p. 2. Mantell's reply to the latter (J, 1 Mar) was not published. GM–BS, 28 Feb and 3 Mar 1845 (Yale); *AJSA*, 49 (Oct 1845), 191. *Vestiges of the Natural History of Creation*, 4th ed. (London, Apr 1845; presentation copy to GM, ATL); J (1 May 1845). Gideon declined J. G. Lockhart's invitation to review *Vestiges* in *Quarterly Review* (J, 24 Feb 1845; RIM–GM, 25 Feb 1845 [ATL]). He met Chambers at a dinner party in London on 12 Mar 1845 (J).

"The work is a tissue of errors all through and will soon fall into oblivion – at least I hope so. Not one eminent man in this country defends any part of it." Writing again eleven days later, he transmitted lengthy excerpts from Sir John Herschel's address, which had just been given. Silliman then responded with thanks on 22 July, Herschel's being "in all respects a sound and judicious view." In his own public lectures, Silliman added, he had "always opposed the doctrine of equivocal generation and Lamarck's absurd theory of transmutation, maintaining that every new organized being or pair of beings was the direct result of creative power and that there is no inherent tendency in matter to produce organized forms, much less life and reason." The *American Journal of Science* remained cool to *Vestiges* thereafter, largely owing to Gideon's influence.

In November 1845, as part of a forthcoming book, Mantell explained his most fundamental disagreement with the transmutationists. "All animals and plants," he asserted, "may justly be regarded as definite aggregations of cells, endowed with specific properties in the different types, and subjected to a never-varying law of development." Though cells formed the basis of all animal and vegetable structures, there was not the slightest reason to assume that those from even the simplest animals or plants were identical. "The cell that forms the germ of each species of organism," he was certain, "is endowed with special properties which can result in nothing but the fabrication of that particular species." Throughout his seriously flawed argument, moreover, the unknown author of *Vestiges* had, in too many instances, "assumed *analogy* to be a proof of identity." Believing that transmutation was therefore impossible, Gideon presumably saw no alternative to a succession of special creations, as Lyell, Owen, Sedgwick, Agassiz, Silliman, and others held.[16]

Toward year's end Chambers published a short volume of *Explanations*, "by the author of *Vestiges of the Natural History of Creation*," in which he responded to the charges of his critics – defending the nebular hypothesis, development, and his religious beliefs, for instance – while reviewing the argument of *Vestiges* as a whole. In general, those who read Chambers' *Explanations* thought better of him as a result. "The author of

16. *Vestiges* controversy. Sir John Herschel (1792–1871; *DNB*), as quoted in GM–BS, 21 June 1845 (Yale); also "Presidential Address," BAAS *Report* for 1845 (London, 1846); Adam Sedgwick, "*Vestiges of the Natural History of Creation*," *Edinburgh Review*, 82 (July 1845), 1–85; compare *Westminster Review*, 44 (Fall 1845), 152–203; *AJSA*, 48 (Apr 1845), 395. The sixth annual meeting of the Association of American Geologists took place at Yale, 30 Apr–6 May 1845; GM's remarks came from GM–BS, 28 Feb & 3 Mar 1845 (Yale). GM–BS, 10, 21 June, 6 July 1845; BS–GM, 22 July 1845 (Yale). GM, *Animalcules*, p. 24n. See also GM, "Dr. Buckland's Organic Theory," London *Times*, 26 June 1845, p. 5 (part of a broader antigeological controversy, 23 June–4 July).

Vestiges," James Forbes wrote privately in response, "has shown himself a very apt scholar, and has improved his knowledge and his arguments so much since his first edition that his deformities no longer appear so disgusting." "It was well that he began to write in the fullness of his ignorance and presumption," Forbes continued, "for had he begun now he would have been more dangerous." Gideon's opinion was similarly improved, particularly since *Explanations* manfully addressed points he had criticized. The author had not meant to assert that the developing mammalian fetus was identical with earlier stages in the development of life but only that it was analogous; nor had he meant to suggest that the cells of plants and animals are alike, but only that the embryos of all animals are, in their earliest stages, indistinguishable. These emendations gave him a much stronger case.

Having read *Explanations,* Mantell (and others) now felt that they had "misapprehended (not misrepresented)" Chambers' opinions. In January 1846, therefore, Gideon boldly supported Chambers to some extent, strongly recommending that everyone interested in transmutation "peruse, with serious attention and unprejudiced mind, the *Explanations* of this able anonymous author." Arguments derived from the fossil record on behalf of the development theory, however, "should not, in the present state of our knowledge of the ancient inhabitants of the globe, be regarded as conclusive." Overhasty generalizations on that topic by many eminent geologists and paleontologists had come to grief, he cautioned, *Vestiges* itself being an example. The writer himself could have provided another.[17]

Mantell's major contribution to the substance of *Vestiges,* his supposed discovery of bird bones in the Wealden, had been announced by him in *South Downs* and *Illustrations,* confirmed by Cuvier and Owen, published by Gideon in the Geological Society's *Transactions,* and summarized by several authors, including himself in *Medals.* Now, however, Owen had changed his mind. Thus, on 11 December he and Gideon spent two hours together at the British Museum comparing "bird" bones with various reptilian specimens as Owen attempted to convince Mantell that his supposed birds had more likely been pterodactyls. Six days later Owen read a paper at the Geological Society "On the Supposed Fossil Bones of Birds from the Wealden," asserting that the specimen Gideon had described

---

17. *Explanations.* Anon. but Robert Chambers, *Explanations: A Sequel to "Vestiges of the Natural History of Creation,"* By the Author of That Work; second edition (London, 1846; GM's copy, ATL). J. D. Forbes–William Whewell, 8 Jan 1846 (St. Andrews University Library; also in *L&L of JDF* [1873], p. 178). GM, *Animalcules,* pp. 95–98; quoting p. 97 twice.

(at Owen's insistence) as the tarsometatarsal bone of a wader was actually the lower portion of a pterodactyl's humerus. Buckland obsequiously agreed. In defense, Gideon pointed out Owen's role in the original identification. The bones, he continued, might well belong to pterodactyls, but so long as available evidence remained imperfect, the question must be considered an open one till more certain data were obtained.

Mantell restated his position more formally in a short paper "On the Fossil Remains of Birds," read at a further meeting of the Geological Society on 7 January 1846. When Owen replied, Gideon once again defended. Relations continued delicate between them until he generously attended Owen's lecture on the osteology of birds in April and the latter, similarly gracious, accompanied him partway home. Mantell then made a special trip to the Geological to withdraw his paper of 7 January replying to Owen, who had "expressed himself chagrined" at some of its remarks. "I would gladly give up any point in science," Gideon declared uncharacteristically, "than hurt the feelings of any man." But he was too late, the paper having already been printed, and with a very incorrect woodcut that Mantell had not even been permitted to see in proof. The unnecessary controversy, a footnote to that regarding *Vestiges*, therefore ramified, further exacerbating Gideon's already difficult relationship with Owen.[18]

House Hunting

When Gideon moved to Clapham Common in 1838, his retinue consisted of his wife, Mary Ann; two daughters, Ellen Maria and Hannah Matilda; two sons, Walter and Reginald; an assistant, Hamlin Lee; and his usual three servants: a cook, a maid, and a coachman-footman. By the next year, his wife, son, daughter, and cook had all deserted, and his second daughter died thereafter. Since his accident in 1841, Mantell's widowed sister Mary West (herself an invalid, eventually alcoholic) had come to live with

18. "Bird" bones. GM, "On the Bones of Birds," *Trans GS,* ns 5 (1840), 175–177. J (11, 17, 19 Dec 1845); RO, "On the Supposed Fossil Bones of Birds from the Wealden," *QJGS,* 2 (1846), 96–102; GM, "Supposed Birds' Bones of the Wealden," *AJSA,* ns 1 (May 1846), 274–275 (=GM–Editors, 20 Dec 1845; BS had retired from the editorship); GM, "On the Fossil Remains of Birds in the Wealden Strata," *QJGS,* 2 (1846), 104–106. J (7 Jan, 2, 8, 23, 25, 30 Apr [quoted] 1846). GM–RO, 27 Apr, 2 May 1846 (BMNH); D. T. Ansted–GM, 29 Apr 1846 (ATL). *Petrif,* pp. 187–193. That authentic birds existed during the Cretaceous is now thoroughly established. But all of GM's specimens (*Illus,* Plate VIII) had come from pterosaurs.

him, but with Ellen Maria returned and then gone again, only Reginald was left.[19] Gideon's medical practice at Clapham, meanwhile, had dwindled. Though fossils had once more begun to fill up otherwise unneeded rooms, Crescent Lodge was becoming an encumbrance.

On 16 August 1844, accordingly, Gideon and his brother Thomas, up from Lewes, went house hunting to the newly built Eaton and adjacent, less expensive Chester squares in London. On the twenty-ninth Gideon engaged (for twenty-one years, but with the privilege of leaving after five) No. 19 Chester Square, a cheerful, unsullied house in Regency style within twenty minutes' drive of his former one and yet (he claimed) "in the heart of the most aristocratic part of London," rather distantly adjoining Belgrave Square and Buckingham Palace. Faced with the upcoming change of habitat and friends, Mantell characteristically despaired. "Another remove," he foresaw despondently, "another career of exertion and anxiety before me – with a shattered constitution, and without one kind heart to rest upon." Yet Gideon relished the activity of packing and devoted most of September to it; on Monday, the twenty-third, he slept at his new home for the first time.[20]

19. GM's immediate family. EMM left because GM had treated her poorly, despite her generous assistance with *Medals*. She moved to London and almost certainly became an illustrator for the publisher John W. Parker (1792–1870; *DNB*), a widower nearly her father's age, whom she married on 12 February 1848. As GM strongly disapproved of this successful union, she did not see her father again until 24 July 1852 (J), following a separation from him of more than eight years.

RNM, expensively apprenticed in 1844 to the noted Isambard K. Brunel (1806–1859; *DNB*), became a civil engineer, working first on Hungerford Bridge, London (and therefore living at home with GM), then in Trowbridge and elsewhere on the Great Western Railway, in the newly made cuts of which he found important belemnites and other invertebrate fossils. After having been laid off because of poor economic conditions in England, he went twice to America and was working at a railroad camp in Kentucky when news of GM's death in 1852 necessitated his return to England. He presided over the auctioning of GM's specimens and effects in 1853. Eventually, RNM accepted a position in India, failed to adjust there, and during the Sepoy Mutiny fell victim to cholera, dying in Allahabad on 30 June 1857, at age thirty. By then WBDM had returned to England (1855–1859), as required by his father's will. In 1859 he went back to New Zealand, taking his father's and brother's effects with him.

MAM visited GM at Chester Square on 11 May 1850 (Reg J; unmentioned in GM's) and was coolly received; they had last seen each other at HMM's funeral ten years before. Her nephew, the dentist Alfred J. Woodhouse (b. June 1824), son of George Edward Woodhouse the younger, was GM's frequent companion during the last four years of his life – the only "family" remaining to him.

20. Because its contents were auctioned in 1853 (catalog, ATL [RNM's copy, annotated]), No. 19 Chester Square can be reconstructed in some detail. At the top of the house, front and back attic rooms were occupied by female servants. Three third-floor rooms had once been children's bedrooms. The two second-floor rooms were Mrs. West's and Gideon's.

His first floor included richly decorated drawing rooms (Belgian carpet, damask curtains, sofa, chairs, busts, best paintings, displays). An equally elegant dining room on the ground floor, similarly appointed, included a fine Spanish mahogany sideboard, with table and chairs for twelve; a large case displayed antiquities. The other major room on this level, Gideon's study, featured an original portrait of Sir Charles Lyell, a bust of Benjamin Silliman, and numerous framed engravings, including heads of Cuvier, Humboldt, Sedgwick, Buckland, Northampton, and the Earl of Egremont, among others. The entrance way and stairs were fully carpeted. Below, in the basement and kitchen, Gideon stored six different sets of china, each extensive; useful crockery, various decanters and glasses, tea urns and pots, a coffee machine, pairs of candlesticks, numerous pans and kettles, and cooking and cleaning equipment. A "butler's pantry" on the same level provided basic sleeping facilities while storing horse collars and harnesses as well.

11

Chester Square

On 3 June 1844, just as *Medals* was being finished, Gideon Mantell dined with Andrew Pritchard, author of *The Natural History of Animalcules* (1834) and *A History of Infusoria* (1842), and Sir John Herschel. When Reginald arrived later on, bringing Gideon's Ross microscope, Herschel saw infusoria for the first time in his life. Pritchard, who sold microscopes, then displayed some fine specimens with a hydro-oxygen apparatus of his own. In March 1845 Gideon and Reginald attended Murchison's Geological Society soiree at Belgrave Square, taking the microscope and some live animalcules fed with carmine, to make them show up. On Easter Sunday they visited J. S. Bowerbank in Islington to see his fossils and pursue microscopics. Gideon also took his microscope to Murchison's next soiree, on 9 April; there was then a good exhibition of infusoria by Reginald and Hamlin Lee in Chester Square on the twentieth. At Murchison's third soiree, on 7 May, three bishops (among other "intelligent persons") examined Gideon's live infusoria. The Reverend J. B. Reade and William C. Williamson (a new friend, who owed his interest in the subject to Mantell) visited Gideon a few days later and probably heard him read a paper on the microscopical examination of chalk and flint to the Geological Society on the fourteenth.[1]

This effort, published quickly in the *Annals of Natural History* that August, soon led Mantell to his next book, *Thoughts on Animalcules*, which he wrote at breakneck speed during the first two or three weeks of November. On 17 December, John Murray (no less) agreed to publish it, paying one hundred pounds for the copyright. Gideon received the first proof on 6 January, and sat up till 2 A.M. on the nineteenth correcting others. Proofs for the last sheet arrived on 17 February, being returned on the

---

1. Microscopics again. J (3 June 1844, 6, 23 Mar, 9, 20 Apr, 7, 14 May 1845); GM, "Notes of a Microscopical Examination of the Chalk and Flint of the South-east of England, with remarks on the Animalculites of certain Tertiary and Modern Deposits," *AMNH*, 16 (1845), 73–88; *AJSA*, ns 2 (Nov 1846), p. 150. James Scott Bowerbank (1797–1877; *DNB*) was professionally a distiller. GM–BS, 10, 21 June, 30 Aug 1845 (Yale); GM–Northampton, 30 Aug 1845 (Castle Ashby).

twenty-sixth (after a reading by John Phillips.) *Animalcules* was then officially published on 18 April, in an edition of two thousand copies. Like *Wonders* and *Medals*, this newest book significantly awakened Victorian readers to aspects of nature they had previously overlooked.

Thoughts on Animalcules; or, A Glimpse of the Invisible World Revealed by the Microscope (dedicated to the Marquess of Northampton, president of the Royal Society) accepted the researches of Ehrenberg and others, but derived largely from Mantell's own. Consisting mostly of descriptions, it was attractively illustrated with twelve colored plates and seven woodcuts. In some "General Remarks" following his survey of the various forms (infusoria, polygastria, monads, vorticellina, rotifera, floscularia, stephanoceros, rotifer or wheel-animalcules, and those with durable cases or shells), Gideon thought it likely that "many of the most serious maladies which afflict humanity are produced by peculiar states of invisible animalcular life" (90), including cholera, influenza, and other epidemic diseases. He attributed this remarkable anticipation of the germ theory of disease to Henry Holland, who had proposed such a hypothesis in 1838. So early as September 1840, Mantell himself had advised Silliman that both cancer and tuberculosis were "probably entozoic," though the microscope had not yet detected any specific organisms.

In an appendix to *Animalcules*, Gideon discussed Holland's idea further, even inching his way toward an admission that the spontaneous generation advocated in *Vestiges*, could it be proven, would "explain many obscure physiological phenomena, and bring the laws of vitality into harmony with those which preside over the inorganic kingdom of nature" (106). For members of the medical profession especially, *Animalcules* proved to be unusually significant. Its more immediate impact, however, was to equate Gideon with microscopics, so that he spent much of his time during the next few months either entertaining prominent visitors with slide displays or using the Ross microscope himself for serious research.[2]

2. GM, *Thoughts on Animalcules* (London, 1846; author's copy at ATL). J (3 Nov, 19 Dec 1845, 6, 19 Jan, 17, 26 Feb, 28 Mar, 17 Apr, 26 May 1846, 4 Jan 1847). GM–BS, 23 Nov 1845 (Yale); JM–GM, 13, 17, 29 Dec 1845 (ATL); GM–CL, 27 July 1849 (ATL). See also WCW–GM letters, 1844–1852 (ATL) – "It was reading your *Medals* that drew my attention to chalk" – and WCW, *On Some of the Microscopical Objects Found in the Mud of the Levant* (Manchester, 1847; GM mentioned, pp. 73, 74, 80, 82, 96, 112–113, and 115). Reviews and comments: *The Lancet*, 2 (1846), 478–479; GM–BS, 21 June 1846 (Yale); *AJSA*, ns 2 (Nov 1846), 149–150; *Westminster and Foreign Quarterly Review*, 46 (1846), 29–60. See also W. A. S. Sarjeant, "The Rediscovery of a Lost Species of Dinoflagellate Cyst," *Microscopy*, 30 (1967), 241–251 (summarizing GM's contributions); and Sarjeant, "Joseph B. Reade (1801–1870) and the Earliest Studies of Fossil Dinoflagellate Cysts in England," *Journal of Micropaleontology*, 1 (1982), 85–93.

Germ theory of disease. *Animalcules*, pp. 90–91; Henry Holland, *Medical Notes and*

The Isle of Wight

Disliking the tasteless ostentation and lingering moral taint of the Royal Pavilion at Brighton, Queen Victoria had purchased Osborne House, adjoining Cowes, in 1845 and began now to replace it with an Italianate mansion, destined to be her summer residence for more than fifty years. Henry Bohn foresaw enough of that to urge from Mantell a book on the geology of Wight, which was one of the few places in England other than Cuckfield where the bones of Wealden saurians could be found. In his letter of 23 November 1845, consequently, Gideon announced to Silliman that he was preparing "a geological ramble round the Isle of Wight." On 21 February 1846 he took a geological model of the Isle with him to Lord Northampton's first Royal Society soiree of the season and spent much of the evening discussing it with Prince Albert, to whom his book would be dedicated.

When the last proof of *Thoughts on Animalcules* passed through Mantell's hands five days later, he was finally able to begin. "I am getting on with the Isle of Wight," he reported to S. P. Wodward on 11 April, "but my health is so very bad that professional duties overpower me, rendering any literary labor very irksome and oftentimes impossible." On 21 May, Gideon sold Bohn the copyright to a book called "Geological Tour Round the Isle of Wight"; its title would change twice more after that.[3]

Temporarily set aside on behalf of Mantell's nostalgic *Ramble* to Lewes, *Wight* next appeared on 6 August, when Gideon approached Bohn once more to complete their agreement. Joseph Dinkel, who had finally parted from Agassiz when the Swiss moved to America, spent three days that month on drawings for the book, and Mantell himself paid the Isle one more visit. On 25 February 1847, after some further delays, he delivered a special copy of *Wight* to Buckingham Palace for Prince Albert, its dedicatee, together with a nicely cased collection of nearly one hundred relevant fossils. Publication then followed on 6 March, when Gideon received a number of copies and gave most of them away. Finally, Bohn paid him

Reflections (London, 1839), Chap. 34. GM–BS, 2 & 26 Sept 1840 (Yale); *Animalcules*, pp. 104–106. Also relevant is Henry Edwards, *Illustrations of the Wisdom and Benevolence of the Deity as Manifested in Nature* (London, 1845; GM's copy, ATL, incl. longhand additions by BS); though his book was dedicated to Mantell, Edwards rejects the germ theory of disease.

3. Isle of Wight. J (3–13 July 1844); GM–BS, 18 July, 24 Aug 1844 (Yale); *AJSA*, 47 (Oct 1844), 402–406; J (30 Mar–2 Apr, 15–18 Sept, 19 Nov, 3 Dec 1845); GM–CK, 4 Apr 1845 (DRD); GM–BS, 6 July, 23 Nov, 30 Dec 1845, 24 Feb 1846 (Yale). J (26 Feb 1846); GM–SPW, 11 Apr 1846 (APS). J (21 May 1846). GM, "Notes on the Wealden Strata of the Isle of Wight," *QJGS*, 11 (1846), 91–96; *London Geological Journal*, 1 (1847), 41–44.

£150 on 5 May; deducting all expenses, the book had brought its author about £100 – "far too little, but ten times what Milton received for *Paradise Lost.*"[4]

Geological Excursions round the Isle of Wight, and along the adjacent Coast of Dorsetshire began haphazardly with touring tips, a pedestrian itinerary, and descriptions of its twenty plates (the last, a colorful geological map). Mantell then reviewed the relevant stratigraphical sequence (an elaborate folding chart), the geology of southeast England, and the geological structure of Wight as a whole. Specific itineraries followed, together with remarks on Cretaceous and Wealden strata. Chapter 11, on fossil reptiles (he avoided "dinosaur"), included a brief history of previous finds and discussed each genus, beginning with *Iguanodon.* After describing a femur and tooth he had acquired in 1845, and remarking on its sacrum, Gideon went on to the "Form of the Iguanodon," a topic he had never before discussed so openly. His comments are of particular interest in that he genuinely thought they might well be his last.

Tempting as restoration of the living animal was, Mantell argued, only a vague idea of its original form and appearance could be derived from present evidence. "In all probability," he supposed, "the entire or a considerable portion of the skeleton of a young iguanodon will sooner or later be brought to light, and yield the information necessary to enable the paleontologist to reconstruct the skeleton, and delineate the form of the living original" (320). *Iguanodon*'s body, Gideon assumed, must have been large and as massive as an elephant's, with limbs proportionately huge. The hind legs in particular, supported by very strong short feet with clawed toes, probably resembled those of a hippopotamus or rhinoceros. The forelegs with their hooked claws appear to have been less bulky, but *Iguanodon* was seemingly quadrupedal. Its teeth, however, masticated tough vegetable food, requiring powerful jaw muscles having no counterpart in modern saurians. The length of the iguanodon, Gideon admitted, had been variously estimated, depending on how one reconstructed its tail; submissive to Owen, he now suggested thirty feet. *Iguanodon,* Mantell concluded, "was a gigantic but inoffensive herbivorous reptile, which lived on the ferns, cycadeae, palms, and conifers that constituted the flora of the country of which it appears to have been the principal inhabitant" (322). He then dispensed quickly enough with *Hylaeosaurus* and other

4. *Ramble:* J (29–30 June 1846); GM, *A Day's Ramble in and about the Ancient Town of Lewes* (London, 1846). *Wight:* Agreement with Bohn, 6 Aug 1846 (ATL); J (6, 21, 24–29, 31 Aug, 3 Sept, 11, 25 Dec 1846; 22, 23 Jan, 2, 15, 18, 25 Feb, 6 Mar, 5 May 1847). The case and specimens given Albert are still preserved at Osborne House, in the Swiss Cottage.

fossil reptiles (including a marine *Cetiosaurus*) to climax his remarks with another evocation of "The Country of the Iguanodon." Comparing this book, excellent in its way, with *The Fossils of the South Downs* helps one to see how both the science of geology and Gideon's place within that science had changed.[5]

A Ramble to Lewes

On 29 June 1846 Gideon went by rail to Brighton and, on a newly opened line, continued to Lewes, where the terminus had replaced Nehemiah Wimble's now-demolished Friars, the site of Mantell's reception by William IV in 1830. St. Mary's Lane, on which he was born, would be renamed Station Street. Following reunions with his brother Thomas and Warren Lee, Gideon inspected Castle Place (still his, but tenantless) and the Priory, then walked through town and up the castle for a lovely panorama. He returned to London the next day, after collecting fossils at Hamsey and other familiar sites. This nostalgic homecoming set him to work on his next book, *A Day's Ramble in and about the Ancient Town of Lewes,* to which Henry Bohn indulgently agreed. By 3 July, Gideon had taken a portion of the manuscript to Richard Clay, and on the ninth printing began. On 6 August, Mantell corrected the last proof and received his traditional first copy from Clay on the eighteenth; it went out a few days later to Ellen Maria. Publication then took place on 1 September.

Originally called *Notes for a Day's Ramble in and about the Ancient Town of Lewes,* Gideon's little book dropped the first two words of its title while in press. It also expanded from 138 pages to 160 through the addition of some "Supplementary Notes" and G. F. Richardson's anonymous poem on Mantell's public excursion to Lewes from Brighton in June 1836. Amazingly enough, *Ramble*'s frontispiece, a drawing of the long-ago demolished dovecote of Lewes Priory, had been sketched by eleven-year-old Gideon for his sister Jemima in 1801, then published forty-five years later in *Archaeologia*. In eleven chapters, beginning with the railway journey from Brighton, Mantell surveyed the topography, geology, and history of his native town, highlighting its Priory and the Battle of Lewes in 1264.

5. GM, *Geological Excursions round the Isle of Wight, and along the adjacent Coast of Dorsetshire* (London, 1847). For useful background (and a high opinion of GM's *Wight* [p. 120]), see John Challinor, "Thomas Webster's Letters on the Geology of the Isle of Wight, 1811–1813," *Proc Isle of Wight Natural History and Archeological Society,* 4 (1949), 108–122; and Wilson, pp. 96–102, 111–115. GM had long admired Webster's work (e.g., GM–BS, 28 Feb & 3 Mar 1845 [Yale]).

Among many points of interest, he noticed Southover church, the so-called Anne of Cleves house, St. Anne's church, Castle Keep, the High Street, the Bull Inn (Paine's home – later a meetinghouse for Dissenters; Thomas Horsfield, pastor), St. Michael's church, impersonal Castle Place, Lewes Castle, and the new church of St. John's sub Castro ("an unsightly edifice, recently erected"). Visiting Fisher Street (as it was now called), Gideon recommended the fossil shop of a Mr. Martin, but said nothing about his father's shoe repair. In Albion Street, however, he paid special tribute to the Lewes Library Society, founded in 1785 and fostered by the late Thomas Woollgar; without its books, Mantell declared, he could never have attempted geological researches. At Cliffe Corner, similarly, he remembered John Button's Academy – and the avalanche of December 1836. In a nearby chalk pit, Gideon "began" his investigation of Sussex fossils.

One archeological digression aside, the last ten pages of text were devoted to geology. Chalk he thought nothing more than aggregated shells, corals, and carbonate of lime; flints originated from hot, silicated waters or vapors periodically ejected from great depths within the earth into ocean basins where chalk was accumulating. These layers gradually consolidated and were then elevated from the bottom of the sea by ubiquitous subterranean forces. Since the Weald is a U-shaped trough filled with sediments, considerable displacement of the originally horizontal strata must have taken place, producing rents and fissures that are now the bold escarpments and deep glens of the South Downs. Some of the chasms, however, were subsequently filled by alluvial sediments partly marine in origin, as the Weald estuary gradually shallowed into a swamp, later drained by human ingenuity. Creation and destruction, in short, were the themes of Gideon's discourse throughout.[6]

Charlesworth's Journal

Two days after *Ramble* began printing, Edward Charlesworth, a staunch foe of Owen, called on Mantell to reenlist his aid toward a long-projected *London Geological Journal*. In February 1845 Gideon had "scribbled many articles for Mr. Charlesworth's journal," but competition from the Geological Society's newly inaugurated *Quarterly Journal* had prevented its coming out. Nonetheless, Charlesworth tried again the following year. He proposed to issue six numbers annually, and the first three actually appeared, beginning in September. "The arrangements for the commencement

6. GM, *A Day's Ramble in and about the Ancient Town of Lewes* (London, 1846; author's copy at ATL); *Archaeologia*, 31 (1846), 431.

of this journal were made in 1844," Charlesworth explained to his few readers, "when the editor's removal from London, in consequence of his receiving a provincial scientific appointment [curator of the York Institute], rendered it expedient that the publication of the work should be postponed." Now his periodical was struggling to emerge, at an expensive three shillings sixpence per copy.

This first issue contained several items of Mantell interest. In "A Few Notes on the Prices of Fossils," Gideon recalled the most significant paleontological auctions or sales he knew of, beginning with that of Colonel Birch's collection of Anning family specimens on 15 May 1820; then the James Parkinson collection, April 1827; the Hawkins collection, evaluated by himself and Buckland, 1834; the Finnell collection of American proboscidean remains, 17 February 1836; the sale of his own collection, 1838; two less distinguished sales in 1840 and 1843; and the second Thomas Hawkins sale (which Gideon attended) of 25 July 1844.

The editor's contribution, ostensibly on *Mosasaurus,* soon degenerated into one of the ill-tempered (though not necessarily wrongheaded) diatribes for which Charlesworth would be famous. A mammalian tooth discovered in the London Clay, for example, had been proclaimed that of an opossum by Owen in 1838 but reappeared in his *British Fossil Mammals* (1846) as a monkey's. The Hunterian professor had been even more egregiously mistaken in a comparable instance, foolishly identifying the nearly complete skeleton of a modern deer as that of an Eocene ancestor – merely on dental evidence. Though well aware of his insufficient data, Charlesworth admonished, Owen "preferred to hazard a guess rather than admit that a tooth does not *always* enable a comparative anatomist to decide upon the genus to which it belongs" (25). The British Association *Report* for that session, moreover, had been rewritten to disguise Owen's blunders, which were subsequently compounded by his angry public letters against Charlesworth, who had seen the truth.

A second major example of Owen's ineptitude was his controversy with Gideon Mantell regarding birds in the Wealden. The latter's just-published paper ("On the Fossil Remains of Birds," 1846) was, for Charlesworth, a model of proper caution. In further portions of his meandering commentary, Charlesworth repeatedly praised Mantell's work while objecting to Owen's professional conduct and British science more generally. Large portions of the BAAS meeting, he declared, might "with perfect justice be denominated 'twaddle.'" Several anonymous and more temperate "Short Communications" by Gideon included one on his newly found fossil reindeer from Wight, dated 2 September; a notice of his Royal Society paper on foraminifera in chalk and flint; a notice of Owen's fossil catalogue (about which Mantell had written Silliman the previous year); and a touching obituary of Etheldred Benett (died 11 January 1845) by Gideon, whose forthcoming *Wight* was also announced.

Charlesworth's second issue appeared in February 1847, with apologies for lateness. Thanked within it for subscriptions help, Mantell also contributed a brief essay on his recently discovered large bivalve *Unio Valdensis*. The major controversy in this issue, however, involved belemnites, about which Owen had published a medal-winning but mistaken paper in 1844. Now the taxonomic fallacy underlying that paper was to be exposed, and the fame of J. Chaning Pearce advanced. Besides promoting the latter's new genus of *Belemnoteuthis,* Charlesworth repeatedly attacked Owen, while emphasizing the dangers of "official" opinion and the desirability of maintaining open scientific debate.

Charlesworth's third issue, of May 1847, was also his last. In an editorial recalling the perfidy of Sir Everard Home (d. 1832), who had first appropriated unpublished anatomical discoveries by his late brother-in-law John Hunter and then burned telltale Hunterian manuscripts placed in his keeping, Charlesworth strongly questioned the propriety of appointing William Buckland, now Dean of Westminster, a trustee of the British Museum. Further discussion of the belemnite question revealed that Pearce had just died of tuberculosis at thirty-six. But two new champions had arisen on behalf of *Belemnoteuthis:* William Cunnington of Devizes, whose manuscript was now in hand; and Reginald Mantell, whose beautiful specimens from Trowbridge provided Cunnington with decisive evidence.

"Birds versus Reptiles," one of several "Short Communications" in this issue, reported that J. S. Bowerbank had utilized his microscope to distinguish birds' bones from reptilian ones, thereby twice confuting Owen, who had on the one hand pronounced some pterodactylian relics to be avian and on the other called Gideon's Wealden bones pterodactylian whereas their structure proved them "undoubtedly referable to birds" (131). Despite occasional acknowledgments of his prestige, the dying journal found almost no issues on which Owen had been right. In thereby opposing the Hunterian professor's scientific dictatorship, Charlesworth temporarily provided the London intellectual community with a badly needed corrective, but his having to do so outside official circles was ominous.[7]

Attempting to professionalize themselves, Victorian scientists in several fields endured a period of stifling authoritarianism during the 1840s;

---

7. *London Geological Journal.* Edward Charlesworth (1813–1893) despised RO more openly than GM ever did. *The London Geological Journal, and Record of Discoveries in British and Foreign Paleontology,* vol. I, nos. 1–3 [all published], 1846–1847, edited by him, is filled with hits at Owen. GM's several well-mannered contributions (pp. 13–17, 36, 40–44, 130–131) include an anonymous obituary [initialed by GM in his own copy, ATL] of Etheldred Benett, p. 40.

Mantell himself experienced outright censorship. For example, the last paragraph of his innovative paper on the soft parts of foraminifera (1846) had predicted that traces of microscopic life would someday be found in so-called Azoic (i.e., lifeless) rocks underlying the Cambrian. "I therefore submit," he had asserted, "that in the present state of our knowledge of the earth's physical history, as derived from paleontological evidence, the period when organic creation commenced must still be regarded as one of those hidden mysteries of nature from which science has not yet withdrawn the veil." But the Council of the Royal Society found these reasonable (and eventually verified) opinions so heretical that they resolved not to publish Gideon's paper unless he agreed to delete them. "For myself, I care not one straw about it," Mantell then advised Lyell, "but I do regret having been compelled to sanction a precedent which may hereafter seriously impede freedom of expression on disputed points, not only in geology, but also in other sciences." Lyell himself had become part of the problem, however, for (as Gideon remarked to Silliman on 21 June) he and Murchison were carrying all before them, effectively dominating the learned societies with their productive researches in America and Russia, respectively. Other, less fortunate researchers naturally felt shut out, and as a result (Gideon observed) "Geology is sadly on the wane here."[8]

Wonders, Sixth Edition

Of those who sought to preserve geology as an open science, no less professional for being popular, Mantell now stood among the foremost. His major contribution toward that end was the extensively rewritten, considerably sophisticated sixth edition of *The Wonders of Geology*, to which he turned in January 1847, one day after completing the index to *Wight*. This virtually new book, well under way by April, began its slow journey through the press in May. "The *Wonders* going on by fits and starts; about one-third printed," Gideon noted on 10 June, adding that the writing was "very hard work indeed." After correcting the last page of Volume 1 on 13 July, he had to re-do about twenty folio pages of manuscript the next month when his manservant lost them en route to the post office. By mid-October, Mantell was "hard at work on the last lecture of *Wonders!*" and after further days of "suffering, anxiety, and profitless toil" wrote the final page of text on the thirtieth, receiving it back in proof on 5 November. He completed the index ten days later. "*Wonders* will be published

---

8. GM, "On the Fossil Remains of the soft parts of Foraminifera," *PT*, 136 (1846), 465–472; GM–CL, 10 Nov 1846 (ATL, APS); GM–BS, 21 June 1846 (Yale).

next week," Silliman was informed on 30 December. But advance copies appeared only on 19 January 1848, when Gideon received three, the first of which he sent to Lyell; publication then took place the next day. This new edition garnered £105 on 1 May, when Bohn paid him for it.[9]

The Wonders of Geology, sixth edition, with unchanged frontispiece and title page vignettes, was dedicated to Benjamin Silliman; at Bohn's suggestion it included Silliman's preface to the American edition. There were still two volumes and eight talks, but lectures four (on the Cretaceous and Wealden), five (Jurassic, Triassic, and Permian), and eight (Devonian, Silurian, and "Cumbrian"; Volcanic and Primary rocks) were now each in two parts. Among Mantell's many changes (since the third edition), his sketch of the nebular theory now included an important footnote tentatively endorsing "the formation of living beings from inorganic elements, . . . a doctrine which would explain many obscure physiological phenomena, and bring the laws of vitality into harmony with those which preside over the inorganic kingdom of nature" (48n), as had been advocated in *Vestiges.* Once again, he straightforwardly preferred uniformitarian to catastrophic forces and excluded the Deluge as a geological agent. Gideon then discussed Agassiz's glacial theory sympathetically, accepting former glaciers of great extent but doubting that the *whole* of Europe had ever been covered with ice, as Agassiz contended, and underestimating the extent to which glaciers could explain the Drift of Great Britain. His human fossils no longer included some Mississippi River footprints discredited by David Dale Owen, a section on geysers added New Zealand to Iceland, and small changes in the conclusion to Lecture 1 made its geology more Lyellian than formerly.

In Lecture 2, Mantell's remarks on the extinction of animals strengthened considerably, reflecting the influence of both *Vestiges* and Sir Roderick Murchison's *Geology of Russia* (1845, with coauthors). Under a new topic, "Law of Extinction" (pp. 125–126), Gideon noted that whole races of animals sometimes disappeared (as in Russia) from strata undisturbed by physical catastrophes. Perhaps, then, species (like the individuals within them) were allotted predestined life spans, with some given only a few hundreds or thousands of years; others, ages; and some few, perpetuity. By way of illustration, his remarks on New Zealand's wingless bird *Apteryx* were now supplemented with others on the moa. A section on fossil elephants (pp. 147–156) had also been enlarged from Murchison, who held that the abundant mammoth carcasses of Siberia must have

<hr />

9. *Wonders,* sixth edition. J (23 Jan, 30 Mar, 8 Apr, 30 May, 10, 26 June, 13 July, 16 Aug, 9, 15, 30 Oct, 5, 15, 25 Nov 1847). GM–BS, 30 Dec 1847 (Yale). J (19, 20 Jan, 1 May 1848).

been transported there by a mighty river – like the iguanodon remains of southeast England. Reviews of other mammalia depended primarily on work by Owen and American paleontologists but included Darwin.

The "Chronological Synopsis" of rocks and strata in Lecture 3 (pp. 200–207) was considerably updated. Working backward, as before, Mantell now distinguished four epochs: Human or Modern, Tertiary, Secondary, and Palaeozoic. The Secondary Epoch divided into Cretaceous, Wealden, Jurassic, Lias, and Triassic formations, while the Palaeozoic included Permian, Carboniferous, Devonian, Silurian, and Cumbrian. His treatment of metamorphic, plutonic, and volcanic rocks (all nonfossiliferous) was fuller, now adding remarks on drifted or erratic boulders, another topic on which Lyell and Murchison had made important contributions. Gideon's discussion of the Tertiary (pp. 213–292) was ambitious; his earlier questioning of Lyell's statistical subdivisions based on alleged percentages of extinct species had now, he believed, been significantly confirmed. Citing his own publications, Gideon endorsed and summarized a stratigraphical analysis of Tertiary deposits proposed by Edward Charlesworth (pp. 224–225).[10]

In accord with Dr. Grant's opinion of 1835, Mantell's brief treatment of fossil monkeys again stressed their anatomical nearness to man. He also reviewed the controversy that had followed his declaring Murchison's fossil fox identical with the recent species. On the basis of what had seemed to Gideon only trivial differences in structure, Owen subsequently assigned the fox to an entirely new *genus*. "It has been shrewdly remarked by a celebrated anonymous author [i.e., Robert Chambers]," Mantell observed no less shrewdly, "that such is paleontological refinement nowadays, that an extra plication of enamel in the tooth of a fossil pachyderm, or an additional notch on the tooth of a carnivore, is sufficient to obtain a specific and even subgeneric name for that animal, and constitute its origin a *separate* creation!" "If this be admitted," he continued presciently, "the fixity of species must soon be deemed a chimera" (264n). As before, Gideon's concluding remarks to Lecture 3 emphasized the extinction of past life and the rapidity of such changes. To it he now added a relevant passage from Playfair on the efficacy of existing causes (pp. 291–292).[11]

10. GM, *The Wonders of Geology; or, A Familiar Exposition of Geological Phenomena*, sixth edition (2 vols.; London, 1848). Re p. 48n, GM had said almost the same thing earlier in *Animalcules*, p. 106 (note 2 above).

11. Fox controversy. RIM, "On the Fossil Fox of Oeningen," *Proc GS*, 1 (1830), 167–169; *Trans GS*, ns 3 (1835), 277–290, and GM, "Anatomical Description of the Fox," pp. 291–292; RO, "On the Extinct Fossil Viverrine Fox of Oeningen," *QJGS*, 3 (1847), 55–60. I wish to thank R. J. G. Savage for his comments on this specimen, which is at BMNH, about 13 million years old (Middle Miocene) and poorly preserved. Though too early

Lecture 4, Part 1, dealing with the Cretaceous, expanded to include Mantell's researches on microfossils. The true nature of xanthidia remained problematical, he believed, despite several important memoirs. Since most of his major publications dealt with Cretaceous strata to some considerable degree, revisions throughout this chapter were frequently extensive. Lyell and indigenous geologists furnished additional information about North America while Murchison was again noteworthy on Russia. Gideon had nothing fundamentally new (i.e., beyond Agassiz) to say regarding fossil fishes, but he recorded several important reptilian finds. In Part 2, on the Wealden, Mantell further defended his priority in establishing its freshwater origin. Besides citing his own publications and Fitton's, Gideon drew on important German ones by Friedrich Roemer and Wilhelm Dunker, two more authors influenced by himself. Some further remarks depended heavily on *Wight*.

Mantell emended his ordering of fossil reptiles in this lecture (pp. 412–449) to reflect Owen's more nearly. He began now with *Plesiosaurus*, then *Cetiosaurus* (still probably marine), and the crocodiles, of which there were several additional types. *Megalosaurus, Iguanodon,* and *Hylaeosaurus* were indeed "dinosaurians," but though Gideon accepted Owen's new word he also regretted it (p. 419n). The same Greek root (*deinos*) occurred also in "*Dinotherium*" and "*Dinornis*" (terrible beast and bird, respectively), and would mislead the unscientific by suggesting affinities among these unrelated creatures. It was preferable to avoid such conundrums. In any event, certain terrestrial crocodile-lizards of the Mesozoic could now be grouped, and if their size had been exaggerated in the past (he did not admit by whom), they were still "sufficiently colossal to satisfy the most enthusiastic lover of the marvellous" (420).

Regarding specific dinosaurs, Buckland received full credit for *Megalosaurus,* a carnivore some thirty feet long. Mantell's section on *Iguanodon* was more elaborate – with fresh details, for example, about how the original series of teeth had been discovered. Noting the resemblance of these teeth to those of the iguana, Gideon himself (in this version of events) had "proposed the name of *Iguanodon*" (423). In affirming a general resemblance between iguanodon and iguana, however, he had never intended to suggest any close analogy; still less to assert, as some writers had supposed, that the iguana is a living miniature of his colossal reptile. His famous "horn" was certainly that, but only tentatively iguanodontian.

Hylaeosaurus, now thought to be twenty or thirty feet long, retained his dermal spines (contra Owen) and was basically unchanged. Mantell's

to be a fox, it is probably ancestral. John Playfair, *Illustrations of the Huttonian Theory of the Earth* (Edinburgh, 1802); *Works,* 1822, I, 117, inaccurately quoted.

long-controversial "tarso-metatarsal bone of a wader," however, was probably from the humerus of a pterodactyl. J. S. Bowerbank reidentified some bones thought by Owen to be from an extinct type of albatross as reptilian, but others still appeared to be avian. At present, Gideon concluded, whether or not birds existed during the Wealden and Cretaceous epochs remained uncertain. He was inclined to believe, however, that their presence in the Secondary formations would sooner or later be satisfactorily established (442).[12] Lecture 4 (and Volume 1) then concluded with "The Country of the Iguanodon" and other passages largely derived from his previous books.

Though primarily devoted to Jurassic strata, Lecture 5, Part 1, at the beginning of Volume 2, continued Mantell's discussion of the Cretaceous and Wealden, reconstructing their flora and fauna insofar as possible while endorsing Lyell's suggestion that the real "Country of the Iguanodon" must have been located far from Sussex. Following a general view of the Oolite and Lias, Gideon described several localities adjacent to the Great Western Railway, including Bradford, Faringdon, Swindon, Stonesfield, Oxford, and Reginald's Trowbridge.[13] He also cited a large number of geologists in this chapter, among them Phillips, Lonsdale, and several Germans. Though he detailed both plesiosaurs and ichthyosaurs, however, Gideon failed to mention J. Chaning Pearce's discovery in 1846 that ichthyosaurs were viviparous.

In Lecture 5, Part 2, on the Triassic and Permian formations, Mantell

---

12. Cretaceous birds controversy. GM, "Supposed Birds' Bones of the Wealden," *AJSA*, ns 1 (1846), 274–275; GM, "On the Fossil Remains of Birds," *QJGS*, 2 (1846), 104–106; *LGJ*, 1 (1846–1847), 27, 130–131. See also Chap. 10, footnote 18 above.

13. Belemnite controversy. *LGJ*, 1 (1846), 65–78; (1847), 127–128, praising RNM; RO, "A Description of Certain Belemnites," *PT*, 134 (1844), 65–85; GM, "Observations on some Belemnites . . . Discovered by Mr. Reginald Neville Mantell, C.E., in the Oxford Clay near Trowbridge in Wiltshire," *PT*, 138 (1848), 171–181; and "Supplementary Observations on the Structure of the Belemnite and Belemnoteuthis," *PT*, 140 (1850), 393–398. William Cunnington's copies (with relevant GM letters) are at Wiltshire Archeological and Natural History Society, Devizes. According to one authority,

> In his 1848 and 1850 papers, Mantell showed a higher degree of accuracy in observation and deduction than Owen. Demonstrating that *Belemnoteuthis* exhibited both phragmocone and guard, he supported Pearce and Cunnington in distinguishing it from 'Belemnites.' The concluding remark of Mantell's 1850 paper ('consequently, the form and structure of the body and arms and other soft parts of the cephalopoda to which the belemnites belonged have yet to be discovered') is still true, as far as I know.

I wish to thank D. Phillips (BMNH) for these comments. See also J (25 Sept 1846); GM–BS, 13 Mar 1848 (Yale); *Petrif*, p. 460; GM, "A few Notes on the Structure of the Belemnite," *The Annals and Magazine of Natural History*, 10 (1852), 14–19; and RO, "On the Structure of the Belemnite," *AMNH*, 10 (1852), 158–159.

depended thoroughly on the researches of others. The Triassic, for example, had been defined almost solely on the basis of German fossil deposits. Among its most interesting genera were *Labyrinthodon,* a supposed giant frog-like creature identified by Owen primarily on dental evidence, and several mysterious beings known only from their footprints. Though Gideon freely admitted that Professor Hitchcock's avian identification of the American ones remained current, he specifically demurred from endorsing it (pp. 556–559). As with the Triassic, the new Permian system (established by Murchison in 1841) was based primarily on distant sites, in Russia and Germany, of which Mantell had no firsthand knowledge. From these strata had come the earliest remains of reptiles yet known, but still earlier ones would probably be found. After characterizing reptiles in general, Gideon considered in turn turtles, crocodiles, ichthyosaurs, plesiosaurs, pterodactyls, and even J. J. Scheuchzer's infamous *Homo Diluvii Testis* before concluding with a general "Review of the Age of Reptiles." In it he emphatically rejected Chambers' hypothesis of a "half-finished planet, unsuitable for warm-blooded animals" (584). The general temperature of the earth and the physical constitution of the sea and atmosphere, Gideon believed, were not essentially different from those of today.

Darwin's "delightful volume" *On the Structure and Distribution of Coral Reefs* (1842) was important to Lecture 6, on corals and crinoidea, which encompassed the development of the nervous system as well. Gideon, moreover, repeated his strictures against *Vestiges;* analogy was not identity (pp. 603–604). Lecture 7, on the Carboniferous, naturally discussed the formation of coal and many fossil plants as well as the nagging problem of prehistoric climates. A retrospect of botanical epochs offered no support for the theory of progressive development (747). Lecture 8, Part 1, summarized the Devonian, Silurian, and Cumbrian periods, before which life records were unknown. In a long footnote, Mantell praised Murchison's *Silurian System* (1839), the publication of which, like its contents, "formed an era in British geology" (767).[14] As for separating the Silurian and Cumbrian systems, Gideon endeavored to remain neutral in what was presently a titanic controversy between Murchison and Sedgwick. Like them, therefore, he called the disputed strata Cumbrian, leaving only the lowermost and nearly barren slate rocks as Cambrian. Unexpectedly, this chapter also included a substantial note on the geology of New Zealand. Finally, Lecture 8, Part 2 (nearly one hundred pages long) dealt solely with volcanic and other nonfossiliferous rocks.

14. GM's copy of this expensive work, given to him by RIM in January 1839, was sold at auction in 1853. For Murchison and Sedgwick, see James A. Secord, *Controversy in Victorian Geology: The Cambrian-Silurian Debate* (Princeton, 1986).

Among his "General Inferences," Mantell concluded that the same "fixed and immutable laws established by Divine Providence for the maintenance and renovation of the material universe" (890) had been at work from the beginning – and that further laws which governed the appearance and extinction of species were perfectly in accord with them. Such a possibility had once attracted him to the developmental hypothesis, as we know, but in the course of writing this book he had resolved to abandon that theory, returning to Lyellian uniformitarianism.[15] Ironically, Gideon then went on to discuss Charles Darwin's remarkable characterization of the Galápagos Islands (from his *Journal of Researches*, 1839; 1845), which "swarm with herbivorous marine reptiles allied to the iguanidae" (893) and a unique flora. Though transmutation was no longer a question about which Mantell was prepared to think further, it continued to influence him unawares.

Iguanodon Reconstructed

Gideon Mantell's involvement with the Isle of Wight, culminating in his book, brought Wealden strata and their saurians into prominence again. As a result, significant new finds soon appeared. Thus, on 24 November 1847 he received some interesting teeth and vertebrae from his collectors, unwisely showed them to Richard Owen, and five days later wrote of them to Silliman: "Some fishermen I set hunting for fossils in the cliffs at Brook Bay in the Isle of Wight occasionally send me specimens," he explained, "and I have this week received several fine vertebrae and some teeth from an old iguanodon." One of these teeth, sketched in his letter, looked so perfect that Owen had thought it worth a fifty-mile journey to see. The tooth's margin, Gideon continued, appeared serrate all round: "Not a point is worn away – a proof that in old age the iguanodon, like the living iguana, crocodiles, etc., had a continual renovation of its dental organs." A young tooth, then, did not necessarily indicate a young animal. This tooth, he surmised, never fully emerged, and was therefore unused. Gideon now renewed his hopes that he might at last acquire an iguanodon's jawbone with teeth attached.

As if in fulfillment of his wish, on 25 March 1848 Mantell received an unexpected specimen from Captain Lambart Brickenden of Warminglid, Sussex. Though personally unacquainted with Gideon, Brickenden was

---

15. Though Lyell was now a major influence on *Wonders,* Mantell's emphasis on "fixed and immutable laws" (p. 890) may also reflect the strong opinions of Charles Babbage, *The Ninth Bridgewater Treatise* (London, 1837), both directly and through *Vestiges.*

not only interested in geology but the current proprietor of those quarries at Whiteman's Green in which *Iguanodon* had originally been discovered. Sent only to be described, his twenty-one-inch specimen was part of an iguanodon's lower jaw! Elated with this good luck, after so many years of waiting, Gideon immediately announced his acquisition to prominent geological friends like Sedgwick, then replied to Brickenden on the thirtieth, expressing great interest in the specimen and wishing to have it in exchange for a set of his own works. On 2 April, Brickenden graciously agreed.

Mantell, meanwhile, had taken his new jaw to the British Museum for comparisons with mandibles from other fossil and recent reptiles. Together with Dr. A. G. Melville, a well-known comparative anatomist, he looked over relics of *Iguanodon* and *Hylaeosaurus* from his former collection. News of the specimen and results of these 1 April comparisons were then relayed to Silliman eight days later. After "thirty" years' search, he had at last found *Iguanodon* bone and teeth unequivocally associated. Unfortunately, the specimen did not contain any used teeth, only seventeen sockets and two germs of teeth. It was also incomplete at the rear, but the front or chin part remained entire, preserving structural modifications of great interest. (Eventually, those modifications would convince Gideon that *Iguanodon* had a long, prehensile tongue and grasping lips.)[16]

When some unnecessary fears of Chartist agitation had subsided, with troops and cannon uneventfully returned to royal garrisons, Mantell received his usual succession of scientific visitors to see the new specimen – Northampton, Lyell, Buckland, and Phillips, among others. On 12 May he and Melville worked out the positioning of teeth within *Iguanodon*'s upper and lower jaws. At the Royal Society six days later Gideon exhibited his iguanodon jaw to a small audience once more distracted from science by uprisings in Paris. Despite recurrent illness, he then completed a technical analysis of *Iguanodon*'s jaws and teeth. Sent and read to the Royal Society on the twenty-fifth, it resolved problems that had puzzled Cuvier

16. *Iguanodon* jaw. Captain (later Major) Richard Thomas William Lambart Brickenden (1809–1900), of Warminglid, Sussex (later Elgin, Morayshire, Scotland), exchanged letters with Gideon from 1848 to 1852 (LB–GM, ATL; GM–LB, BMNH). Dr. Alexander Gordon Melville (dates unknown) co-authored *The Dodo* (London, 1848) with Hugh Edwin Strickland; he was later professor of zoology at Queen's College, Galway (*Petrif*, p. 250n). J (25 Mar, 1, 5, 6 Apr 1848). GM–AS, 25 Mar 1848 (ATL); GM–WB, ca. 18 Apr 1848 (RS). LB–GM, 14, 20 Mar, 2 Apr, 31 May, 5, 19 June, 2, 29 Oct, 7, 21 Dec 1848; 18, 26 Nov, 26 Dec 1850 (ATL); GM–LB, 30 Mar, 5, 14 Apr, 1 June, late Oct, 31 Oct, 15 Dec 1848 (BMNH). LB–GM, 18 June 1848 (BMNH). GM–BS, 9 Apr, 6 June, 27 Oct 1848 (Yale). Leonard Horner–Charles Bunbury, 15 May 1848 (in Katherine M. Lyell., ed., *Memoirs of Leonard Horner* (2 vols.; London, 1890), II, 128–129.

and others since 1823. Still obsessed with size, however, Gideon concluded that *Iguanodon*'s jaw may have been as much as four feet long; Owen had limited its entire head to two and a half.

In this same paper, Mantell also described part of an upper jaw found by himself at Cuckfield in 1838, which Melville now analyzed in conjunction with the lower. Remarkably (as Silliman had already been informed on 9 April), the jaws of the iguanodon differed from those of the iguana and every other known reptile, living or fossil, its nearest analogue being found among herbivorous mammalia, like the sloth. In *Iguanodon*, Gideon determined, we can see how a gigantic terrestrial reptile successfully maintained its basic anatomical structure despite small, simple modifications required by its size and exclusively vegetarian diet. Some further remarks clearly presupposed a concept known sixty years later as the ecological niche. *Iguanodon,* he concluded, "occupied the same relative station in the scale of being, and fulfilled the same general purpose in the economy of nature, as the mastodons, mammoths, and mylodons of the Tertiary period and the large pachyderms of modern times" (198). For this superb paper, Gideon received the Royal Society's gold medal, the highest recognition British science could bestow.

In a startling finale, Mantell reidentified a specimen formerly presented to the Royal Society (in 1841) as part of the lower jaw of a young iguanodon. While similar, it now appeared sufficiently distinct to deserve a separate designation: *Regnosaurus Northamptoni.* The Regni had been an ancient Sussex tribe, so for Gideon the generic name meant "Sussex saurian." It was not of his own invention, however, having been suggested to him by Conybeare (for *Iguanodon*) in November 1824. Soon to retire as president of the Royal Society, Northampton had been Mantell's friend and consistent supporter for years. Though possibly a stegosaur (and if so, another major first by Gideon), *Regnosaurus* is still known only from this one specimen, which has been variously identified. Nevertheless, it was Gideon's fifth dinosaur.[17]

---

17. An attempted Chartist rally failed to materialize on 10 April, in part because Gideon and other aging reformers of his generation were becoming more conservative. J (8, 9, 10, 16, 20 Apr, 12, 18, 25 May 1848). GM, "On the Structure of the Jaws and Teeth of the Iguanodon," *PT,* 138 (1848), 183–202; *Lit Gaz,* 17 June 1848; *AJSA,* ns 6 (Nov 1848), 429–431. While the phrase "ecological niche" was a twentieth-century invention, the concept itself predates 1848; Buckland (*G&M,* I, 216) and Owen (1842, p. 197), for example, were somewhat familiar with it. WDC–GM, ca. 23 Nov 1824 (ATL). *Regnosaurus* (see also *Petrif,* pp. 333–335) was equated with *Hylaeosaurus* by Rodney Steel in 1969; John H. Ostrom later suggested that it might be a sauropod instead. More recently, George Olshevsky has called it "a rare Early Cretaceous stegosaur" and "by a wide margin the first stegosaur ever to be scientifically described."

Stirred to further productivity by his striking success, Mantell wrote a more general notice of organic remains recently discovered in the Wealden formation. Sent to the Geological Society on 31 May, it was read two weeks later at a "very full" meeting of 14 June. In this paper he reviewed several of his own finds from the Isle of Wight, the new jaw, and *Regnosaurus* but emphasized spectacular finds (some named for himself) made in the Wealden of Germany by Dr. Wilhelm Dunker. Besides *Regnosaurus,* the established genera now included *Plesiosaurus, Megalosaurus, Iguanodon* (of which numerous specimens had recently been discovered at Brook Bay and Hastings), *Hylaeosaurus,* and the crocodile *Goniopholis.* Dunker added *Macrorhynchus,* a gavial, from Germany. Less certain were bones from Wight and elsewhere attributed to *Cetiosaurus, Poekilopleuron,* and *Streptospondylus.* Another possible saurian, *Pholidosaurus,* had been proposed by Von Meyer. With more and more of its great creatures acknowledged, the Wealden was coming to be regarded (in Silliman's apt words) as Europe's Westminster Abbey of prehistoric life.

In both papers Mantell further refined his concept of the iguanodon, a gigantic herbivorous reptile bulky and massive as an elephant, its exclusively vegetable diet requiring a large abdomen. *Iguanodon*'s limbs must have been proportional to its body, the hinder ones resembling those of the hippopotamus and rhinoceros, with strong, short feet and broad, horny toes; the less bulky forelegs were adapted for seizing plants and pulling down branches. *Iguanodon* masticated powerfully, living on conifers, arborescent ferns, and cycads. Like the sloths, moreover, it possessed a large, prehensile tongue and fleshy lips capable of being protruded and retracted. Gideon's bold but well-considered speculations animated *Iguanodon* with a degree of physiological and behavioral specificity unmatched by any other saurian then known.[18]

Once apprized of Mantell's find, Owen deliberately sought to minimize its importance. Thus, when Gideon exhibited his fossil jaw to the Geological Society on 31 May, his rival claimed to have a more perfect one, which he would bring to a subsequent meeting! This crafty, well-staged announcement not only shocked and embarrassed Mantell (who had selflessly bestowed some valuable moa bones upon Owen less than six months before) but also jeopardized the publishability of his Royal Society paper. Not read fully at the next evening's meeting on 1 June, it was only "read in," a previous paper having taken too much time. Owen appeared that night also, with a drawing others were not permitted to study. "The most

18. GM, "A brief Notice of Organic Remains recently Discovered in the Wealden Formation," *QJGS,* 5 (1849), 37–43. GM–Sir H. De la Beche, 13 June 1848 (Nat. Mus. Wales); J (1, 14 June, 25 July 1848).

extraordinary incident has occurred," Mantell then wrote Silliman on 6 June; "another specimen about half the size of mine has been found at Horsham. Like mine, it is the right dentary bone, and has five or six successional teeth in place, but no mature ones. It is imperfect at the symphysis [chin], but more entire posteriorly than mine." Where had the evil magician obtained it?

Owen's specimen belonged to the same George B. Holmes of Horsham who had previously contributed finds to Owen's discussion of *Iguanodon* in 1841. As a neighbor and fellow collector, however, Brickenden had seen Holmes's jaw and was able to supply Mantell with some particulars; it was from a very young iguanodon and only one-third the size of his own specimen. Owen, for his part, was likewise anxious to learn more about Gideon's. With that purpose in mind, he sent Holmes to meet Mantell at the British Museum on 31 July. As this obvious ploy failed to deceive, Owen then wrote directly to solicit additional information. "The designing effrontery of this request," Gideon protested, "is too obvious even to me!" He resolved to avoid Owen in the future, but regretted having to become so reserved and selfish as persons he despised.[19]

His previous collaboration with Melville having worked well, Mantell initiated another in November. They visited W. D. Saull's collection on the fifteenth, to see his Wealden bones, and the British Museum two days later, "with a view of determining the true character of the vertebral column of the iguanodon." Several more such conferences followed, as Gideon probably sought to undermine Owen's assured quadrupedal identification of *Iguanodon*. On 11 December, in particular, he and Melville spent two hours at Horsham looking over Holmes's Wealden fossils. Though not permitted to make either drawings or notes, Gideon analyzed the collection shrewdly, convincing himself that many of Owen's reconstructions in 1841 had been based on imperfect data.

In January 1849 Mantell kept George Scharf busy for a week drawing vertebrae and other bones for associated memoirs on the osteology of the iguanodon that he and Melville completed at 11 P.M. on the fourteenth. Though the Royal Society received both parts the next day, they were not read until 8 March – a meeting Owen did not attend. Northampton, Murchison, and others then praised Gideon freely and the library table was covered with his specimens, attracting almost all the fellows.

---

19. GM–BS, 6 June 1848 (Yale); J (31 July, 1, 3 Aug 1848). G. B. Holmes–RO, 1 Aug 1848 (BMNH). RO–GM, 1 Aug 1848 (ATL), with draft reply. G. B. Holmes–GM, 3 Aug, 11 Nov 1848 (ATL). See John A. Cooper, "The Life and Work of George Bax Holmes (1803–1887) of Horsham, Sussex: A Quaker Vertebrate Fossil Collector," *Archives of Natural History*, 19 (1992), 379–400.

Mantell's paper deliberately echoed his 1841 essay on saurian osteology, augmenting and correcting it as needed. In large part, however, it was also Gideon's long-delayed reply to Owen's famous memoir of the same year. In the published version (1842), Owen had emphasized that all three of his "dinosaurians" exhibited sacra composed of five fused vertebrae. But Mantell had borrowed from Saull – and then further developed – the same fine *Iguanodon* sacrum in matrix that Owen had described. When more fully exposed, it proved to consist of *six* vertebrae rather than five. Owen had also transferred to his misidentified *Cetiosaurus* some angular caudal vertebrae that Melville now restored to *Iguanodon*. As before, in *Wight*, Gideon no longer attributed a "horn" to his revised iguanodon, but continued to insist (contra Owen) that *Hylaeosaurus* had spines along its back. Further remarks served not only to contradict Owen at other points but to establish the reasonability, competence, and sometimes brilliance of Mantell's long-continued involvement with *Iguanodon*. Arguing masterfully from persuasive evidence, he was again solving problems of fifteen and even twenty-five years' standing. In conclusion, Gideon affirmed that except for its cranium, sternum, forearms, wrists, and ankles, *Iguanodon* was now fully ascertained, its pectoral arch, arm, and vertebral column having been determined for the first time in the present paper. This new evidence confirmed his earlier inferences; he continued to believe, moreover, that *Iguanodon*'s forelimbs were long, slender, and prehensile while the hind ones were strong and massive, like those of the hippopotamus. With this, Gideon Mantell explicitly concluded his twenty-five-year effort to reconstruct *Iguanodon*, of whose former existence a few isolated, water-worn teeth had once been his sole evidence.[20]

20. J (15, 17, 22 Nov, 2, 7, 9, 11, 19, 21, 22 Dec 1848; 6, 8, 11, 12, 14, 15 Jan, 8, 9, 10 Mar 1849). GM, "Additional Observations on the Osteology of the Iguanodon and Hylaeosaurus," *PT,* 139 (1849), 271–305 (incl. A. G. Melville, "Notes on the Vertebral Column of the Iguanodon," pp. 285–300); *Lit Gaz,* 7 Apr 1849; *AJSA,* ns 7 (May 1849), 439–441.

12

Petrifactions and Their Teachings

Despite Gideon Mantell's medal and his self-proclaimed withdrawal from paleontological research, dinosaur hunting was far too interesting and rewarding ever for him to abandon. On 31 July 1849, therefore, Mantell visited J. S. Bowerbank (in Highbury Grove, Islington) to see bones of a "young iguanodon" recently obtained from the Isle of Wight. The specimen consisted of some badly crushed vertebrae, together with other bones, but from it Gideon attempted to reconstruct the neck structure of an immature iguanodon. He added illustrations and remarks concerning it to Plate XXIX of his just-given paper before publication.[1]

Pelorosaurus

Less than two months later, on 21 September, George Fowlestone of Ryde (a professional lapidary) brought Mantell the head of an enormous *Iguanodon* tibia, fifty-eight inches in circumference, and portions of two unknown bones. Four days after that Gideon was off to Oxford, where he spent hours in Buckland's museum trying to sort out *Iguanodon* from *Cetiosaurus* and an undetermined alternative. On the twenty-seventh he visited Malling Hill, near Lewes, to see a fine humerus from Cuckfield collected two years earlier by Peter Fuller, a miller. Some fifty-four inches long, it was similar to, but distinct from, the equivalent bone in *Iguanodon*. Gideon returned a month later with Dinkel for a drawing. Combining the new humerus with some further bones in his collection not

---

1. On Plate XXIX of his "Additional Observations" paper, Mantell depicted "cervical vertebrae of a very young Iguanodon" (also *Petrif,* pp. 264–265). Much later, in 1974, these bones were reassigned by Peter M. Galton to the possibly warm-blooded dinosaur *Hypsilophodon* (a genus established on other evidence by Thomas Henry Huxley in 1870). *Hypsilophodon* is indeed "like a very much smaller and lightly built Iguanodon" (Charig, p. 118) but lacks the thumbspike that Gideon would never assign properly. Though *Hypsilophodon* would have been Mantell's seventh dinosaur, he failed to recognize it as a separate genus, having only a few vertebrae on which to base his identification.

otherwise assignable, he now proposed the existence of yet another saurian.

Mantell had discovered his sixth and largest dinosaur, which as of 4 November (when Buckland dropped by to see the evidence) he was calling *Colossosaurus*. Fuller then wrote Gideon to accept his offer of seven guineas for the humerus, provided that copies of *Wonders* and *Medals* be included also; his bone arrived at Chester Square on the twelfth. Writing to Silliman three days later, Mantell revealed that he had named his new find *Pelorosaurus*, from the Homeric word for monster. Though he would never have enough evidence to know it, Gideon had discovered the first brachiosaurid, one of a family of ponderous quadrupeds as big as he first thought *Iguanodon* to be – in fact, the largest dinosaurs of all.

Mantell began writing his memoir on *Pelorosaurus Conybeari* the next day; Dinkel then finished its necessary drawings. Though received by the Royal Society on 2 November, the new paper and its dinosaur languished till 14 February 1850, when Gideon also presented his second on belemnites. Having long thought that an enormous, previously unidentified lizard remained within the Wealden strata, Mantell declared in his Valentine's Day paper, he had at last obtained sufficient evidence to announce the existence of a terrestrial reptile contemporary with *Iguanodon* and at least as big. The great humerus he now displayed had been found by Peter Fuller in 1847; it was from the same remarkable quarry in which Gideon had originally obtained his iguanodon teeth and jaw. Once reassembled from its various pieces, the humerus proved to be four and a half feet long. As comparisons soon established its uniqueness from any previously known, Mantell proposed his new genus, to which he attributed not only the humerus but also several vertebrae found by himself many years since. They had originally been identified as iguanodontian by Mantell, then as cetiosaurian by Owen, and next by Mantell and Melville as from a new species of cetiosaur, but were now seen to be different again. Gideon also discussed the general similarity of terrestrial flora and fauna during Oolitic, Wealden, and Cretaceous times. This realization, he said, had taken him to Oxford, where he found additional pelorosaur remains and had a useful discussion with Buckland. Extrapolating from its humerus alone (in Cuvier's manner), Gideon estimated the length of *Pelorosaurus* at eighty-one feet. He would subsequently confirm this new dinosaur with further discoveries, including a scapula in 1851 and some splendid foreleg bones (from a second species) the next year.[2]

2. *Pelorosaurus*. J (31 July 1849); GM, "Additional Observations," pp. 302–303. J (21, 25, 27 Sept, 27 Oct, 4, 6 Nov 1849). H. E. Strictland–GM, 2 Oct, 5 Nov 1849 (ATL). P. Fuller–GM, 7 Nov 1849 (in J, ATL). J (9, 12 Nov 1849). Brachiosaurids now include

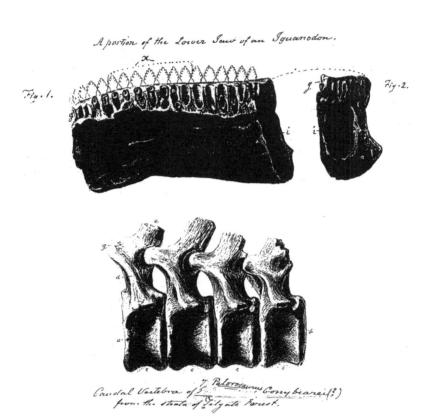

A portion of the Lower Jaw of an Iguanodon.

Fig. 1.   *Fig. 2.*

Caudal Vertebræ of Ĩ. Pelorosaurus Conybearii(?)
from the strata of Tilgate Forest.

Fig. 13. One of a series of four consecutive *caudal* vertebræ of a Saurian ; ~~probably of~~
~~the Iguanodon~~ *the Cetiosaurus.*

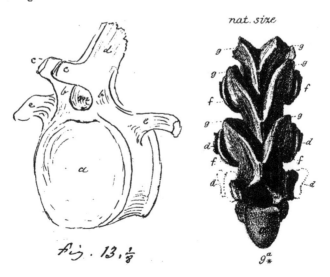

nat. size

Fig. 13. ⅔

9a
⁕

A Pictorial Atlas

Soon after announcing *Pelorosaurus,* Mantell began work on his next book, *A Pictorial Atlas of Fossil Remains,* the compiling of which had been suggested to him by Henry Bohn. This bulky folio reprinted seventy-four full-page colored plates from Parkinson's *Organic Remains* (1804–1811) and Edmund Artis' *Antediluvian Phytology* (1838), a book on fossil botany, while updating their captions. With considerable assistance from John Morris (1816–1886), author of *A Catalogue of British Fossils* (1843), Gideon attempted to reidentify each depicted specimen. He and Morris met for that purpose on 2 April 1850, subsequently exchanging notes. The manuscript and additional woodblocks were sent to Bohn precisely one month later. Gideon then received a proof copy of the entire

Pelorosaurus (80 feet), *Brachiosaurus* (75–90 feet), and *Supersaurus* (over 100 feet). *Pelorosaurus* derived its name from the following remarks: "The word Colossus signifies simply a statue. . . . Although therefore the word Colossosaurus might be indulgently accepted, it would after all . . . mean only the Statuesaurus. Now, there is a Greek word, and a well known one, *Pelor,* signifying "monster" and as such it is applied more than once by Homer to the monstrous Cyclops. . . . If you call your beast Pelorosaurus, you will mean the monster lizard and to all classical scholars will convey the idea of something unusually gigantic." (The Řev. Charles Pritchard [of Clapham Common, F. R. S.]-GM, ca. 12–15 Nov 1849 [ATL]). Also GM–BS, 30 Oct, 9, 15, 23 Nov 1849 (Yale); GM–LB, 16 Nov 1849 (BMNH); *Petrif,* pp. 330–333.

J (16, 20, 22 Nov 1849). GM, "On the Pelorosaurus," *PT,* 140 (1850), 379–390. Mantell had discovered and depicted significant pelorosaurian remains so early as 1826 (*Illus,* Plate XV). Specimens he then identified as iguanodontian were subsequently reidentified by Owen as cetiosaurian (*Geol SE Engl,* p. 296; GM, 1839 catalog, #544–550; Owen, 1842, pp. 94–98). In 1848 (*PT,* p. 297) Gideon then proposed them as a new species of *Cetiosaurus.* Scapula: J (12, 14 June 1851). Foreleg bones: J (18, 25, 26, 27 Feb, 4, 5 Mar, 8 May, 29 July 1852); GM–BS, 1 Apr 1852 (Yale). GM's "second species" of *Pelorosaurus* may, however, be an entirely different dinosaur altogether. See *The Dinosauria,* p. 398 and Table 16.1. If so, it would have been his eighth.

Figure 12.1. Four new dinosaurs (opposite): *Top: Regnosaurus,* drawing by George Scharf as "A portion of the Lower Jaw of an Iguanodon" (*Philosophical Transactions of the Royal Society,* 131 [1841] Plate V) but reidentified by GM in 1848. *Second from top: Pelorosaurus,* drawing by Joseph Dinkel (*Philosophical Transactions of the Royal Society,* 140 [1850], Plate XXII). *Bottom left: Cetiosaurus,* drawing by Mantell as "probably of the Iguanodon" (*Philosophical Transactions of the Royal Society,* 131 [1841], Plate IX; inscribed caption, ATL) but reidentified by GM, following Owen, in 1844 (*Medals,* II, 726–728). *Bottom right: Hypsilophodon,* identified by Mantell as "a very young Iguanodon" (drawing by J. Dinkel, *Philosophical Transactions of the Royal Society,* 139 [1849], Plate XXIX; *Petrif,* pp. 264–265) but reidentified by Galton in 1974.

book on 19 July – and (from an entirely different source) two further fragments of iguanodon jaw!

Written probably in May and June, the final section of *Atlas* comprised
almost three dozen pages of "Supplementary Notes" in no particular order, representing Mantell's opinions on whatever paleontological topics
moved him to speak out: fossil bears and caverns (paying tribute to Buckland), the belemnite controversy, fossil birds (in New Zealand, the
Wealden, and America), paleobotanical classification, cephalopods, coal,
corals, Cuvier, sloths (citing Darwin), microfossils, the Irish elk, and (inevitably) saurians. In no area was the progress of paleontological knowledge more evident than in this last. "Although when Mr. Parkinson's work
was published, many fossil bones and teeth of reptiles had been discovered in various parts of England," Gideon stressed, "yet the abundance
and variety, and the extraordinary modification of form and structure of
this class of vertebrated animals, which prevailed throughout the Secondary geological formations, were not for a moment suspected" (192).

Mantell's note on fossil reptiles ("cold-blooded oviparous quadrupeds")
specified that they had lived throughout the entire Mesozoic and as far
back as the Carboniferous. He then saw them gradually declining in numbers and species at the close of the Cretaceous, until replaced in the Tertiary by warm-blooded mammals and birds. An extended note on *Iguanodon* became deeply personal. Written fairly soon after 31 May (there is
a reference to the Horsham jaw), it detailed Gideon's struggles to overcome the scientific opposition of first Cuvier and then Owen before
achieving a final synthesis with Melville. At the end he added in proof a
note on his acquisition of two further fragments of iguanodon jaw, those
given him on 19 July by Samuel Beckles.[3]

His Last Book

Mantell's occasional but flattering references to the British Museum in *Atlas* foreshadowed his last book, *Petrifactions and Their Teachings; or, a*

3. GM, *A Pictorial Atlas of Fossil Remains* (London: Bohn, 1850); author's copy, ATL. H.
Bohn–GM, 8, 21 Oct 1849 (ATL), recording changes in the title. J (3 Apr, 2 May, 29 June,
19, 30 July, 14, 17, 26, 27 Sept, 23 Oct 1850). GM–WBDM, 29 Oct 1850 (ATL). Gideon
himself bought a number of Parkinson's specimens at auction in Oct 1827. For John Morris, see *DNB* and *Hist GS,* pp. 159–161. Though probably written in August, Gideon's
loyal dedication to William Buckland had been tactfully backdated to January because
the Dean of Westminster (who, so late as November 1849, vigorously defended GM's
right to the Royal Medal) was afflicted in February 1850 with incurable mental debility.
Buckland, confined for a time in a Clapham Common asylum, died in 1856.

Hand-Book to the Gallery of Organic Remains of the British Museum (1851), which – continuing some remarkably sustained creativity – he began as soon as the *Atlas* was done. Thus, on 23 November, Gideon visited the British Museum on Great Russell Street (the Natural History Museum, of course, did not then exist) to make notes on its fossil gallery. By 9 January 1851 he was lecturing on "Petrifactions and Their Teachings" to the Whittington Club in London. Joseph Dinkel began his drawings for the book in April, and Mantell spent several more days that month at the Museum, noting its principal fossils. Throughout the latter part of May – and well into July – he was "hard at work night and day" on the book, while Dinkel completed its numerous wood engravings. Frequently interrupted, and set aside entirely during the latter part of August, *Petrifactions* was seriously resumed the following month. Gideon sent the manuscript of Chapter 4 to his printer, Richard Clay, on the eighth, and some further pages on the twelfth, then spent another week in the Museum looking over iguanodon, hylaeosaurus, and New Zealand birds' bones.

Both the long hours of note-taking and the continued sitting at his desk that followed inflicted terrible suffering, but Mantell devoted himself to this book as if it were his final testament. On 25 September, after reinspecting Room VI at the Museum, he visited Bohn, whose procrustean insistence forced him to limit the completed volume to five hundred pages, which meant throwing out half of what he had written about rooms V and VI. Gideon sent thirty more pages of manuscript to Clay on the twenty-ninth, revisited the Museum again, and came to final agreement with Bohn on 6 October. Having written his dedication and preface on the eleventh, he began the index four days later, completing it on the seventeenth. A preliminary first copy arrived in four more days, and the book was out by month's end, though published officially on 1 November. Bohn paid him £105 for it on the twenty-first.

The Museum's collection of organic remains, situated on its far north side, extended from east to west in a series of six rooms. Within them, fossils were arranged by taxonomic affinity rather than geological period. The first room, therefore, housed fossil plants. Room II, in relative disarray, included moa bones and miscellaneous fauna; III and IV, reptiles; V, fishes (with one gigantic Irish elk); and VI, mammalia. Mantell's handbook, accordingly, had six subdivided chapters, one for each room. Each began with a diagram of the room and a synopsis of its contents, then gave more detailed attention to whatever fossils and minerals most seemed to warrant it. Since the whole book extended to 496 pages, just short of the limit imposed by Bohn, it would have taken a great deal of time and stamina to follow Gideon's commentary throughout.

Not surprisingly, Mantell's extensive remarks on rooms III and IV, both devoted to fossil reptiles, dominated the book. Together, they comprised the longest, fullest, and most systematic exposition of his views regarding archosaurs that Gideon would ever write. Being essentially his final one as well, these two extended chapters are of some importance. In reading them, moreover, one detects the pride, irony, and nostalgia of a successful but disappointed man reviewing his life's work. (The opinions, as we would expect, are similar to those in *Atlas* but a good deal fuller.) Throughout this section, Gideon conducted a running battle with Owen, responding to almost every disputed point involving saurians that their various quarrels had raised.

Beginning with its Part 2, Mantell's Chapter 3 was an orderly and lucid treatise on fossil reptiles – his *replacement* for Owen's of 1842. Citing Cuvier's vague pronouncement that there had been an Age of Reptiles, Gideon systematically discussed its duration, then the nature of fossil evidence. After reviewing fossil turtles, he characterized saurian teeth and vertebrae. Both, and the sacra of dinosaurs in particular, resembled those of the larger mammals. Part 3 described the Swanage Crocodile and other crocodilians, frogs and salamanders, pterodactyls, and *Mosasaurus*. Utilizing Gideon's frontispiece of 1827 and title of 1833, Part 4 nostalgically surveyed the geology of southeast England. Within it, for the first time, Mantell specifically located his iguanodon quarry at Whiteman's Green. "From that quarry, long since filled up and the area covered by pasturage and gardens," he recalled, "I collected the first and most interesting fossil remains of the iguanodon, hylaeosaurus, pelorosaurus, and other stupendous creatures whose existence was previously unknown and unsuspected" (204). For several paragraphs thereafter, Gideon's account became unabashedly autobiographical, stressing his eventual discovery of the fluviatile origin of the Wealden formation. He then listed the various strata and traced them from London to the coast, concluding with a review of Wealden fauna and flora.

Part 5, comprising ninety pages, dealt solely with *Iguanodon,* the discovery of which Mantell attributed entirely to himself. He described the iguana and its teeth, then the jaws and dentition of *Iguanodon,* Saull's iguanodon sacrum, and the six caudal vertebrae found by Trotter. Gideon noted that he had also acquired more than thirty iguanodon vertebrae during the past year; should Providence grant him life and health, he hoped to continue his investigations of saurian osteology. As for other portions of the skeleton, Gideon remained doubly mistaken about his famous "horn," which he thought perhaps not iguanodontian but certainly a horn. (In 1851, however, Owen had tentatively supposed it digital instead.) Following an extended analysis of the Maidstone Iguanodon, which was majestically displayed in a hexagonal floor case beneath the central north window, Mantell attempted to restore the living animal. Even more than formerly, he stressed its sloth-like characteristics. *Iguanodon*, Gideon suggested, had been the most mammalian of all reptiles, just as the sloth was the most reptilian of mammals. He still believed that *Iguanodon* might have been sixty to seventy feet long. Part 6 described *Hylaeosaurus* (triumphantly establishing the reality of its dorsal spines), *Megalosaurus, Pelorosaurus,* and *Regnosaurus,* then the country of the iguanodon.[4]

Mantell and Owen

Because of his discoveries and their significance, Gideon Mantell enjoyed both honors and friends during his later years, including recognition and respect throughout Europe and the United States. His lectures and books were continually in demand and did much to overcome lingering prejudices that had long retarded popular acceptance of geology and its findings. It is a tragedy that during the same years he was also menaced by an unscrupulous, unnecessarily competitive adversary who did what he could to deprive Gideon of still further rewards. Since Gideon's quarrels with Richard Owen have never been fully chronicled, it is appropriate to summarize their relationship, which began sometime before 5 December 1832 (their earliest-known association), when Owen attended the Geological Society to hear Mantell announce his discovery of *Hylaeosaurus*. In his *Geology of South-East England* (1833), Gideon complimented Owen's paper on the nautilus. Surviving correspondence between them dates from

---

4. GM, *Petrifactions and Their Teachings* (London: Bohn, 1851). Reviewed in *Lit Gaz,* 15 Nov 1851, pp. 767–769: "In many respects it is a personal narrative of his own researches, and those of his worthy sons. . . . The name of Mantell will be forever associated with the history of the British Museum" (767). Other reviews in *AJSA,* ns 13 (May 1852), 407–409; *The British and Foreign Medical Review* (Jan 1852), p. 249; and *The Atlas* (clipping untraced). "Lignographs of Petrifactions and Their Teachings," proof sheets annotated by GM (Nov 1851) and WBDM (Apr 1858), ATL. J (23 Nov 1850, 9 Jan, 1, 11, 18, 24, 25, 26, 29 Apr, 16, 17, 21, 22 [quoted], 26, 30 May, 3, 5 June, 3, 15, 25 July, 27 Oct, 21 Nov 1851).

The chief interruptions were Benjamin Silliman's visits to GM and the Great Exhibition. Silliman visits: J (11–31 Mar 1851). BS–GM, 20 Feb, 15 Apr, 23 May, 15 July, 13 Aug 1851 (all from the Continent; ATL). J (16–31 Aug 1851). BS, *A Visit to Europe in 1851* (2 vols.; New York, 1853), I, 75–109; II, 385–446. BS, "Notes on a Tour to Europe in the Spring and Summer of 1851" (five-volume manuscript, Yale), I, 79–127; V, 70–172. GM–RNM letters (ATL) reiterate details. The Great Exhibition: J (1 May–11 Oct 1851, *passim*); GM–RNM, 2, 16 May, 12 Oct 1851 (ATL); GM–WBDM, 21 Oct 1851 (ATL); GM–BS, 21 Oct, 11 Nov 1851 (Yale).

1835, the year in which Owen confirmed the supposedly avian nature of Gideon's Wealden "bird" bones. So long as Mantell remained at Brighton, their interactions were minimal. After December 1832, no further mentions of Owen appear in Gideon's journal till 28 May 1841, when Mantell attended a lecture of Owen's on fossils. Yet Owen had been among those members of the Geological Society who in 1838 urged the British Museum to complete its purchase of Gideon's museum, and they had dined together at Owen's home in November 1840, sharing their common interest in microscopics. Both, moreover, had been elected to the prestigious Athenaeum Club that year, and on the same day.

By 1841, however, Mantell and Owen were locked in a serious contest for credibility, largely because the latter had so tactlessly dismissed several of Gideon's pioneer conjectures. Thus, the first of four major clashes between them that year resulted from Owen's blunt pronouncement (in *Odontographia*) that the teeth of iguanodons and iguanas were entirely distinct. A second quarrel followed when Gideon presented the Royal Society a bone he took to be a part of an iguanodon's lower jaw, Owen disagreeing. (Mantell himself later reidentified this specimen as *Regnosaurus*.) The third disagreement concerned fossil turtles, as Owen significantly reassigned one of Gideon's. Then, on 2 August, Owen read his historic paper on dinosaurs to the British Association at Plymouth. Full of backhanded insults disguised as science, this momentous publication particularly exasperated Mantell, who struggled to believe that Owen's numerous "corrections" of his work represented nothing more than advancements of knowledge that he, a votary, ought properly to applaud. At last, he wrote a complaining letter to *Literary Gazette*. Two months after Owen's paper was presented, and perhaps to some extent because of it, Gideon suffered his crippling carriage accident.[5]

Nonetheless, Mantell's general attitude toward Owen's work continued positive, and in May 1842 he attended five of Owen's lectures on the nervous system at the College of Surgeons. Some private letters of his written early in 1843 criticized one instance of Owen's thinking as absurd, but Gideon's direct communications with the fecund Hunterian professor re-

---

5. Owen. In Chapter 8 I traced Owen's career through 1835. In 1836 he was named Hunterian professor of comparative anatomy and physiology at the Royal College of Surgeons. Two years later he won the GS Wollaston medal; the RS medal followed (for a mistaken paper about belemnites) in 1846. Owen helped found the Royal Microscopical Society (1839; president, 1840–41). In 1842 he became joint conservator of the Hunterian Museum and was awarded a civil list pension of two hundred pounds; he became sole conservator in 1849. His major publications have been noted within my text. (For additional details and his subsequent reputation, see *DNB, DSB*, the biographies by Owen and Rupke, the work of Adrian Desmond, Charles Darwin's autobiography, and Lynn Barber, *The Heyday of Natural History* [London, 1980], Chap. 12.) GM–RO letters are at BMNH and APS; RO–GM, at ATL.

mained gracious, friendly, and admiring. There were difficulties even so. "You misunderstood me," Mantell assured Silliman in 1843, "if you suppose I have had any personal quarrel with Owen. No such thing, we are upon good acquaintance terms, because I have not attacked his unwarrantable conduct, nor shall I." He was too ill to care "one straw" about worldly reputation. "But should I live," Gideon added, "then I shall assert my rights." He did live.

In *Medals* (1844), Mantell fully appreciated Owen, renaming *Hylaeosaurus armatus* in his honor as *H. Owenii* and deferring to him in several contexts. Writing privately to Silliman (28 February 1845), he characterized everything Owen had thus far published as "very excellent." While openly admiring Owen's work, however, Mantell continued to be wary of the man. His distrust grew considerably in December 1845, when Owen abruptly reversed an opinion of ten years' standing and endeavored to convince Mantell that his Wealden bird bones were in fact pterodactylian. Owen's paper to that effect, on the seventeenth, proved less than accurate regarding discussions between them a decade ago. "It is deeply to be deplored," Gideon asserted a few hours later, "that this eminent and highly gifted man can never act with candor or liberality." His reply to Owen, written two days after, was read to the Geological Society on 7 January 1846, eliciting further antagonism from both men.

On 5 March, however, Owen sent a copy of his newly published *History of British Fossil Mammals and Birds* to Mantell, who appeared briefly but honorably within it as an "original and successful explorer of the Wealden" (xv). Replying graciously the next day, Gideon thanked Owen "most warmly" for his invaluable present and congratulated him on the completion of this "imperishable monument of . . . genius, talents, untiring industry, and successful research." He regretted that circumstances would not permit him to prepare a volume on the flora and fauna of the country of the iguanodon, but promised to send Mrs. Owen a copy of *Animalcules* when it appeared. Shortly thereafter he did so, characterizing the book as "a very humble attempt to familiarize the general reader with a few principles of the science which Professor Owen has so greatly advanced." In April, Mantell sat through a further series of Owen's lectures. Following the last of them, on the twenty-fifth, Owen accompanied him partway home. In consequence of their reconciliation, Gideon attempted unsuccessfully to withdraw his reply to Owen regarding Wealden birds' bones, only to be informed by the secretary of the Geological Society that he was too late. When the reply appeared, some of Owen's unreliable cordiality evaporated.[6]

---

6. Cretaceous birds. J (10, 17, 19, 26, 28 May 1842). GM–RIM, 20 Feb 1843 (APS); GM–BS, 30 Mar 1843 (Yale); GM–RO, 5 Feb 1844 (BMNH); *Medals*, II, 734–735. RO–GM, 15 July 1844 (ATL). GM–BS, 28 Feb 1845 (Yale). RO, "On the Supposed Fossil

As Owen's paper of 1843 on *Dinornis* circulated throughout New Zealand, a feverish hunt for additional moa bones began, resulting in a series of important finds. Caught up in that enthusiasm, Walter Mantell dropped his father in England a brief note in 1846 affirming that he would himself be going into the interior. Beginning in January 1847, Walter reconnoitered known moa bone sites on the North Island and then went exploring, living with the Maori to enlist their aid while hoping to secure at least one moa that was still alive. He returned moaless to civilization on his father's birthday, 3 February, but in good spirits and with a collection of about eight hundred specimens, including many bones and some fragments of egg shell. The moa's egg was then entirely unknown in Europe.

When Walter's unique and valuable specimens of moa egg shell arrived from New Zealand in September 1847, Gideon immediately shared them with Owen, who replied with unfeigned gratitude. On receiving two boxes of Walter's moa finds in December, Mantell washed and sorted the bones in his drawing room, then donated them to Owen, who visited Chester Square on 2 January 1848 for that purpose and presented a Zoological Society paper on the moa little more than a week later. Appearing the same month, Gideon's *Wonders,* sixth edition, tentatively accepted Owen's position regarding Wealden "bird" bones and referred them to pterodactyls instead. His paper of 2 February, "On the Fossil Remains of Birds," paid Owen an immortal compliment for having deduced the essential nature of the moa correctly from a single fragment of bone. Sir Robert Peel then invited both men to a dinner party for sixteen on 26 February, and a very pleasant evening resulted. Gideon subsequently attended Owen's first two lectures in March.[7]

Bones of Birds from the Wealden," *QJGS,* 2 (1846), 96–102; J (17 Dec 1845); GM, "On the Fossil Remains of Birds in the Wealden Strata," *QJGS,* 2 (1846), 104–106. J (7 Jan, 5 Mar 1846); GM–RO, 6 Mar 1846 (BMNH); GM–Mrs. RO, 15 Sept 1846 (CUL). J (2, 8, 23, 25, 30 Apr 1846); GM–RO, 27 Apr 1846 (BMNH); D. T. Ansted–GM, 29 Apr 1846 (ATL); GM–RO, 2 May 1846 (BMNH). See also Chapter 11, note 12 above. In *A History of British Fossil Mammals, and Birds* (London, 1846), pp. 545–558, RO's supposed "Bird of the Chalk" (*Cimoliornis*) is in fact a pterosaur (Figs. 230, 231). All of his Eocene specimens, however, are legitimately avian. I wish to thank Dr. Peter Wellnhofer for providing this information.

7. Moa remains. J (11 June 1846); WBDM, moa notebook, 1847 (ATL); WBDM–GM, 3 Feb [GM's birthday], 20 June 1847 (ATL); J (4, 22–23 Sept 1847); GM–RO, ca. 24 Sept 1847 (APS); RO–GM, 25 Sept 1847 (ATL); GM-derived, "Discovery of the Eggs of the Moa or Dinornis of New Zealand," *Athenaeum,* 25 Sept 1847, p. 1013; GM–BS, 2 Nov 1847 (Yale). Owen generally: J (9, 11, 30 Mar, 8, 15 Apr 1847). GM–RO, ca. 24 Sept 1847 (APS); RO–GM, 25 Sept 1847 (ATL); GM–RO, 14, 25 Dec 1847 (BMNH); RO–GM, 25 Dec 1847 (*Petrif,* p. 487). J (2, 11 Jan 1848); GM–RO, 7 Jan 1848 (BMNH). J (26 Feb, 14, 18 Mar 1848).

By 3 Aug 1848, however, Gideon had good reason to believe that his high praise of Owen for identifying *Dinornis* from a single bone had been decidedly inappropriate (J).

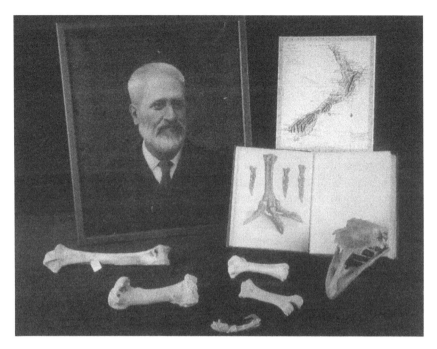

Figure 12.2. New Zealand composite. *Top left:* "Walter Baldock Durrant Mantell, 1886" (H. M. Gore, portrait in oils). *Top right:* Map of the coast of New Zealand, Plate 25 in Sidney Parkinson, *Journal of a Voyage to the South Seas* (London, 1784). Gideon Mantell, *A Pictorial Atlas of Fossil Remains* (London: Bohn, 1850), showing the frontispiece (tinted lithograph) by Joseph Dinkel, "Bones of the right foot of the moa, or extinct colossal ostrich-like bird of New Zealand found . . . at Waikaouaiti . . . by Walter Mantell." *Front foreground:* Maori trawling hook (mother-of-pearl and plaited flax). All items Alexander Turnbull Library except *Foreground:* Moa bones, including skull (collection of the Museum of New Zealand, Te Papa Tongarewa).

Yet after hearing Mantell's presentation on Reginald's belemnites at the Royal Society on 23 March (which politely refuted Owen's mistaken but medal-winning paper of 1844), Owen responded hysterically with snide ridicule and vicious personal attack. Though defended graciously by Buckland and Northampton – and vindicated by posterity – Gideon went home that night with deep regrets. "He was a man for whose genius I have the highest respect," Mantell told Silliman, "whose friendship I have tried to preserve, to whom I gave up all the treasures poor Walter sent me, kept my [drawing] room in a state of discomfort for weeks to suit his convenience . . . and yet because I dissented from his views in a single instance, could attack me in the most virulent and illiberal manner." After this, he concluded, "I can never put confidence in the apparent civility of Professor

Owen, and I must, to my great regret, keep aloof from him." This episode, rather than any previous, was the effective beginning of Gideon's profound disillusionment regarding Owen.

He attended Owen's further lecture of 22 April nevertheless. Nothing more took place between them until the disgraceful affair of the iguanodon's jaws on 31 May. Twice provoked, Mantell and Melville then retaliated with a fairly thoroughgoing demolition of Owen's dinosaurian criteria of 1842 and Gideon's triumphant reconstruction of a nearly complete (if still slightly conjectural) iguanodon. The identification of *Pelorosaurus* corrected yet another significant mistake by Owen, who had meanwhile begun to adulterate his science with metaphysics, which did nothing to enhance his professional reputation.[8]

Mantell's announcement of *Pelorosaurus* in November 1849 coincided with increased recognition of his later work, though it was never so freely bestowed as he would have liked. In particular, the Geological Committee of the Council of the Royal Society met that month to recommend Professor J. D. Forbes (who had written importantly on glaciers) for that year's Royal Medal. Gideon was the only other candidate. While not a member of the Committee, Owen had nonetheless appeared at both its meetings (as Mantell learned from Murchison) to ridicule his rival's work. After hearing Owen also, the full Council likewise supported Forbes. Unofficially informed of that decision, Gideon remonstrated against it to Buckland, who replied with kindly support. Mantell then wrote to the President and Council, requesting further consideration. They agreed to meet on 26 November for that purpose, which gave both sides a few days to organize. Significantly, this forthcoming meeting was the first at which Lyell would be present, having been called away from the others by his father's funeral.

At the fourth and final meeting, Owen appeared once again to argue against Mantell on procedural grounds (where his case was strongest); to assert the alleged superiority of physical geology to paleontology; and then to attack the substance of Gideon's recent papers, that on the iguan-

---

8. Other dealings. J (23 Mar 1848); GM–RO, 23 Mar 1848 (BMNH); Charles Darwin–Joseph Hooker, 10 May 1848. J (2 Apr 1848). GM–BS, 9 Apr 1848 (Yale). For the iguanodon jaws incident, see Chapter 11, note 16 above; for *Pelorosaurus*, note 2 above. In marked contrast to Gideon's generosity, when Owen received a shipment of NZ bird bones in July 1848, GM was not even allowed to see them! (GM–BS, 31 July 1848 [Yale]). Owen's bizarre metaphysical beliefs came to light most obviously in such publications as RO, *On the Archetype and Homologies of the Vertebrate Skeleton* (London, 1848) and *On the Nature of Limbs* (London, 1849). Gideon privately denounced the latter as transcendental quackery (J, 9 Feb 1849). For Thomas Henry Huxley's evaluation, see Owen, *Life*, II, 312–319.

odon jaw in particular. Dr. Mantell, he argued, was nothing more than a collector of fossils, furnishing specimens to distinguished scientists like himself. He also ridiculed Gideon's inference that *Iguanodon* had cheeks and prehensile lips. To all this, Lyell replied coolly and devastatingly, quoting praise of Gideon's researches from Owen's own works, Cuvier's famous retraction, and a powerful letter to the Council from Buckland declaring that any *one* of Mantell's recent papers – whether on *Iguanodon,* belemnites, or foraminifera – was worthy of the highest honor the Royal Society could bestow. In the final vote, Gideon carried all but two. Though Lyell and Murchison leaked the news to him immediately, he was officially informed at a meeting of the Geological Society on the thirtieth, where Owen sat opposite, "looking the picture of malevolence." At the Royal Society meeting on 6 December, which Owen did not attend, Mantell was congratulated all round – publicly for his medal, and privately for his victory over Owen. When the latter then came forward on 13 December, at a meeting of the Royal Society's Council (both being members), Gideon bowed stiffly and declined his proferred hand. "I will not abet such consummate hypocrisy," he affirmed thereafter.[9]

The next significant tiff between them took place on 27 February 1850, when Mantell read a paper on Walter's latest moa specimens to the Geological Society. Knowing that he would do so, Owen approached the Council of the Zoological Society the previous day and arranged to substitute a hasty fourth paper by himself on *Dinornis* for one on another topic he had been scheduled to give that night. In it, he described the feet of the genus – thereby anticipating Mantell, who was known to have had a spectacular pair specially mounted for display. Gideon read his "Notice of the Remains of the Dinornis" as scheduled on the twenty-seventh. Having come loaded with moa bones, Owen then commented "in his usual deprecating manner" on Mantell's presentation. Both Edward Forbes (who had received New Zealand specimens of his own a few days before) and Lyell supported Gideon and Walter, while Gideon himself replied to Owen rather sharply.[10]

Beginning the following October, just after Mantell had received Walter's box containing *Notornis,* a rare contemporary New Zealand bird, a further but even more ludicrous quarrel ensued. On the twenty-fourth, at

9. Royal Medal. J (18, 29 Oct, 16–30 Nov, 2, 6–13 Dec 1849). CL–GM, 29, 30 Nov 1849 (ATL); GM–BS, 14 Dec 1849 (Yale). RIM–AS, 23 Nov 1849, 17 July 1851 (CUL); AS–RIM, 23 Nov 1849, 15 July 1851 (GS); see also Lyell, *LLJ,* II, 105–113. Gideon actually received the medal on 4 Jan 1850 (J). See pp. 230–232 above.

10. J (27 Feb, 4 Mar 1850); see GM, *Atlas,* frontispiece. GM, "Notice of the Remains of the Dinornis" *QJGS,* 6 (1850), 319–343; RO, "On Dinornis (Part IV)," *Trans ZS,* 4 (1850), 1–20. *Athenaeum,* 2 Mar 1850, p. 237. GM–BS, 17 May 1850.

a meeting of the Royal Society, Owen congratulated Gideon for his new bird and asked to reprint just one lithograph (Saull's *Iguanodon* sacrum) from Mantell's osteological paper of 1849 – but the number was soon raised to nine. In actuality, however, Owen had already petitioned the Society's Council to the same effect on the twenty-third; it approved his request the next day. Gideon brought up the matter again at the Council's next meeting, on the thirty-first, only to discover that everyone who voted to approve Owen's request had done so believing that the plates in question were his own! On 14 November, at a third Council meeting, Owen defended his claim on the grounds that the specimens depicted in Mantell's plates had been *described* by him in his *Report on British Fossil Reptiles,* II (1842). "The poor man," Gideon commented with mock solicitude, "must be demented." He next wrote the Society a letter, which was taken up by the Council at its meeting of the twenty-eighth. According to Mantell, Owen then "made a jesuitical apology, and was thoroughly humbled; not a shadow of a reasonable excuse could he offer." Gideon replied forcefully, exposing the "injustice and misrepresentation" of Owen's original application. Every one of the Council members present, he believed, "seemed to be convinced that Owen for once had been caught and exposed in his duplicity."[11]

In May and June 1851, while at work on *Petrifactions,* Mantell unsuccessfully opposed further Geological Society grants to Owen, whom he accused in private of monopolizing money, specimens, and discoveries. Before the end of July, Owen's treatise on Cretaceous reptiles appeared, full of statements to which Gideon took offense. It began, for example, with a discussion of fossil turtles that revived Owen's differences with Mantell in 1841 and charged him with a further error. In discussing *Mosasaurus* Owen perforce admitted that Gideon had found the first example in England (just as he had found the first turtle remains in the English Chalk). Similarly, Mantell had first noticed some teeth that Owen the next year (1840) described as belonging to a new mosasauroid reptile, *Leiodon;* Edward Charlesworth had added an important jawbone fragment in his *London Geological Journal* of 1846. Acceding to another item in the same journal, Owen agreed that James Scott Bowerbank had first established the remains of *Pterodactyl* in the Chalk, correcting a misidentification by Owen. But Gideon, who had collaborated with Bowerbank,

11. Lithographs. Walter's box arrived on the eighteenth. J (24, 28 [incl. 23, 24], 31 Oct, 14, 28 Nov 1850). GM–BS, 27 Nov, 6 Dec 1850 (Yale); GM–LB, 23 Nov, 12, 18 Dec 1850 (BMNH). RS minutes and other printed documents annotated by GM, incl. GM–RS, 31 Oct 1850; RO–RS, 14 Nov 1850; GM–RS, 27 Nov 1850 (RS, ATL). LB–GM, 26 Dec 1850 (ATL), with comment on Owen. See page 235 above.

was ridiculed for his statements regarding these specimens in the *Wonders* of 1848. Owen also denied that Bowerbank's microscopic examination of specimens in 1845 had been decisive. Owen's final topic, *Iguanodon,* included a long series of hits against Gideon, who wrote an unavailing letter of remonstrance to the president and council of the Palaeontographical Society.

Even worse, there was a strong likelihood that Owen would be named to fill the British Museum post of Charles König, who had died unexpectedly at the end of August. Mantell revealed his accumulated anger regarding Owen in a letter to Reginald on 9 October. Responding to the aggressions made against him, Gideon wrote:

> I have not spared him [Owen] in this new work [*Petrifactions*], and as this will be the first time anyone of my standing has reprobated his conduct, I expect the most vicious conduct from this unscrupulous man. He is now attempting to get the late Mr. König's place in the British Museum and thus prevent Mr. Waterhouse (O's supposed friend!) from having the promotion he has been expecting. If Owen gets it (and I dare say he will), my collection and health will soon be deprived of an ally and I shall be obliged to absent myself from the British Museum, as I have done from the College of Physicians.

Such, then, was the Owen-Mantell relationship as of the writing of *Petrifactions,* a book which, not surprisingly, was replete with comments on Owen, some of them highly complimentary, others ironic and biting.[12]

---

12. J (27 May, 3, 17 June 1851). GM–LB, 27 or 28 May 1851 (BMNH). RO, *A Monograph on the Fossil Reptilia of the Cretaceous Formations* (London, 1851–1854); Part I, pp. 1–118, plates I–XXXVII, VIIA, IXA. For GM, see esp. pp. 1–4, 30–31, 36, 40–42, 48–49, 82–83, 105–118; tabs 1, 2, 8, 33–37. *Iguanodon* was the only dinosaur Owen discussed (pp. 105–118; see also supplement II, Mar 1861).

   J (25 July, 6 Dec 1851); *Petrif,* Chapter 3. GM–RNM, 9 Oct 1851 (ATL). Ironic references to Owen: *Petrif,* pp. 162–163, 168n, 191–192n, 193n, 226, 229, 240, 257, 260, 285–286, 286n, 287, 295, 299, 307n, 321–322, 331, 333n, 334, 335, 460; others, pp. 259, 276, 282n, 349, and 358n. Owen had also published *A History of British Fossil Reptiles,* I (London, 1849), incl. *Iguanodon, Megalosaurus,* and *Hylaeosaurus,* and (as crocodiles, but see also the Errata) *Cetiosaurus* and *Pelorosaurus;* completed in four vols., 1884, with references to GM throughout.

   One should also mention Frederick Dixon, *The Geology and Fossils of the Tertiary and Cretaceous Formations of Sussex* (London, 1850), which was completed by Owen after the author's death in 1849. Some twenty-five references to GM appear within it, esp. p. 6: "In Sussex Dr. Mantell has been most conspicuous in the advancement of geological knowledge, and as the historian of the Wealden formation and discoverer of Iguanodon, his name is placed among the first geologists of any age or country."

The *Telerpeton* Controversy

On 19 October 1851, the day after *Petrifactions* was completed, Lyell called on Mantell to discuss a blatant attack on himself published by Owen in *Quarterly Review* the previous month. While criticizing his Anniversary Address of 21 February 1851 as president of the Geological Society, Owen had unwisely given way to intemperate vituperation against Lyell, whose genial if sometimes distant personality and highly regarded character were virtually sacrosanct. Recognizing Owen's blunder for what it was, Gideon wrote to Silliman:

> I presume some of the libraries in your city take *Quarterly Review*. If so, pray get it and read the article on Lyell's views of animal creation. It was written by Owen and is a most unfair attack. I am heartily glad of it, for it will tend to open the eyes of Lyell and others of his stamp to the real character of the man. The cause of Owen's doing this was Lyell's refusal to unsay what he had said in favor of Mr. Waterhouse of the British Museum as successor to poor Mr. König. Owen wishes to get this place. . . . If he does (which is very likely, for such un-scrupulous men usually succeed), he will tyrannize over the British Museum as he does now over the College of Surgeons. Lyell is ex-ceptionally annoyed . . . ; it serves him right for not more openly op-posing this unprincipled varlet.

Uncharacteristically nettled by Owen's wholesale condemnation of his an-tiprogressionist views, which denied any form of organic "development" or evolution, Lyell took the unusual step of composing a reply. On the evening of the twenty-eight, he showed the draft to Mantell, who gloated afterward that "Lyell is not accustomed as I am to unjust attacks and mis-representations!"

The immediate result of Owen's bad judgment regarding Lyell was that Sir Charles (since 1848) and Gideon became, for several months, collab-orators – to an extent reminiscent of their earliest work together during the 1820s. On 10 December, Lyell began (or at least dated) a "Postscript to the Third Edition" of his *Manual of Elementary Geology,* the most sig-nificant feature of which was a sustained refutation of the biological the-ory of progressive development. Written to substantiate a position Lyell had advocated consistently since 1832, it also defended his Anniversary Address against Owen's strictures. Though many families of animals and plants could not be found early in the geological record, Lyell empha-sized, it was "no part of the plan of Nature to hand down to after times a complete or systematic record of the former history of the animate world" (14). One should not, therefore, accept the absence of, say, birds

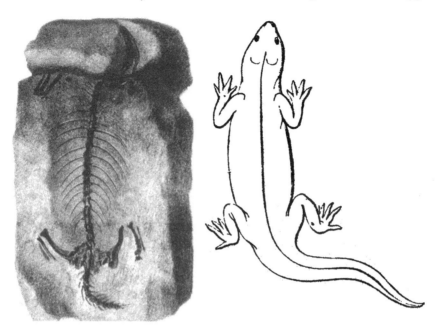

Figure 12.3. *Left: Telerpeton,* the original specimen (drawing by James Lee: *QJGS,* 8 [1852], Plate IV; Lyell, *Manual,* fourth edn [1852], p. x; *American Journal of Science,* ns 13 [1852], p. 279); *right:* and as restored (GM–BS, 11 October 1852; Yale).

or reptiles at a certain time merely because no fossils of them had been found.[13]

On the other hand, some new discoveries strongly suggested that reptiles were demonstrably older than had been formerly believed. Mantell's friend Captain Lambart Brickenden, for example, had recently discovered a trackway of thirty-four turtle footprints near Elgin, in Morayshire, Scotland, where he was now living. They appeared to be from the Old Red Sandstone (which is to say, Devonian strata); if so, then turtles were much older than had previously been thought. Even more remarkably, Patrick Duff, author of *Sketch of the Geology of Morayshire* (1842) and Brickenden's brother-

13. J (19 Oct 1851); RO, "Lyell on Life and Its Successive Development," *Quarterly Review,* 89 (1851), 412–451 (reviewing *Principles of Geology,* eighth edition; *Manual of Elementary Geology,* third edition; and esp. Lyell's Anniversary Address to the Geological Society, as reported in *QJGS,* 7 [1851], xxv–lxxvi). GM–BS, 21 Oct, 11 Nov 1851 (Yale); J (28 Oct 1851); CL–GM, 28 Oct, 22 Nov 1851 (ATL). Lyell had also replied privately to Owen (CL–RO, 9 Oct 1851 [BMNH]). CL, "Postscript to the Third Edition [of] Lyell's *Manual of Elementary Geology,*" 16 pp., Dec 1851 (ATL, with annotations by GM); rept. in CL, *Manual,* fourth edition (London, 1852), pp. vii–xxii; Gideon received his copy of the fourth edition on 16 Jan 1852 (J).

in-law, obtained a saurian skeleton from the same strata, at Spynie quarry, near Elgin. This was the famous "Elgin reptile," which appeared most opportunely for both Lyell and Mantell.

On learning of the specimen Brickenden realized its unique scientific importance. If from the Old Red as well, Duff's fossil would take its place as the oldest known example of reptilian life. On 22 November, accordingly, Brickenden expectantly sent Mantell the manuscript of an intended publication on "The Fossil Footprints of Moray," in which he also announced "the late discovery of a reptilian" in the same Devonian rocks, illustrating both. Receiving these interesting materials on the twenty-fifth, Gideon responded immediately, advising his friend that the reptile should be more fully described and available for display when the paper was read; he asked that it be entrusted to his care.

Mantell visited Lyell the next evening, taking Brickenden's drawings with him. During the previous month, with Gideon's help, Lyell had been composing his letter of remonstrance to the editors of *Quarterly Review* regarding Owen's lengthy denunciation of his nonprogressionist views in the tardy September issue. With this newly discovered fossil, however, a superior rejoinder became possible. Heretofore, almost the only evidence available to substantiate Lyell's stubborn a priori antievolutionary assumption had been the minute mammalian fragments from Stonesfield found long ago. Now there was an entire reptilian skeleton, virtually complete and stratigraphically anomalous – perhaps the first of many such revelations. Understandably then, as Gideon recorded, "Lyell was greatly pleased with this discovery."

On the twenty-seventh Lyell sent Mantell a proposed notice of the reptile, which he would delay the next edition of his *Manual of Geology* to include. After reading Brickenden's manuscript the next day, he requested further details via Gideon, and on the twenty-ninth suggested *Elginosaurus* as a name. Writing Brickenden then, Mantell sought to confirm that the strata involved were indeed Devonian, underlined Lyell's fascination with the specimen, and insisted that it be sent to himself for describing. Brickenden, meanwhile, had already responded faithfully to Gideon's letter of the twenty-fifth by approaching Duff, who was eager to obtain metropolitan recognition for his discovery and cooperated readily. Only six or seven inches long from tip to tail, the small but important reptilian was soon en route to London, where Duff's brother, Dr. George Duff, received it on 1 December at his home in Hanover Square. Dr. Duff apparently notified Owen that the specimen had arrived. Learning of it from him, Lyell visited Dr. Duff that afternoon to see the reptile for himself.

From Dr. Duff's it was only a short drive to Chester Square. Having seen the specimen himself, Lyell arranged a showing for Mantell the next morning. By viewing it then, he would forestall Owen, who was scheduled to inspect the same fossil at noon. On receiving Lyell's note, however, Gideon came immediately. Lyell then called at Chester Square that evening to confer regarding strategy. Writing Brickenden later that night, Mantell emphasized that a technical description and detailed lithograph of the new fossil should accompany the "Footprints" paper. "The reptile," he declared, "is evidently a batrachian," which he proposed to name *Telerpeton Elginese* (Greek for remote or most ancient reptile from Elgin). Lyell happily acceded to Gideon's new designation on the fourth.

Dr. Duff apparently took *Telerpeton* to the College of Surgeons on 2 December, where Owen saw it and expressed great interest. Duff then wrote his brother Patrick that day, noting the ambitions of Mantell, Lyell, and Owen toward the specimen. Would he object to Mantell's borrowing and describing the find? There being no binding previous commitment, the fossil arrived at Chester Square on the seventh (from Lyell's, where drawings had been made). Gideon then spent most of the eighth, ninth, and tenth examining it. On the eleventh he compared the Elgin reptile with recent specimens of lacertae and batrachians at the British Museum. Unable to wait longer, Lyell completed his remarkably current "Postscript" on the twelfth. Featuring lithographs by James Lee, it both depicted and described Brickenden's chelonian footprints, Duff's reptile, and some probably amphibian fossil eggs which his third edition had ascribed to gastropods. Published within a few days as part of this six-penny separate, Lyell's discussion of *Telerpeton* was the earliest to appear in print.

A critical question for Lyell, and one that had recurred insistently throughout a frenzied triangular correspondence among Mantell, Brickenden, and himself, was whether or not Elgin sandstone genuinely belonged to the Old Red. Brickenden (and later Murchison) stoutly insisted that it did. The precise identity of *Telerpeton* also remained in doubt. In any case, Lyell affirmed, the reptile was either a freshwater amphibian or a small lizard. He further stated that Mantell had completed a detailed osteological account and was responsible for its name.

Gideon, meanwhile, had submitted Brickenden's portion of their joint paper on *Telerpeton* to the Council of the Geological Society on 3 December, and completed his own within a week. William Hopkins assured him on the ninth that it would be read at the next meeting. Still uncertain as to whether the Elgin specimen was amphibian or reptilian, however, Mantell visited the British Museum on the eleventh (as we know) to compare it with recent skeletons; Joseph Dinkel finished his drawings of the specimen the same day. The reptile, Gideon then assured Brickenden on the twelfth, "is a very primitive one," its ribs being saurian but its pelvis, vertebrae, and tail all more nearly amphibian. It was, therefore, "either a batrachian with saurian affinities or the reverse," but in either case the first Devonian reptile – which is to say the oldest anywhere.

On Wednesday the seventeenth Mantell exhibited *Telerpeton* at a Council meeting of the Geological Society and was to have presented his and Brickenden's paper on it that evening; Dr. Duff and Henry Bohn came to hear it as his guests. Because a very long paper by President Hopkins ran overtime, however, that on the Elgin reptile was postponed. Owen, who had seen the specimen earlier, carefully inspected it and Gideon's drawings nonetheless. He then left the room in a towering rage. Three days later a letter by Owen appeared in *Literary Gazette,* claiming that Patrick Duff had sent the reptile to London for naming and description by *him.* He also attempted to supersede Gideon's name for the creature with his own: *Leptopleuron lacertinum* ("slender-ribbed reptile"). Whereas Mantell had remained equivocal regarding the new animal's specific character, Owen briefly but straightforwardly identified it as a lizard.

Shocked by his rival's effrontery once again, Gideon protested that same day to Lovell Reeve (the editor of *Literary Gazette*) about exposing "the base feelings of a highly gifted man," but his letter was not published. He also wrote Brickenden, enclosing a note for Duff, in hopes of ascertaining whether or not Owen's unlikely claim of priority contained any possible degree of truth. To Gideon's surprise, it did. As a chagrined Duff admitted four days later:

> Professor Owen's statement is substantially correct. I did not address him directly but sent him first a copy of the Elgin *Courant* [for 10 October] which contained the first announcement of the fossil. I afterwards sent to my friend Dr. Duff a drawing of it and requested him to submit it to Professor Owen in the hope that he would notice it.
>
> Subsequently, my friend Captain Brickenden called on me and stated that the fossil had attracted your attention and that you thought a notice of it in conjunction with his paper on fossil footprints in the Old Red or Morayshire would suit well. I at once determined to consign the specimen to the charge of Dr. Duff and expressed a wish that he would show it to any of the leading geologists in London who might desire to see it. Dr. Duff wrote me on the 2nd of this month that you, Sir Charles Lyell, and Professor Owen were seeking to have possession of it and that I must decide who was to have the priority. I immediately wrote him that if he had not promised it to Professor Owen you should have the first turn of it, in order that you might describe it in Captain Brickenden's paper, and I understand that you got possession of it accordingly.

Dr. Duff received this letter in London on the twenty-sixth and immediately forwarded it to Chester Square. "My course is clear," Gideon then wrote Lyell the following night; "to take no notice whatever of the published letter, and if after my paper is read, Owen disputes my priority, then to refer him to Captain Brickenden and Mr. Patrick Duff." Despite some unfortunate confusion, the specimen was legitimately his.

Priority for *Telerpeton* had already been established through Lyell's "Postscript," which was out by mid-December. Speaking on geology at Ipswich on the seventeenth, moreover, Lyell had exhibited Gideon's cast of *Telerpeton* and summarized his equivocal identification of its nature, Owen's more definite opinion not yet having appeared. On 3 January 1852, however, *Literary Gazette* reported Lyell's remarks as if he had *compared* Mantell's opinion with Owen's, thereby establishing some claim to equal priority for the latter. Believing a response necessary, Lyell wrote the journal on 7 January 1852 and succeeded in obtaining a correction. Meanwhile, as had become known, Owen was unexpectedly passed over at the British Museum in favor of Waterhouse.

At the Geological Society's Council meeting on 7 January, Mantell also responded to Owen's *Literary Gazette* opinions of 20 December, laying before President Hopkins the letters he had received from Brickenden and Dr. Duff supporting his claim to the specimen. Hopkins, meanwhile, had done some investigating on his own, garnering written evidence from Sedgwick, Murchison, Dr. Duff, and Owen himself. The latter defended his claim to priority and contritely attributed his admittedly absurd behavior at the meeting of 17 December to "painful surprise." "I know that private slander has been rife," he concluded with unconscious irony, "but it must ultimately recoil upon its source."

Brickenden's "Notice of the Remains of Reptiles in the Old Red or Devonian Formation of Scotland" (his "Footprints" paper, retitled by Mantell) was then read at the evening meeting. Gideon followed, reading a dual contribution. The first part described *Telerpeton,* which he continued to think either a lizard or an amphibian. The second discussed Lyell's problematic fossil eggs (and possible freshwater deposits in Scottish Paleozoic formations). At the end of a good debate, in which Lyell, Murchison, and others participated (Owen being conspicuously absent), President Hopkins ruled to warm applause that Dr. Mantell's paper on *Telerpeton* had been published, though not presented, on 17 December and therefore took precedence over any other description of the Elgin fossil. Gideon then wrote joyously to Brickenden on the eighth and to America on the thirteenth, including with the latter an account of Devonian fossil reptiles written two days before. Owen replied with further letters in both *Literary Gazette* and the *Athenaeum,* but to no immediate avail. In the last of their important battles, Gideon had defeated him once again.[14]

14. *Telerpeton.* In part because it derived from a significant stratigraphical mistake, the *Telerpeton* controversy has attracted considerable scholarly attention. See, for example, Michael J. Benton, "Progressionism in the 1850s: Lyell, Owen, Mantell and the Elgin

Fossil Reptile *Leptopleuron (Telerpeton),*" *Archives of Natural History,* 11 (1982), 122–136, with its helpful geological map (we differ on more than details, however); and Chapter 4, note 4 above.

LB–GM, 4 July 1850 (ATL); GM–LB, 28 Oct 1850 (BMNH). J (July 1850–Oct 1851, *passim*). The fossil reptile's discovery was announced in the Elgin (Scotland) *Courant* of 10 Oct 1851. LB–GM, 22 Nov 1851, with manuscript, "The Fossil Footprints of Moray," and inaccurate annotation by GM (ATL); GM–LB, 25 Nov 1851 (BMNH); J (26 Nov, 19, 28 Oct 1851); LB–GM, 27 Nov 1851 (ATL); CL–GM, 27, 28, 29 Nov 1851 (ATL); GM–LB, 29 Nov 1851 (BMNH); LB–GM, 1, 6, 10, 16 Dec 1851 (ATL); J (1 Dec 1851); GM–LB, 1, 7, 12 Dec 1851 (BMNH); CL–LB, 1, 5 Dec 1851 (BMNH); G. Duff–GM, 2 Dec 1851 (ATL); CL–GM, 3, 4, 5, 9 Dec 1851 (ATL); GM–CL, 10 Dec 1851 (APS); J (4, 6, 7, 8, 11 Dec 1851). CL, "Postscript" (for price: GM–BS, 21 Jan 1852 [Yale]). J (3, 7, 11 Dec 1851). W. Hopkins–GM, 9 Dec 1851 (ATL). GS (OM), 17 Dec 1851; J (17, 20 Dec 1851).

RO, "Vertebrate Air-Breathing Life in the Old Red Sandstone," *Lit Gaz,* 20 Dec 1851, p. 900; annotated copy at ATL: "Prof. Owen was at the Geological Society on the 17th when my paper was commenced and he examined my drawings, etc. and then sent this letter to Lovell Reeve!" According to Owen's letter, dated 15 Dec, "Mr. Duff, the proprietor of the very remarkable fossil recently discovered in a sandstone of the Devonian system of rocks at Elgin, transmitted me a drawing of it, with the request that I would undertake its examination, to which having gladly acceded, the specimen was brought to me by a friend of Mr. Duff's." The issue is further clarified by a hasty note (dated 5:30 P.M. Monday [1 Dec]) from Lyell to Mantell: "I met Owen today, who told me he had got a promise from Dr. Duff that he would bring the Elgin reptile to the College of Surgeons tomorrow about noon. So I came here to Dr. Duff, who tells me that if you call before 11 A.M. . . . he wishes you to see it first. Pray come, so that I may cite you" (ATL). See also LB–GM, 25 Dec 1851 (ATL).

GM–Lovell Reeve, 20 Dec 1851 (Alan G. Nicholson); GM–LB, 20 Dec 1851 (BMNH); P. Duff–GM, 27 Dec 1851 (ATL). London *Times,* 23 Dec 1851, p. 8. Also GM–LB, 22, 27 Dec 1851 (BMNH); CL–LB, 24 Dec 1851 (BMNH); LB–CL, 26 Dec 1851 (BMNH). GM–CL, 25, 27, 29 Dec 1851 (ATL; APS). Anon., "Sir Charles Lyell on the White Chalk, and on Progressive Development," *Lit Gaz,* 3 Jan 1852, pp. 16–17; CL, "The New Fossil Reptile," 10 Jan 1852, p. 41, with conciliatory editorial reply.

J (7 Jan 1852); GS (OM), 7 Jan 1852; *Lit Gaz,* 24 Jan 1852, pp. 90–91. LB, "Notice of the Discovery of Reptilian Foot-Tracks and Remains in the Old Red or Devonian Strata of Moray," *QJGS,* 8 (1852), 97–100; GM, "Description of the *Telerpeton Elginese,*" *QJGS,* 8 (1852), 100–109 (with inscribed separate combining both at ATL); *ENPJ,* 52 (1852), 353–355; *AJSA,* ns 13 (May 1852), 278–281. The evidence before President Hopkins included LB–GM, 27 Nov 1851; and P. Duff–GM, 24 Dec 1851; also RIM–AS, 22 Dec 1851 (RO's "towering rage" on 17 Dec); AS–Hopkins, 29 Dec 1851; Dr. Duff–Hopkins, 7 Jan 1852; and RO–Hopkins, 6, 7 Jan 1852 (all four CUL).

For additional perspectives on the *Telerpeton* controversy, see AS–RIM, 19 Nov 1852 (GS); RIM–AS, 22 Nov 1852; J. S. Henslow–AS, 27 Nov 1852; J. D. Hooker–AS, 14 Dec 1852 (all three CUL); RIM–LB, 10 Sept 1855; P. Duff–LB, 13 Oct 1856; and P. Duff–LB, 6 Oct 1858 (all three BMNH). James A. Secord called several of these letters to my attention. GM–LB, 8 Jan, 8 Apr 1852 (BMNH); GM–BS, 13 Jan, 1 Oct 1852 (Yale). *Lit Gaz,* 17 Jan 1852, pp. 64–65; *Athenaeum,* 17 Jan 1852, p. 85. WCW–GM, 9 Jan 1852 (ATL), with comment on Owen. J (24, 27 Jan 1852). With *Telerpeton* out of the way, Mantell, Lyell, and their friends next turned to another enigmatic Elgin fossil, *Stagonolepis,* a Triassic thecodont not then identified as such. By 1860, *Telerpeton* was seen to be Triassic also, and Owen's name for it eventually prevailed.

Epilogue: Norwood Park

Though Richard Owen continued to harass him throughout much of 1852, a more serious opponent for Gideon Mantell was persistent bad health. His long professional round to Stockwell, Clapham, and Kilburn on 19 January, for example, resulted in a night of torment as Gideon's nerves shot unpredictable writhings throughout his body. He tried cold hydrocyanic liniment and hot fomentations in vain, as well as calomel and gamboge. On 3 February, his sixty-second birthday (which he thought the sixty-first), Gideon was unwell all day. Often now, he would alternate days of activity – usually including scientific meetings, to which he was normally accompanied by Alfred Woodhouse – with days after of convalescence and drugs. "I have been suffering and existing," he wrote Silliman on 26 February, "that is all." By month's end, however, he was noticeably better and often in public.

On 5 March, Mantell lectured triumphantly at the Royal Institution on the structure of the iguanodon. A few days later neuralgia seized him once again, and he passed another night in agony, with prussic acid, fomentations, calomel, opium, and hot brandy all unavailing. Yet he appeared at a scientific soiree the next evening, participated regularly in others, attended meetings and lectures, and in April even gave four lectures of his own at Leeds. While active among those who were seeking to preserve the Crystal Palace from destruction, Gideon also worked on a second (and potentially more profitable) edition of his *Medals*. After attending the Council and regular meeting of the Geological Society on 19 May, however, he came home alone in great pain. That night, in a futile effort to achieve sleep, he consumed a full ounce of self-prescribed opiate – thirty-two times the normal dosage.[1]

"Life runs on," Gideon philosophized to his son Walter in New Zealand in a letter of 5 June: "In the hustle and turmoil of this great Babylon, I

---

1. J (19, 20 Jan, 3, 4, 5, 7, 10, 11, 20 Feb 1852); GM–BS, 26 Feb 1852 (Yale). J (23 Feb, 5, 12, 13, 18, 19, 20, 24 Mar 1852). Leeds: J (3–15 Apr 1852). Opium: J (19 May 1852); I have accepted "thirty-two times" from Curwen, p. 289n.

find it impossible to do anything satisfactory. I am never quiet, never idle, always suffering mentally or bodily, striving against physical infirmities in vain, and growing more and more dissatisfied with the little I can or do effect." Though hard at work on his new edition of *Medals,* Mantell had begun to find the effort too much for him. "My memory fails me most dreadfully," he confessed to his other son Reginald in America on 24 June; "I cannot remember anything that has happened but a few days or weeks before without notes!"[2]

Three days later Gideon learned that Lord Derby, the minister, had offered him a royal annuity of one hundred pounds, in recognition of his scientific labors. "It is a miserable pittance from the Crown of the British Empire!!" he then complained to Walter, yet as a compliment of sorts Mantell received it thankfully. He said as much to Silliman on 16 July, adding that the news had come to him quite unexpectedly in a very kind note from the Earl of Rosse, current president of the Royal Society, at whose suggestion the pension had been bestowed.

Mantell responded to this fairly minimal honor with renewed ambitions. "I am still dreaming of a history of the Wealden," he revealed to Silliman in the same letter. "I shall not die content unless I do publish a complete work on my discovery of the freshwater character and fossils of that formation." His collection of Wealden fossils (now including insect remains) was increasing daily. "What a glorious volume the fauna and flora of the country of the iguanodon would make!" Gideon envisioned, – "if I had but the means wherewithal to do it." He thought of projecting a one-volume work costing perhaps two guineas; with two hundred subscribers, there would be no loss.[3]

One of his last major public concerns was the now-dismantled Crystal Palace, which a well-supported private company had bought and planned to reerect at Syndenham, a few miles south of London. At a meeting of the Crystal Palace Company on 10 August, the board of directors had resolved (perhaps at the suggestion of Prince Albert) "that a Geological Court be constructed [at the new site] containing a collection

---

2. GM–WBDM, 5 June 1852 (ATL); J (5, 6, 7, 14, 15, 16 June 1852). J's Volume 4, the typescript version, and Curwen's edition all end at 14 June. For the text of Volume 5, discovered by DRD in 1982, see Sharon Dell, ed., "Gideon Algernon Mantell's Unpublished Journal, June–November 1852," *Turnbull Library Record,* 16 (1983), 77–94. GM–RNM, 24 June 1852 (ATL).

3. Pension, 4 Aug 1852 (ATL); J (27 June 1852); GM–WBDM, 2 Aug 1852 (ATL); GM–BS, 16 July 1852 (Yale); announcements in *Lit Gaz,* 24 July 1852; and *AJSA,* ns 14 (Nov 1852), p. 290. Murchison had suggested the pension idea to Rosse (RIM–AS, 22 Nov 1852 [CUL]). J (15 Sept, 21 Oct 1852). On the latter date Gideon collected twenty-one pounds as a first quarterly payment (minus expenses); he never received another.

of full-sized models of the animals and plants of certain geological periods, and that Dr. Mantell be requested to superintend the formation of that collection." They then wrote him on the twelfth to introduce Benjamin Waterhouse Hawkins, a fellow of the Linnaean Society recommended by several persons (including Charles Darwin) to model the prehistoric creatures. Gideon had attended groundbreaking ceremonies at Syndenham on the fifth, but he disliked the idea of life-size models – as opposed to more scientific displays – and abruptly withdrew from the project. By so doing, he left the restoration of *Iguanodon* and other dinosaurs to Owen.[4]

Early in September, after a final visit to the Isle of Wight, Mantell did what he could toward the new edition of *Medals,* but for all his effort it went on slowly. "Another week," he wrote despondently on the sixteenth, "and though continually at home and unremittingly employed, have done but little in anything. In truth, I am used up." Four days later he bade good-bye to his uncomprehending brother Joshua at Ticehurst. On 28 September, Gideon made his last living pilgrimage to Hannah Matilda's grave. His final return to Lewes and Brighton took place on the twenty-ninth, when he visited a local collector to see some recently discovered Chalk fishes, took tea with friends, and called on a patient at Brighton.[5]

In a last major letter to Walter, Gideon turned against much that had previously been dear to him. "Selfishness, frivolity, and excitement are in the ascendent," he affirmed.

> There is no such thing now as an Englishman's fireside. Eternal hustle, movement with the greatest rapidity, constant change are characteristic of society. The facility of travelling, and the speed with which persons can be transported from one end of Europe to the other, have completely metamorphosed the English character. I believe I am the only man in my station who has remained in England three months together the last few years.

---

4. Crystal Palace. J (2, 3, 17, 27 Apr, 1, 28 May, 22 June, 11, 18 July, 5, 13, 18, 19 Aug 1852). Crystal Palace Company: Extract from Minutes of a Meeting of the Board of Directors, 10 Aug 1852 (ATL); GM–WBDM, 11 Aug 1852 (ATL); Crystal Palace Company–GM, 12 Aug 1852 (ATL). Had the models reflected his understanding rather than Owen's, Victorian conceptions of the major dinosaurs would have been rather different. For the Syndenham restorations (which still survive), see *The Illustrated London News,* 23 (31 Dec 1853), 559–560; 24 (7 Jan 1854), 22; Owen, *Life,* I, 397–399; and Adrian J. Desmond, *The Hot-Blooded Dinosaurs,* Chapter 1. Despite Owen's control, the Waterhouse Hawkins restorations preserve *Iguanodon*'s "horn" and *Hylaeosaurus*'s dorsal spines.

5. Wight: J (2–4 Sept 1852). *Medals:* J (26 Jan, 13, 21, 29 May, 5, 7, 11, 14, 16, 21, 28, 29 June, 10, 17, 19, 22, 26, 29 July 1852 – mostly about illustrations); GM–BS, 24 Aug 1852 (Yale). J (31 Aug, 16, 20, 28, 29, 30 Sept, 1 Oct 1852).

Railways, moreover, had also revolutionized the English mind; novels and nonsense were superseding everything of more sterling character.[6]

Mantell's last letters to Silliman, written on 24 August and 11 October 1852, dealt mostly with Owen's recent attacks upon himself in *Quarterly Review* and the *Annals of Natural History*. Despite persistent illness, Gideon presided at the opening meeting of the West London Medical Society on 15 October, reading an address as well. He was then kept busy correcting proofs, the printing of *Medals* having begun. On 3 November, Gideon managed to attend both the Council and general meetings of the Geological Society. He even saw patients in Clapham on the fifth and seventh.

Mantell's last journal entry, written the next day, recorded a letter from Reginald. That evening, the eighth, Gideon lectured heroically (but necessarily from a chair) to the opening meeting of the Clapham Athenaeum, all the while enduring nervous shocks that he described afterward as universal toothaches. Those who saw him beforehand remarked his languid and careworn appearance. The lecture itself, on petrifactions, went well, but immediately thereafter Mantell complained of extreme exhaustion and had to be rushed home. He spent all the next day doubled over in constant pain. Hoping to obtain relief through sleep, Gideon retired at the extremely uncharacteristic hour of 8 P.M. Falling on his way upstairs and unable to rise, he crawled to bed on hands and knees. After two large draughts of opiate Gideon at last fell asleep around half past one. In the morning he was unconscious, noticeably worse, and obviously dying. What Gideon Mantell had described in a journal entry the previous month as the "sad enigma" of life ended for him at 3 P.M. on Wednesday, 10 November 1852.[7]

6. GM–WBDM, 23 Sept 1852; a "Friday morning" sequel (24 Sept 1852) was Gideon's last letter to his elder son (both ATL). In particular, a well-written American story on the slave trade, called *Uncle Tom's Cabin,* had been enormously fashionable, selling a hundred thousand copies in England alone. Anxious to proceed with *Medals,* Gideon visited his printer, Richard Clay, on 2 October, only to learn that Henry Bohn had diverted all his available paper to three precipitous editions of Mrs. Stowe's novel (J [2 Oct 1852]).

7. For RO and GM in 1852, see esp. Anon. but RO (Owen, *Life,* I, 373–374), "Professor Owen – Progress of Comparative Anatomy," *Quarterly Review,* 90 (Mar 1852), 362–413, an important (but scarcely humble) self-analysis that reviews fifteen of Owen's publications, hits at GM on belemnites (p. 383), and disputes Colenso's role in NZ bird bone discoveries (404–405n). Mantell responded with "A few Notes on the Structure of the Belemnite," *AMNH,* ns 10 (July 1852), 14–19. Owen then published an anonymous letter by "The *Quarterly Reviewer*" in the same periodical ("On the Structure of the Belemnite," *AMNH,* ns 10 [Aug 1852], 158–159). When they refused to publish his reply to that, Gideon wrote the editors on 16 Aug 1852 to "retire from the unequal conflict" (transcribed fully in J, 18 Aug 1852; and GM–BS, 24 Aug 1852 [Yale]). "After the *Telerpeton* affair, I am surprised at nothing" (GM–BS, 11 Oct 1852 [Yale]). See also RO, "On a New Species of Pterodactyle," *AMNH,* ns 10 (Nov 1852), 378–392. GS (OM), 3 Nov 1852; J (3, 5, 7, 8, Nov 1852). Details of Gideon's last days and hours are from AJW–BS, 18

A round of necessary duties then devolved on his survivors. "Do you attend Mantell's funeral?" P. J. Martin inquired of the Reverend Charles Pritchard in a note. "If you do, I have a strong inclination to meet you there."

> When I saw him last, [Martin continued] I jested with him about leaving the honor of the extraordinary tumor on his spine, post mortem, to *Owen*. In the same humor, he replied: 'No, no! Owen shall not have the picking of my bones! I mean to depute the task to my friend Hodgkin.' I presume [Martin concluded pragmatically] they will anatomize the part in question and bury the rest of him.

Gideon's body was disposed of entirely as he wished. During March 1851 he had entrusted Alfred Woodhouse with a letter to be opened after his death. In it, Mantell asked that a postmortem examination be made; if the pathological changes that had taken place within his body were of medical interest, relevant parts of himself should be given to the Hunterian Museum at the College of Surgeons. His funeral, moreover, was to be as plain as possible, taking place in the early morning at Norwood Cemetery, where he wished to be buried alongside his beloved child Hannah Matilda.

Conducted by Dr. Thomas Hodgkin and William Adams (the latter, assistant surgeon to the Royal Orthopedic Hospital), the postmortem examination took place on 13 November. According to Adams' report, Gideon had been a "tall, well-developed, muscular man" (168). Since his spine was of primary interest, the body was first opened from behind, dissection then progressing toward the abdominal cavity. No tumor, abscess, cyst, or morbid growth of any kind was found, but Hodgkin and Adams soon exposed very pronounced lateral curvature in the lumbar spine, which had been severely twisted as well. A portion including three lower dorsal and all the lumbar vertebrae was then removed for separate examination.

Remarkably, though Mantell had consulted the most eminent physicians and surgeons in London, none of them suspected lateral curvature of the spine or realized that the hard nodules felt in his lumbar region were not tumors but transverse processes of his twisted vertebrae, a condition that had been accurately described from other examples so early as 1824. Hodgkin and Adams subsequently presented their "Case of Distortion of the Spine" to the Medico-Chirurgical Society on 27 June 1854, exhibiting the removed portion. As specimen #4808, "the lower part of a spinal

Nov 1852, 8 Jan 1853, and BS papers (Yale). "Sad enigma": J (27 Oct 1852; re G. B. Sowerby the elder [d. 1854]).

column affected with osteo-arthritis," it then remained on exhibit at the Hunterian Museum for almost ninety years, until destroyed by enemy action in May 1941, a casualty of World War II.[8]

The first obituaries appeared on 13 November. According to quite a personal one in the *Sussex Express,*

> The last interview we had with him was about a month ago [probably on 29 September]. His conversation was cheerful and his mental powers were not only unimpaired, but lively as ever. He was entertaining new objects for literary enterprise, and appeared confident of accomplishing the undertakings he projected. Even as he then spoke, however, we felt that the unerring, though silent, hand of death was laid upon him. He is now no more, and we feel in his death that we have lost no ordinary man.

One of quite another tone appeared the same day in *Literary Gazette.* Full of misinformation and snide deprecations, it described Gideon as "fluent and eloquent in speech, full of poetry, and extremely agreeable in manners to all who manifested an admiration of his genius." Dr. Mantell, it was conceded, had "done much after his kind for the advancement of geology, and certainly more than any man living to bring it into attractive and popular notice." But while Gideon had been a most successful lecturer and a prominent author of popular works on geology, he unfortunately lacked "exact scientific (and especially anatomical) knowledge, which compelled him privately to have recourse to those possessing it." In his "popular summaries of geological facts," moreover, Mantell "was too apt to forget the sources of information which he had acknowledged in his original memoirs." The history of *Iguanodon* provided a remarkable instance of this, for what reader of the *Medals of Creation* would suspect that to "Cuvier we owe the first recognition of its reptilian character, to Clift the first perception of the resemblance of its teeth to those of the iguana, to Conybeare its name, and to Owen its true affinities among reptiles, and the correction of errors respecting its bulk and alleged horn"? Having been denied the pleasure of conducting his autopsy that day, Owen did what he could to cut Gideon up in print.[9]

---

8. P. J. Martin–Rev. C. Pritchard, 14 Nov 1852 (ATL). PJM last met GM on 17 June 1852 (J). GM–AJW, 8 Mar 1851 (Spokes, p. 248); GM–RNM, 8 Mar 1851 (never mailed, but deposited with AJW [ATL]); Dr. Thomas Hodgkin and William Adams, "Case of Distortion of the Spine, with Observations on Rotation of the Vertebrae as a Complication of the Lateral Curvature," Royal Medical and Chirurgical Society of London, *Medico-Chirurgical Transactions,* 37 (1854), 167–180; Spokes, pp. 252–257 (quoting p. 256); Curwen, pp. 292–293. Inquest: Stephen Williams–WBDM, 13 Nov 1852 (ATL). Funeral: AJW–BS, 8 Jan 1853 (Yale).

9. Obituaries and Owen. *Sussex Agricultural Express,* 13 Nov 1852, p. 3; Anon. but RO,

His deceased rival was eulogized more satisfactorily thereafter. In a perceptive review of Gideon's career, the *Athenaeum* pointed out on 20 November that "as a man of science Dr. Mantell did not take the position of a great generalizer or of the discoverer of new laws." But "he was naturally an enthusiast, and with such quick observation that he would have distinguished himself in almost any branch of science. The accident of his position made him a geologist; the requirements of the science made him a great one." After reviewing Gideon's major publications and a few of his better-known contributions, the reviewer concluded that these "discoveries and researches vindicate for Dr. Mantell a high place amongst scientific geologists – a higher place, we think, than some of his contemporaries were disposed to allow him" (1271).

At a somber meeting of the Clapham Athenaeum that night Gideon's loss was deeply felt. Outraged by Owen's malicious obituary in *Literary Gazette*, council members voted to buy space in the same periodical for a brief reply. Appearing on the twenty-seventh, it expressed "deep regret at the death of their late distinguished friend and able coadjutor, Dr. Mantell," whose lectures before their group had long been one of its chief ornaments and attractions. "No one who has enjoyed the advantage of hearing him," it was suggested, "can ever forget the singular ability, the felicitous illustrations, and the energetic eloquence which characterized all his discourses." Three days later Earl Rosse borrowed these same words in eulogizing Mantell before the Royal Society. He also presented unique details of Gideon's early years, as recalled by his brother Thomas, and listed his Royal Society papers.

For William Hopkins, president of the Geological Society, Gideon seemed "a memorable instance of a man of genius, constantly and diligently occupied with the practice of a laborious profession, nevertheless reaching great eminence as a man of science" (xxii). In a concise but well-informed review, Hopkins praised Mantell not only for his Wealden and dinosaurian researches but also for his work with ventriculites, belemnites, foraminifera ("He was an expert microscopist"), *Telerpeton,* and the flightless birds of New Zealand. As a popular expounder of geological facts, moreover, Gideon was unequalled and as a lecturer, without a rival. A subsequent obituary by the Linnaean Society began with Mantell's childhood education; it then traced his many felicitous publications, from 1813 onward.

Literary Gazette, 13 Nov 1852, p. 842. Owen's mean-spirited jibes at GM cost him the presidency of the Geological Society of London; see Adrian Desmond, *Archetypes and Ancestors: Palaeontology in Victorian London, 1850–1875* (Chicago and London, 1982), p. 208n13.

Additional memorials appeared in the *Illustrated London News, Gentleman's Magazine, The Cottage Gardener,* and other publications. Though several of them were vague or in error regarding basic facts, the tone throughout was one of deep respect for a brilliant, hard-working man who had risen by his own efforts from unpromising beginnings. This was, for example, the theme of a generally well-informed obituary in *Gentleman's Magazine,* which thought his "a striking instance of a rise in life amidst great difficulties." Based on the *Athenaeum's* account, that in the *Illustrated London News* remembered Gideon not only for his geological. works but as a surgeon and lecturer as well.

Though naturally less incisive, the most heartfelt obituary was Benjamin Silliman's; signed by him, it appeared in the *American Journal of Science* in May 1853. "A number of years ago," Silliman recalled,

> Dr. Mantell sustained a severe injury on the spine, in consequence of a fall from his carriage, and an incurable tumor arose, which by its pressure upon the nerves of the spinal cord produced at first temporary paralysis and, subsequently through life, frequent and intense neuralgic suffering attended by great emaciation. Still, his powerful and enthusiastic mind rose above his sufferings, although they often deprived him of sleep. He wrote several of his works while he was a martyr to pain. At the same time he continued his professional visits, and at the bedside of his patients and when in society at home or abroad, he assumed a degree of cheerfulness which might have led anyone to suppose that he was in perfect health (148–149).

Silliman then recorded some particulars of Gideon's last week sent him by Alfred Woodhouse. As a personal friend, he continued, Mantell had been "most interesting and estimable," outstanding for candor, kindness, and scientific justice, especially to original discoverers; no British intellectual, moreover, excelled him in his liberality toward American science and letters. "Exact and thorough scientific knowledge, the enthusiasm of a discoverer, and the rich but chastened diction of a poet," Silliman concluded, "were never more remarkably united than in him" (150).[10]

10. Brief notice, the *Athenaeum,* 13 Nov 1852, p. 1244; *Sussex Advertiser,* 16 Nov 1852, p. 5; *Athenaeum,* 20 Nov 1852, pp. 1270–71; *Literary Gazette,* 27 Nov 1852, p. 1; "Address of the Right Honourable the Earl Rosse . . . Read at the Anniversary Meeting of the Royal Society on Tuesday, November 30, 1852" (London, 1853); William Hopkins, "Anniversary Address," *QJGS,* 9 (1853), xxii–xxv; *Proc LS,* 53 (1853), 235–237; *Illustrated London News,* 4 Dec 1852, p. 501; *Gentleman's Magazine,* 38 (Dec 1852), 644–647 (quoting p. 644); *The Cottage Gardener, and Country Gentleman's Companion,* 16 Dec 1852, pp. 199–200; *AJSA,* ns 15 (May 1853), 147–150. See also Anon. but Stephen Williams, "A Reminiscence of Gideon Algernon Mantell, Esq . . . To which is Appended

an Obituary by Professor Silliman of New York [*sic*]" (London, 1853; for authorship, AJW–WBDM, 31 Jan 1853 [ATL] and Reg J, 5, 10 Feb 1853).

There were also private reminiscences, incl. BS, "Death of Dr. Gideon Mantell," Private Journal No. XIV, pp. 391–397 (Yale); CL–LB, 20 Apr 1853 (BMNH): "Mantell's death is indeed very sudden, and leaves a great blank which I feel much"; AS–RIM, 19 Nov 1852 (GS); RIM–AS, 22 Nov 1852 (CUL); Dr. J. C. Warren–BS, 14 Jan 1853 (in Fisher, *Life*, II, 231–232). See also BS–GM, 30 Aug 1847 (Yale). GM's final will of 19 Nov 1849 is at PRO, London.

Stratigraphic Table

The following table summarizes (and somewhat oversimplifies) the growth of Gideon Mantell's stratigraphic understanding, as reflected in his major expressions of it:

1822	*The Fossils of the South Downs*
1827	*Illustrations of the Geology of Sussex*
1833	*The Geology of the South-East of England*
1838	*The Wonders of Geology*
1839	*The Wonders of Geology,* third edition
1844	*The Medals of Creation*
1848	*The Wonders of Geology,* sixth edition
1851	*Petrifactions and Their Teachings*

Fossils and locations cited for each stratigraphical unit (arranged from present to past) are those having particular reference to Mantell. The brief historical notes, however, reflect modern understanding, whether of historical geology (what happened in the Jurassic) or of the history of geology (who discovered it). All estimates of geological time are modern.

Divisions of geological time are represented as follows: ERAS, PERIODS, *Epochs,* Formations. Except for "Tertiary" (and sometimes "Secondary"), the formational names were older and in more common use than those of epochs, periods, and eras. As stratigraphical nomenclature became more exacting, however, some formations expanded into epochs (e.g., Oolite, Lias). Now, of course, there are endless subdivisions.

According to Mantell,

> *Strata* are sedimentary deposits that have been formed in the beds of lakes, rivers, and seas, and have subsequently been displaced and elevated above the water by physical causes. A series, or group of strata, is termed a *formation;* and the fossil remains found in one series or formation differ more or less completely from those of another (1851, p. 4n).

But Mantell's idea of what constitutes a formation differs from one book to another.

FORMATIONS ABOVE THE CHALK (Mantell, 1822); later (for others), CENOZOIC (J. Phillips, 1841). 65 million years ago – present.

GM Formations (and specimens)	GM Fossils (and specimens)	GM Localities	Modern Understanding; Historical Notes
	Mankind, modern mammals, flowering plants, grasses	England, New Zealand, France, Italy, India, Burma	Era of Recent Life; Age of Mammals
MODERN OR HUMAN EPOCH (1848); later, QUATERNARY (Morlet, 1854)	Present-day life	Present-day world	Age of Mankind Almost no valid human fossils were known, and the antiquity of mankind was not generally acknowledged.
Alluvial Formations (Mantell, 1822, ch. 18); later, Recent (Lyell, 1833), Drift (Mantell, 1844 only)	Surface deposits	Sussex	Historical times; modern animals and plants. In southeast England, the Flandrian Transgression (rise of sea level); creation of the present-day English Channel; sea cliffs; Weald drainage patterns.
Alluvium (1822, ch. 20); later, Drift (1848) – "the effect of causes still in action"	Humans, mammals, birds, mollusks, animalcules	Human barrows, urns, graves, Sussex; moa beds, New Zealand	Rivers, lakes
Diluvium (1822, ch. 19); later, Neuer Pliocene (Lyell, 1833), Pleistocene (Lyell, 1839), Post Pliocene (Mantell, 1851). "Called also the Quaternary or Diluvian period; these deposits cannot be definitely separated from those of the Modern or Human epoch" (1851, p. 4n).	Mammoth, mastodon, Irish "elk," cave bear, carnivorous mammals, sloths (Megatherium)	Elephant beds, Brighton (1833); Kent's Cavern, Torquay; Val di Noto and Palermo, Sicily; Ava (Burma); Siwalik Hills, India	Deluge era (Buckland, 1823); later, the Ice Age (Agassiz, 1840). Denudation of the Weald; coombes. Cuvier, Recherches (1812) on large mammals.

	GM Fossils (and specimens)	GM Localities	Modern Understanding; Historical Notes
TERTIARY (Arduino, 1760; Brongniart, 1810; Mantell, 1838)	Primates, elephants, other large mammals, birds	Castle Hill, Newhaven; Isle of Wight; London, Hampshire, and Paris basins; Auvergne, France; Subapennines, Italy; Virginia, USA	Lyell's three-part division of the hitherto unitary Tertiary period into Eocene, Miocene, and (newer and older) Pliocene was based more on statistics than on stratigraphy. Though Mantell adopted Lyell's terminology to some extent, he remained skeptical about the soundness of Lyell's methods and the reality of his rather arbitrary periods.
Pliocene (Lyell, 1833; Mantell, 1838)	Bovids, whales, other mammals, shells	English Crag, Kent, Surrey Chiltern Hills	Submergence of the Weald. Early cattle, sheep, goats, antelopes; grasslands, forests.
Crag (Pliocene and Pleistocene)	Shells, occasional bones	East coast of England	Shell banks deposited in very shallow water.
Miocene (Lyell, 1833; Mantell, 1838)	Murchison's "fox"; other mammals, shells	Southeast England; Val d'Arno, Italy; Oeningen, Germany; Bordeaux, France	Alps rise; folding and erosion in southeast England. Mammalian domination; diversity peaks. Early deer, horses, camels, elephants.
Oligocene (Beyrich, 1854)	Freshwater shells	Isle of Wight	Mammalian diversity – dogs, cats, pigs, rodents; social insects

Formation	Fossils	Locations	Notes
Tertiary Formations (1822, ch. 14); later, *Eocene* (Lyell, 1833; Mantell, 1838)	Mammals, reptiles, birds, sharks (teeth), fishes, crabs, lobsters, insects, mollusks, nummulites (large foraminifera); casts from Cuvier	London, Hampshire, and Paris basins; Castle Hill, Newhaven; Isle of Sheppey; Isle of Wight; Monte Bolca, Italy; Aix, France	Tectonic activity; marine incursion into southeast England; Hebridean volcanics. Mammalian ecological expansion; early primates, early elephants, early marine mammals (first whales), bats, birds, crocodiles, modern fishes, nummulites, flowering plants.
London Clay (1822, ch. 17)	Reptiles, fishes, birds, shells	Greater London, Sussex, Hampshire	Fossils (scanty) suggest a tropical climate not far from land.
Plastic Clay (1822, ch. 16)	Marine fossils, lignite	Castle Hill, Newhaven; Paris Basin	The Argille Plastique of Cuvier and Brongniart (1812), intermediate between Chalk and London Clay
Druid Sandstone (1822, ch. 15, then dropped)	None (and not a regular stratum)	Sussex, Berkshire, Wiltshire	So called because Stonehenge and similar monuments are made of it.
Paleocene (Schimper, 1874)	Bivalves	London Basin	Shorelines fluctuate; land mammals proliferate. Earliest primates.

At the end of the Mesozoic Era, between the Cretaceous and Tertiary periods, a great wave of extinctions occurred. As a result, the dinosaurs, plesiosaurs, ichthyosaurs, mosasaurs, and pterosaurs all disappeared, as did the ammonites and (a little later) the belemnites.

SECONDARY FORMATIONS (Mantell, 1822); later, MESOZOIC (J. Phillips, 1841). 250 to 65 million years ago.

	GM Fossils (and specimens)	GM Localities	Modern Understanding: Historical Notes
	Rise of reptiles, mammals, birds	Southeast England; Isle of Wight; Dorsetshire; Wiltshire; foreign countries.	Era of Middle Life; Age of Reptiles (see *Wonders*, 6th, p. 560n)
CRETACEOUS (D'Halloy, 1822; Mantell, 1844; *Chalk Formations* (1822, ch. 9); The Chalk	Dinosaurs, large reptiles, turtles, birds (pterosaurs), small mammals, modern fishes, ammonites, mollusks, echinoderms, siliceous sponges, flowering plants.	Southeast England; Isle of Wight; Wiltshire; Germany; France; New Jersey, USA	The Chalk Period. GM's most important contributions to knowledge (vertebrate and invertebrate paleontology) were derived from specimens found in Cretaceous strata.
Upper Chalk (1822, ch. 13); Chalk with Flints	*Mosasaurus*, turtles, fishes, mollusks (*Inoceramus*), crinoids, sponges; *Marsupites*, *Ventriculites* (GM's *Alcyonium* of 1814)	South Street and Offham pits; South Downs; Isle of Wight; Wiltshire; Brighton to Dover cliffs	The Weald totally submerged; all fossils marine. Topped by a plane of erosion. The famous English Chalk: White Cliffs of Dover, North and South Downs; Salisbury Plain; The Needles, Isle of Wight.
Lower Chalk (1822, ch. 12); Chalk without Flints	Cephalopods, gastropods, brachiopods, terebratula	Southerham and South Street pits; Lewes	Lacks flints common in the Upper.
Grey Chalk Marl (1822, ch. 11); later (1827), Chalk Marl	*Ammonites Mantelli*, *Scaphites*, *Turrilites*, "juli of the larch" (later, coprolites), belemnites, brachiopods, corals, sponges	Hamsey and Offham pits; Eastbourne (all three Sussex)	Deposited in originally shallow, murky water that deepened and clarified over time. Richly fossiliferous.

Formation	Fossils	Locality	Description
Firestone, or Merstham Beds (1827); later (1844), Malm Rock, Upper Green Sand, or Glauconite; in 1851, combined with Chalk Marl as Upper Green Sand	Marine fossils & sponges	Western Sussex; Wiltshire, Dorsetshire	Greensand of some kind overlies the Gault in Wiltshire but underlies it in Sussex.
Blue Chalk Marl (1822, ch. 10); later (1827), Galt, or Folkstone Marl	Fishes, crustaceans, ammonites, hamites, belemnites	Western Weald (Bignor, Sussex); Ringmer, Hamsey (both near Lewes)	A shallow-water marine environment (now Gault). Richly fossiliferous.
Green Sand [Formation] (1822, chs. 4, 8); later (1827), Shanklin Sand; later, Lower Green Sand (Fitton, 1824; Murchison, 1826; Mantell, 1833); "Kentish Rag"	Maidstone Iguanodon; mollusks, ammonites, brachiopods, corals, sponges	Maidstone, Offham, Hurstpierpoint, Eastbourne, Faringdon	Widespread marine incursions.
Wealden Formation (P. J. Martin, 1828; Mantell, 1833). For Mantell in 1833, the Wealden Formation included the Weald Clay, Hastings Beds (Tilgate Beds and former Iron Sand), and Ashburnham Beds. In 1838 he added the Purbeck Beds, which later assimilated the Ashburnham Beds.	*Megalosaurus, Iguanodon, Hylaeosaurus, Cetiosaurus, Pelorosaurus,* Swanage Crocodile, plesiosaurs, freshwater shells (*Unio Valdensis*), starfishes, sea urchins, trees and plants	"Tilgate Forest" (Whiteman's Green); Brook Point, Isle of Wight; Swanage	Weald partially submerged, possibly a delta or estuary (but freshwater). The formation with which Mantell is most closely associated.
Weald, or Oak Tree Clay (1822, ch. 7); later (1827), Weald Clay	Brackish water fossils.	Sussex and Kent	Stiff blue clay, shale, sandstone, and shelly freshwater limestone (Sussex Marble).

	GM Fossils (and specimens)	*GM Localities*	*Modern Understanding; Historical Notes*
Hastings Formation or Beds (Fitton, 1824; Mantell, 1827), including	Swamp fauna; dinosaurian remains	Hastings, Weald	Sandstone and clays alternating; nodular iron ore; deltaic but deepening.
Tilgate Beds (1822, ch. 6, as Tilgate Limestone	Dinosaurian remains; freshwater fossils	Whiteman's Green, Weald	Fawn-colored sandstone, deltaic or lacustrine.
Iron Sand (1822, ch. 5, then dropped); see GM, "On the Iron-Sand Formation of Sussex," *Trans GS*, 2 (1826), 131–134; read 21 June 1822	Few fossils	Uckfield, Hastings, Newick	The Rocks, Uckfield.
JURASSIC (Brongniart, 1829; Mantell, 1848)	Ichthyosaurs, plesiosaurs, dinosaurs, fishes, ammonites, belemnites, crinoids, gastropods, corals	Dorsetshire, Oxfordshire, Hampshire, Wiltshire, Yorkshire; Jura Mountains, France; Solnhofen, Germany	The first major age of dinosaurs; extensive continental and marine exposures. Early stratigraphical investigations by Smith, Buckland, Conybeare. Later (in 1860), the first Mesozoic bird, *Archeopteryx*.
Purbeck Beds (1838, including some Portland)	Fossil forest (cycads); marine and freshwater fossils	Isle of Purbeck; Lulworth Cove	Lagoons and shallows.
THE OOLITIC SYSTEM (1838, p. 437, simplified); OOLITIC OR JURASSIC (1851)	Dinosaurs, marine reptiles, ammonites, gastropods, corals, cycads	Bath district, Cotswolds, Wiltshire	"Eggstone"; limestones composed of small spherules resembling the roe of fishes (and other rocks associated with them).
Upper Oolite Portland Oolite	Saurian bones; huge ammonites; gastropods, bivalves	Isle of Portland; Swindon, Wiltshire	Lagoons and shallows.

Kimmeridge Clay	Isle of Portland; Swindon	Ichthyosaurs, plesiosaurs	Deepening seas; North Sea oil. A dark clay, quarried for brick-making at Swindon.
Middle Oolite Coral Oolite (Coral Rag)	Oxfordshire, Wiltshire, Yorkshire	Corals	Shallower water; remains of coral reefs.
Oxford Clay	Oxfordshire, Yorkshire, Dorsetshire	Marine reptiles, ammonites	Marine environment of fluctuating depth.
Lower Oolite Cornbrash Great Oolite	Dorsetshire, Yorkshire Bath, Cotswolds	Ammonites, brachiopods Brachiopods, mollusks.	Marine limestones. A limestone series; the inferior is always beneath.
Inferior Oolite	Bath, Cotswolds	Ammonites, brachiopods, sea urchins, corals	
Stonesfield Slate	Stonesfield, near Oxford; Cotswolds	*Megalosaurus*, *Cetiosaurus*, small mammals	Sandstone, limestone, but not slate; thinly bedded, easily split (used for roofing); Buckland 1824; the small mammal jaws had major impact on debates about the history of life.
THE LIAS (1838, p. 438); LIASSIC Upper Lias Shale; later; Upper Lias	Lyme Regis, Dorset, to Whitby, Yorkshire	Ichthyosaurs, plesiosaurs, ammonites	"Layer rock" deposited in shallow sea. Classic source of superbly preserved fossil marine reptiles; Mary Anning, Col. Birch, Thomas Hawkins.

	GM Fossils (and specimens)	GM Localities	Modern Understanding; Historical Notes
TRIASSIC (Alberti, 1834, based on German exposures; Mantell, 1851)	*Telerpeton, Stagonolepis*; trifid footprints (later datings).	Elgin, Scotland; the Connecticut River Valley, Massachusetts (now regarded as early Jurassic); English Midlands (fuller extent not then realized).	Rise of the archosaurs: ichthyosaurs, crocodiles, pterosaurs, early dinosaurs. Both the Elgin and Massachusetts exposures were central to later controversies about dating.
THE SALIFEROUS, OR NEW RED SANDSTONE SYSTEM (1838, p. 409); later, *Trias* and PERMIAN *New or Upper Red Marl or Sandstone* (Muschelkalk and Keuper) *Upper New Red, or Trias* (1844), adding Bunter; later (1848), *Triassic or New Red Formation*, TRIAS SYSTEM *Lower New Red* (1844); later (1848), PERMIAN SYSTEM (Paleozoic) Magnesian Limestone and Zechstein	New Red Sandstone (see below)	Devonshire, northern England, Scotland	The distinction between Triassic and Permian remained unclear until the late 1840s, largely because the British New Red Sandstone included both.

"The separation of the strata now termed Permian from the Triassic group, with which they were formerly classed, was first proposed by Sir Roderick Murchison, and is based on the fact that the fossils hitherto discovered are entirely distinct from any that occur in the Trias and subsequent formations; it is, therefore, inferred that after the deposition of the so-called Permian strata, a complete change took place in the faunas and floras of the lands and seas, and the Trias is regarded as the dawn of a new system of organic beings" (1851, p. 5n).

PALEOZOIC (J. Phillips, 1841; Mantell, 1844). 590 to 250 million years ago.

	GM Fossils (and specimens)	GM Localities	Modern Understanding; Historical Notes
	First vertebrates, amphibians, reptiles; no mammals, birds	Wales, Lake District, Derbyshire, Yorkshire	Era of Ancient Life At the end of the Paleozoic, there occurred the greatest episode of extinction in the history of life.
THE PERMIAN SYSTEM (Murchison, 1841; Mantell, 1848, p. 533; formerly, for him, Lower Saliferous); Permian Formation (1851)	Earliest reptiles, fishes, brachiopods, corals (but no trilobites); *Labyrinthodont* (mysterious footprint maker, now known to be an amphibian)	Russia; by report through Murchison (1841, *Philosophical Magazine*)	Diversification of reptiles, ammonites; first belemnites, last trilobites. Murchison, 1845 (*Geology of Russia*).
New Red Sandstone (Permo-Trias)	Fossil reptiles; *Telerpeton*	County Durham; English Midlands; Elgin, Scotland	A continental deposit of wind-blown sand, its age long controversial (because securely dated fossils were unavailable).
THE CARBONIFEROUS, OR COAL SYSTEM (Conybeare, 1822; Mantell, 1838, p. 521); Carboniferous Formation (1851)	Fishes, mollusks, zoophytes, and especially plants; trilobites rare, amphibians undetected	Midlands, Wales, Derbyshire, Yorkshire; Coalbrook Dale, Shropshire; Pennsylvania, USA.	Coal Age forests, shallow seas. Early amphibians, reptiles, and insects.

	GM Fossils (and specimens)	GM Localities	Modern Understanding; Historical Notes
Coal Measures	Coal trees, ferns; *Stigmaria, Sigillaria* (fossil roots and stems, respectively). Earliest land life (then known)	Midlands, South Wales	A major source of the Industrial Revolution. Fossils depicted: Artis, 1825; Mantell, 1850.
Mountain Limestone (Millstone Grit) Old Red Sandstone (1838, as Carboniferous)	Marine fossils	Derbyshire, Yorkshire, Somersetshire; Wales, Scotland, Ireland; Matlock and Crich Hill, Derbyshire; the Avon at Clifton.	The cave districts of Derbyshire, Yorkshire, and Somerset.
THE OLD RED SANDSTONE, OR DEVONIAN SYSTEM (Sedgwick and Murchison, 1839; Mantell, 1839, p. 612); Devonian (or Old Red) Formation (1851)	Armored fishes, trilobites, corals, cephalopods, mollusks, brachiopods, crinoids	Devonshire, Wales, Scotland; the latter by report through Hugh Miller (1841)	Land plants and animals; a few Silurian pioneers excepted, earlier periods were wholly marine.
Old Red Sandstone		Devonshire; Scotland; Shropshire; Wales.	The continental rocks of the Devonian; conglomerates, shales, and (often red) sandstones; freshwater and marine fossils. Graptolites extinct. Miller, *The Old Red Sandstone* (1841).
THE SILURIAN SYSTEM (Murchison, 1835; Mantell, 1838, p. 605); Silurian Formation (1851) *Upper* (1851) *Lower* (1851)	Trilobites, brachiopods, mollusks, corals, graptolites	Wales, Shropshire, Lake District; by report through Murchison (1839); Dudley, Shropshire (trilobites); Malvern Hills, Worcestershire; foreign localities.	Murchison, 1839 (*The Silurian System*). Under the old Wernerian classification, both Devonian and Silurian strata were known as "Transition" rocks.

ORDOVICIAN (Lapworth, 1879)	Trilobites, brachiopods, mollusks, crinoids, graptolites	Wales, Lake District; by report through Sedgwick and Murchison (as Lower Silurian or Upper Cambrian)	Known to GM as "Cumbrian" or "Lower Silurian"; disputed between Sedgwick and Murchison. *Wonders*, 6th, p. 797.
THE CUMBRIAN OR CAMBRIAN SLATE SYSTEM (Sedgwick, 1835; Mantell, 1838, p. 607); later, THE CUMBRIAN OR SCHISTOSE SYSTEM (1848); Cumbrian Formation (1851)	"The Silurian system [as defined by Murchison] is succeeded by a vast series of strata of a slaty character, which are destitute of any distinct assemblage of organic remains, although fossils [not specified] occur in some of the uppermost rocks" (1848, p. 794).	North Wales; by report through Sedgwick (1835)	The earliest evidence of life then known; championed by Sedgwick.
PRIMARY ROCKS (unfossiliferous) Precambrian periods and fossils were unknown.	Granite, basalt.	England, Scotland; the Alps; Mount Vesuvius, Etna.	Primitive rocks. The earliest major classification of rocks by age (Arduino, 1760) had them as Primary, Secondary, and Tertiary, an assumed order of creation having nothing to do with fossils. The term "Primitive" eventually replaced Primary because from the time of James Hutton (1795) onward, it was increasingly realized that rocks of all types, including granite and basalt, were being created and destroyed all the time. Lithological chronology therefore fell out of favor, to be replaced by the stratigraphical chronology represented in this table. There was, as yet, no secure way of dating most nonfossiliferous rocks.

Index

Excepting dates, editions, and occasionally chapters, all numbers refer to pages. Entries are arranged alphabetically by key word, general preceding specific. Within each entry, the arrangement of subentries is usually chronological – in the case of persons, from birth to death; in the case of books, from front to back. Advancements of knowledge also appear in historical order; bones, on the other hand, are not anatomically arranged. As throughout my text, dinosaurs and the strata that contained them are stressed. The front matter and Stratigraphic Table are not indexed.

Milton Keynes UK
Ingram Content Group UK Ltd.
UKHW041519181024
449640UK00003B/10